Mathematics for Economics and Finance

Accessible, concise and interactive, this book introduces the mathematical methods that are indispensable in economics and finance.

Fully updated to be as student-friendly as possible, this edition contains extensive problems, worked examples and exercises (with full solutions at the end of the book). Two brand new chapters cover coupled systems of recurrence/differential equations and matrix diagonalisation, respectively. All topics are motivated by problems from economics and finance, demonstrating to students how they can apply the mathematical techniques they learn.

For undergraduate students of economics, mathematics or both, this book will be welcomed for its clarity and breadth and the many opportunities it provides for readers to practise and test their understanding.

Martin Anthony is Professor of Mathematics at the London School of Economics and Political Science, where he has been Head of Department and Vice-Chair of the Academic Board. He is the author of five books. Martin has been heavily involved in LSE's distance learning initiatives through the University of London International Programme and is a recipient of multiple LSE Teaching Excellence Awards.

Norman Biggs is Professor Emeritus in the Department of Mathematics at the London School of Economics and Political Science, where he was Head of Department and Vice-Chair of the Appointments Committee. He is the author of thirteen books, including *Discrete Mathematics* (Second Edition, 2002). He previously served as General Secretary of the London Mathematical Society.

Mathematics for Economics and Finance

Methods and Modelling

Second Edition

MARTIN ANTHONY
London School of Economics and Political Science

NORMAN BIGGS
London School of Economics and Political Science

CAMBRIDGE
UNIVERSITY PRESS

CAMBRIDGE
UNIVERSITY PRESS

Shaftesbury Road, Cambridge CB2 8EA, United Kingdom

One Liberty Plaza, 20th Floor, New York, NY 10006, USA

477 Williamstown Road, Port Melbourne, VIC 3207, Australia

314–321, 3rd Floor, Plot 3, Splendor Forum, Jasola District Centre, New Delhi – 110025, India

103 Penang Road, #05-06/07, Visioncrest Commercial, Singapore 238467

Cambridge University Press is part of Cambridge University Press & Assessment, a department of the University of Cambridge.

We share the University's mission to contribute to society through the pursuit of education, learning and research at the highest international levels of excellence.

www.cambridge.org
Information on this title: www.cambridge.org/highereducation/isbn/9781108472395
DOI: 10.1017/9781108560177

When citing this work, please include a reference to the DOI 10.1017/9781108560177

First published 2024

Printed in the United Kingdom by TJ Books Limited, Padstow Cornwall

A catalogue record for this publication is available from the British Library.

Library of Congress Cataloging-in-Publication Data
Names: Anthony, Martin, author. | Biggs, Norman, author.
Title: Mathematics for economics and finance : methods and modelling / Martin Anthony and Norman Biggs.
Description: 2nd edition. | New York, NY : Cambridge University Press, 2024. | Includes index.
Identifiers: LCCN 2023056068 (print) | LCCN 2023056069 (ebook) | ISBN 9781108472395 (hardback) | ISBN 9781108459433 (paperback) | ISBN 9781108560177 (ebook)
Subjects: LCSH: Economics, Mathematical. | Finance–Mathematical models.
Classification: LCC HB135 .A62 2024 (print) | LCC HB135 (ebook) |
DDC 330.01/51–dc23/eng/20240104
LC record available at https://lccn.loc.gov/2023056068
LC ebook record available at https://lccn.loc.gov/2023056069

ISBN 978-1-108-47239-5 Hardback
ISBN 978-1-108-45943-3 Paperback

Additional resources for this publication at www.cambridge.org/AnthonyBiggs2ed

For Colleen and Alistair, and in memory of my parents [M.A.]
For Christine and Juliet [N.B.]

Contents

Preface to Second Edition *page* xiii
Preface to First Edition xv

1 Mathematical Models in Economics 1
1.1 Introduction 1
1.2 A Model of the Market 1
1.3 Market Equilibrium 3
1.4 Excise Tax 4
1.5 Comments 5

2 Mathematical Terms and Notations 11
2.1 Sets 11
2.2 Functions 12
2.3 Composite Functions 13
2.4 Graphs and Equations 15

3 Sequences, Recurrences, Limits 23
3.1 Sequences 23
3.2 The First-Order Recurrence 23
3.3 Limits 26
3.4 Special Cases 27

4 The Elements of Finance 35
4.1 Interest and Capital Growth 35
4.2 Income Generation 36
4.3 The Interval of Compounding 37

5 The Cobweb Model 43
5.1 How Stable Is Market Equilibrium? 43
5.2 An Example 44
5.3 The General Linear Case 45
5.4 Economic Interpretation 47

6	**Introduction to Calculus**	51
6.1	The Rate of Change of a Function	51
6.2	Rules for Finding the Derivative	53
6.3	Marginal Cost as a Derivative	54
6.4	The Derivative of a Composite Function	55
6.5	The Derivative of an Inverse Function	56
7	**Some Special Functions**	61
7.1	Powers	61
7.2	The Exponential Function and Its Properties	62
7.3	Continuous Compounding of Interest	64
7.4	The Logarithm Function	65
7.5	Trigonometrical Functions	67
8	**Introduction to Optimisation**	73
8.1	Profit Maximisation	73
8.2	Critical Points	74
8.3	Optimisation in an Interval	76
8.4	Infinite Intervals	78
9	**The Derivative in Economics I**	85
9.1	Elasticity of Demand	85
9.2	Profit Maximisation Again	87
9.3	Competition versus Monopoly	87
10	**The Derivative in Economics II**	93
10.1	The Efficient Small Firm	93
10.2	Startup and Breakeven Points	94
11	**Partial Derivatives**	101
11.1	Functions of Several Variables	101
11.2	Partial Derivatives	101
11.3	The Chain Rule	103
12	**Applications of Partial Derivatives**	110
12.1	Functions Defined Implicitly	110
12.2	The Derivative of an Implicit Function	112
12.3	Contours and Isoquants	113
12.4	Scale Effects and Homogeneous Functions	115
13	**Optimisation in Two Variables**	123
13.1	Profit Maximisation Again	123
13.2	How Prices Are Related to Quantities	123
13.3	Critical Points	126

13.4	Maxima, Minima and Saddle Points	127
13.5	Classification of Critical Points – Introduction	129
13.6	The Classification of Critical Points in General	131
14	**Vectors, Preferences and Convexity**	137
14.1	Vectors and Bundles	137
14.2	Prices and Budgets	138
14.3	Preferences, Utility and Indifference Curves	139
14.4	Linear and Convex Combinations	142
14.5	Choosing Optimal Bundles	144
15	**Matrix Algebra**	152
15.1	What is a Matrix?	152
15.2	Matrix Multiplication	153
15.3	How to Make Money with Matrices	154
16	**Linear Equations I**	160
16.1	A Two-Industry 'Economy'	160
16.2	Linear Equations in Matrix Form	161
16.3	Solution of Linear Equations by Row Operations	162
16.4	The Echelon Form in General	164
17	**Linear Equations II**	173
17.1	Consistent and Inconsistent Systems	173
17.2	The Rank of a Consistent System	175
17.3	The General Solution in Vector Notation	176
17.4	Arbitrage Portfolios and State Prices	179
18	**Inverse Matrices**	190
18.1	The Square Linear System	190
18.2	The Inverse of a Square Matrix	191
18.3	Calculation of the Inverse	193
18.4	The Inverse of a 2×2 Matrix	195
18.5	IS–LM Analysis	196
19	**The Input–Output Model**	204
19.1	An Economy with Many Industries	204
19.2	The Technology Matrix	205
19.3	Why Is There a Solution?	207
20	**Determinants**	212
20.1	Determinants	212
20.2	The Determinant as a Test for Invertibility	216
20.3	Cramer's Rule	217

20.4 The Inverse of a 3 × 3 Matrix via Determinants 217

21 Constrained Optimisation 224
21.1 The Elementary Theory of the Firm 224
21.2 The Method of Lagrange Multipliers 225
21.3 The Cost Function 227
21.4 The Efficient Small Firm Again 228
21.5 The Cobb–Douglas Firm 230

22 Lagrangeans and the Consumer 238
22.1 Lagrangeans: A More General Formulation 238
22.2 The Elementary Theory of the Consumer 238
22.3 The Price Ratio and the Tangency Condition 240
22.4 The Consumer's Demand Functions 241
22.5 The Indirect Utility Function 243

23 Second-Order Recurrence Equations 251
23.1 A Simplified National Economy 251
23.2 Dynamics of the Economy 252
23.3 Linear Homogeneous Recurrences 253
23.4 Non-homogeneous Recurrences 256

24 Macroeconomic Applications 264
24.1 Recurrence Equations in Practice 264
24.2 Oscillatory Solutions 265
24.3 Business Cycles 265
24.4 Improved Models of the Economy 268

25 Areas and Integrals 274
25.1 The Consumer Surplus 274
25.2 The Concept of Area 276
25.3 Anti-derivatives and Integrals 277
25.4 Definite Integrals 279
25.5 Standard Integrals 280

26 Techniques of Integration 289
26.1 Integration by Substitution 289
26.2 Definite Integrals by Substitution 291
26.3 Integration by Parts 292
26.4 Partial Fractions 293

27 First-Order Differential Equations 301
27.1 Continuous-Time Models 301
27.2 Some Types of Differential Equations 303

27.3 Separable Differential Equations 303
27.4 A Continuous-Time Model of Price Adjustment 305
27.5 Cash Flows under Continuous Compounding 307
27.6 Integrating Factors for Linear Equations 308

28 Second-Order Differential Equations 319
28.1 Market Trends and Consumer Demand 319
28.2 Linear Equations with Constant Coefficients 320
28.3 Solution of Homogeneous Equations 321
28.4 Non-homogeneous Equations 323
28.5 Behaviour of Solutions 325

29 Coupled Systems and Diagonalisation 334
29.1 Population Modelling 334
29.2 Coupled Systems of Recurrence Equations 334
29.3 Reduction to Second-Order Recurrence 335
29.4 Eigenvalues and Eigenvectors 336
29.5 Diagonalisation of a Square Matrix 339

30 Applications of Diagonalisation 347
30.1 Using Diagonalisation: Powers of Matrices 347
30.2 Solving Recurrence Systems Using Matrix Powers 348
30.3 Solving Recurrence Systems Using Change of Variable 349
30.4 A Population Model 350
30.5 Systems of Differential Equations 352

Appendix A Solutions to Exercises 361

Appendix B Answers to Problems 408

Index 416

Preface to Second Edition

In this new edition, we have expanded the text in two main ways: extension of topics covered and inclusion of significantly more learning support material.

We have included two new chapters concerning diagonalisation of matrices and its applications: these provide links between linear algebra, recurrence equations and differential equations. We have also expanded the treatment of determinants, matrix inverses and first-order differential equations.

We have considerably expanded the provision of end-of-chapter material that consolidates learning. The first edition had exercises, with some brief answers to most of these provided at the end of the book. Now, we have both exercises and problems at the end of each chapter. At the end of the book, we provide extensive solutions to the exercises and brief answers to (most of) the problems. This will, we hope, in conjunction with examples throughout the text and the fully worked examples in each chapter, enable the book to be used even more actively by readers, by providing various opportunities to practise methods and test their knowledge.

It has been many years since the publication of the first edition, and we are grateful for the extensive feedback from colleagues and readers since then (much of which has resulted in corrections and improvements to the text).

We are very grateful to our colleague Dr James Ward of the Mathematics Department at the London School of Economics for his help in many aspects of preparing this edition, including his extensive work on the exercises, problems and solutions, his improvement of the figures and conversion of the underlying files into LaTeX, and his many and useful comments and observations derived from his extensive experience of teaching this material.

We thank Anna Scriven and Roger Astley of Cambridge University Press for their support and patience.

London, October 2023.

Preface to First Edition

This book is an introduction to calculus and linear algebra for students of disciplines such as economics, finance, business, management and accounting. It is intended for readers who may have already encountered some differential calculus, and it will also be appropriate for those with less experience, possibly used in conjunction with one of the many more elementary texts on basic mathematics.

Parts of this book arise from a lecture course given by the authors to students of economics, management, accounting and finance, and management sciences at the London School of Economics.

We thank Duncan Anthony, Reza Arabsheibani, Juliet Biggs and, particularly, Graham Brightwell for their invaluable comments on various drafts of the book. The final draft was read by Dr Stephen Siklos of Cambridge University, and his pertinent comments resulted in a number of improvements. We are also grateful to Roger Astley of Cambridge University Press for his efficient handling of the project, and to Alison Adcock for her help in preparing the manuscript.

London, October 1995.

1 Mathematical Models in Economics

1.1 Introduction

In this book we use the language of mathematics to describe situations which occur in economics. The motivation for doing this is that mathematical arguments are logical and exact, and they enable us to work out in precise detail the consequences of economic hypotheses. For this reason, mathematical modelling has become an indispensable tool in economics, finance, business and management. It is not always simple to use mathematics, but its language and its techniques enable us to frame and solve problems that cannot be attacked effectively in other ways. Furthermore, mathematics leads not only to numerical (or *quantitative*) results but, as we shall see, to *qualitative* results as well.

1.2 A Model of the Market

One of the simplest and most useful models is the description of *supply and demand* in the market for a single good. This model is concerned with the relationships between two things: the *price* per unit of the good (usually denoted by p) and the *quantity* of it on the market (usually denoted by q). The 'mathematical model' of the situation is based on the simple idea of representing a pair of numbers as a point in a diagram, by means of coordinates with respect to a pair of axes. In economics it is customary to take the horizontal axis as the q-axis, and the vertical axis as the p-axis. Thus, for example, the point with coordinates $(2000, 7)$ represents the situation where 2000 units are available at a price of \$7 per unit.

How do we describe *demand* in such a diagram? The idea is to look at those pairs (q,p) which are related in the following way: if p were the selling price, q would be the demand, that is the quantity which would be sold to consumers at that price. If we fill in on a diagram all the pairs (q,p) related in this way, we get something like Figure 1.1.

We shall refer to this as the *demand set* D for the particular good. In economics you will learn reasons why it ought to look rather like it does in our diagram, a smooth, downward-sloping curve.

Suppose the demand set D contains the point $(30, 5)$. This means that when the price $p = 5$ is given, then the corresponding demand will be for $q = 30$ units. In general,

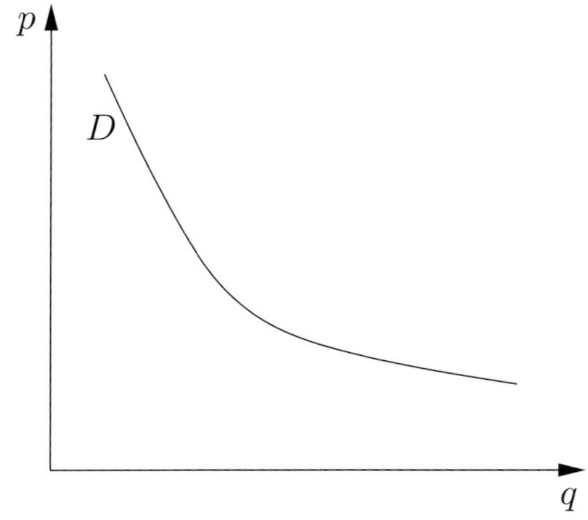

Figure 1.1 The demand set.

provided D has the 'right shape', as in Figure 1.1, then for each value of p there will be a uniquely determined value of q. In this situation we say that D determines a *demand function*, q^D. The value written $q^D(p)$ is the quantity which would be sold if the price were p, so that $q^D(5) = 30$, for example.

Example Suppose the demand set D consists of the points (q,p) on the straight line $6q + 8p = 125$. Then for a given value of p we can determine the corresponding q; we simply rearrange the equation of the line in the form $q = (125 - 8p)/6$. So here the demand function is

$$q^D(p) = \frac{125 - 8p}{6}.$$

For any given value of p we find the corresponding q by substituting in this formula. For example, if $p = 4$ we get

$$q = q^D(4) = \frac{125 - 8 \times 4}{6} = \frac{93}{6}$$

as the quantity. ■

There is another way of looking at the relationship between q and p. If we suppose that the quantity q is given, then the value of p for which (q,p) is in the demand set D is the price that consumers would be prepared to pay if q is the quantity available. From this viewpoint we are expressing p in terms of q, instead of the other way round. We write $p^D(q)$ for the value of p corresponding to a given q, and we call p^D the *inverse demand function*.

Example Taking the same set D as before, we can now rearrange the equation of the line in the form $p = (125 - 6q)/8$. So

$$p^D(q) = \frac{125 - 6q}{8}$$

is the inverse demand function. ■

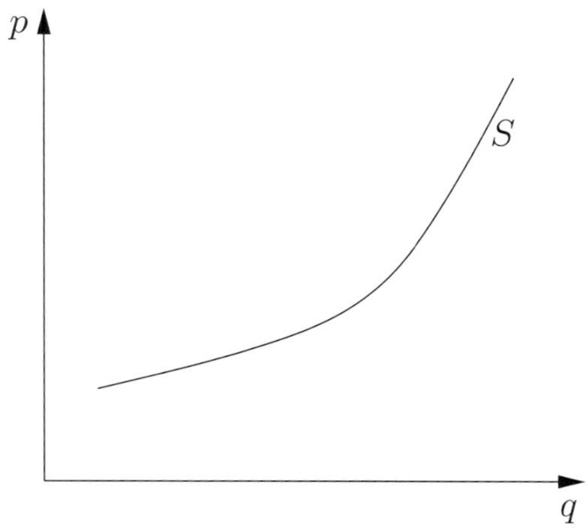

Figure 1.2 The supply set.

Next we turn to the supply side. We assume that there is a *supply set* S consisting of those pairs (q, p) for which q would be the amount supplied to the market if the price were p. There are good economic reasons for supposing that S has the general form shown in Figure 1.2.

If we know the supply set S we can construct the *supply function* q^S and the *inverse supply function* p^S in the same way as we did for the demand function and its inverse. For example, if S is the set of points on the line $2q - 5p = -12$, then solving the equation for q and for p we get

$$q^S(p) = \frac{5p - 12}{2} \quad \text{and} \quad p^S(q) = \frac{2q + 12}{5}.$$

1.3 Market Equilibrium

The usefulness of a mathematical model lies in the fact that we can use mathematical techniques to obtain information about it. In the case of supply and demand, the most important problem is the following. Suppose we know all about the factors affecting supply and demand in the market for a particular good; in other words, the sets S and D are given. What values of q and p will actually be achieved in the market?

Figure 1.3 makes it clear that the solution is to find the intersection of D and S, because that is where the quantity supplied is exactly balanced by the quantity required.

The mathematical symbol for the intersection of the sets S and D is $S \cap D$, and economists refer to $E = S \cap D$ as the *equilibrium set* for the given market.

Fortunately, there is a simple mathematical technique for finding the equilibrium set; it is the method for solving 'simultaneous equations'.

Example Suppose the sets D and S are, respectively, the sets of pairs (q, p) such that

$$q + 5p = 40 \quad \text{and} \quad 2q - 15p = -20.$$

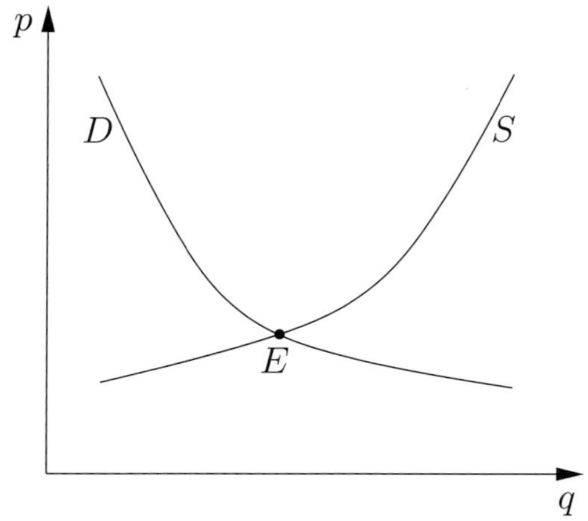

Figure 1.3 The equilibrium set $E = S \cap D$.

Then a point (q^*, p^*) which is in the equilibrium set $E = S \cap D$ must, by definition, be in both S and D. Thus (q^*, p^*) satisfies the two equations

$$q^* + 5p^* = 40 \quad \text{and} \quad 2q^* - 15p^* = -20.$$

The standard technique for solving these equations is to multiply the first one by 2 and subtract it from the second one. This tells us that

$$2(q^* + 5p^*) - (2q^* - 15p^*) = 2(40) - (-20)$$

which, on simplification, means $25p^* = 100$, so that $p^* = 4$. To find q^*, we can note that $q^* + 5p^* = 40$ now implies $q^* = 40 - 5p^* = 40 - 5(4) = 20$. In other words, the equilibrium set E contains the single point $(20, 4)$. ■

It is worth remarking that in this example we get a single point of equilibrium because we took the sets D and S to be straight lines. It is possible to imagine more complex situations, such as that we shall describe in Example 2.5, where the equilibrium set contains several points, or no points at all.

1.4 Excise Tax

Using only the simple techniques developed so far we can obtain some interesting insights into problems in economics. In this section we study the problem of excise tax. Suppose that a government wishes to discourage its citizens from drinking too much whisky. One way to do this is to impose a fixed tax on each bottle of whisky sold. For example, the government may decide that for each bottle of whisky the suppliers sell, they must pay the government \$1. Note that the tax on each unit of the taxed good is a fixed amount, *not* a percentage of the selling price.

Some very simple mathematics tells us how the selling price changes when an excise tax is imposed.

Example In the previous example the demand and supply functions are given by

$$q^D(p) = 40 - 5p \quad \text{and} \quad q^S(p) = \frac{15}{2}p - 10,$$

and the equilibrium price is $p^* = 4$. Suppose that the government imposes an excise tax of T per unit. How does this affect the equilibrium price?

The answer is found by noting that, if the new selling price is p, then, from the supplier's viewpoint, *it is as if the price were* $p - T$, because the supplier's revenue per unit is not p, but $p - T$. In other words, the supply function has changed: when the tax is T per unit, the new supply function q^{S_T} is given by

$$q^{S_T}(p) = q^S(p - T) = \frac{15}{2}(p - T) - 10.$$

Of course the demand function remains the same. The new equilibrium values q^T and p^T satisfy the equations

$$q^T = 40 - 5p^T \quad \text{and} \quad q^T = q^{S_T}(p^T) = \frac{15}{2}(p^T - T) - 10.$$

Eliminating q^T we get

$$40 - 5p^T = \frac{15}{2}(p^T - T) - 10.$$

Rearranging this equation, we obtain

$$\left(5 + \frac{15}{2}\right)p^T = 50 + \frac{15}{2}T,$$

and so we have a new equilibrium price of

$$p^T = 4 + \frac{3}{5}T.$$

The corresponding new equilibrium quantity is

$$q^T = 40 - 5p^T = 20 - 3T.$$

For example, if $T = 1$, the equilibrium price rises from 4 to 4.6 and the equilibrium quantity falls from 20 to 17. Unsurprisingly, the selling price has risen and the quantity sold has fallen. But note that, although the tax is T per unit, the selling price has risen not by the full amount T, but by the fraction $3/5$ of T. In other words, not all of the tax is passed on to the consumer. ■

1.5 Comments

1. Economics tells us why the supply and demand sets ought to have certain properties. Mathematics tells us what we can deduce from those properties and how to do the calculations.

2. Mathematics also enables us to develop additional features of the model. In the case of supply and demand, we might ask questions such as the following:
   - What happens if conditions change, so that the supply and demand sets are altered slightly?
   - If the equilibrium is disturbed for some reason, what is the result?
   - How do the suppliers and consumers arrive at the equilibrium?

A typical instance of the first question is the excise tax discussed above. In this book we shall develop the mathematical techniques needed to deal with many other instances of these questions.

Worked Examples

Example 1.1 If $x - 2y = 3$ and $3x + 5y = 20$, what are x and y?

Solution

We eliminate y from the simultaneous equations. (An alternative, and equally valid, first step would be to eliminate x.) To do this, we multiply the first equation by 5 and the second by 2, obtaining the two equations

$$5x - 10y = 15,$$
$$6x + 10y = 40.$$

Adding these, we have

$$(5x - 10y) + (6x + 10y) = 15 + 40, \quad 11x = 55,$$

so that $x = 5$. Given this, we can use the first equation, $x - 2y = 3$, to determine y: $y = (x - 3)/2 = (5 - 3)/2 = 1$. Therefore, $x = 5$ and $y = 1$.

Example 1.2 Suppose that the supply and demand sets, S and D, for a particular market are described as follows: S consists of the pairs (q,p) such that $2p - 3q = 12$ and D consists of the pairs (q,p) such that $2p + q = 20$. Determine the supply function $q^S(p)$, the inverse supply function $p^S(q)$, the demand function $q^D(p)$ and the inverse demand function $p^D(q)$. Sketch S and D and determine the equilibrium set $E = S \cap D$. Comment briefly on the interpretation of the results.

Solution

The supply function is obtained by expressing quantity in terms of price for points in the supply set. We have $2p - 3q = 12$, and rearranging this gives $q = \frac{2}{3}p - 4$. Thus

$$q^S(p) = \frac{2}{3}p - 4.$$

Similarly,

$$q^D(p) = 20 - 2p.$$

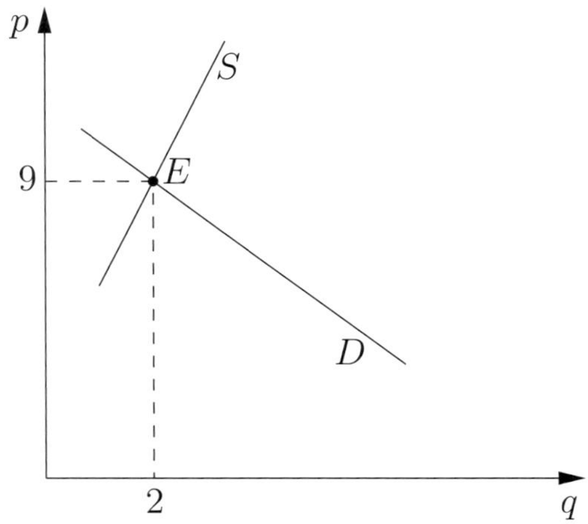

Figure 1.4 The demand set and the supply set for Example 1.2.

To obtain the inverse supply and inverse demand functions, we express p in terms of q on the supply and demand sets. Thus, the inverse supply function is

$$p^S(q) = 6 + \frac{3}{2}q$$

and the inverse demand function is

$$p^D(q) = 10 - \frac{1}{2}q.$$

The sets S and D are sketched in Figure 1.4.

A point (q^*, p^*) lies in $E = S \cap D$ if and only if the point satisfies both the equation of the supply set and the equation of the demand set. Thus, (q^*, p^*) must be such that

$$2p^* - 3q^* = 12$$

and

$$2p^* + q^* = 20.$$

Subtracting the first equation from the second gives $4q^* = 8$, or $q^* = 2$. Substituting this value in the first equation gives $2p^* - 6 = 12$, so that $p^* = 9$. Thus E consists of the single point $(2, 9)$ and when the market is in equilibrium, the selling price will be 9 and the quantity sold will be 2.

Example 1.3 Suppose that, in the market described in Example 1.2, an excise tax of 2 per unit is imposed. Determine the new equilibrium market price and quantity.

Solution

In the presence of an excise tax of 2, the effective price from the supplier's point of view is not the market price, but the market price less 2. Thus, if p^T is the new equilibrium market price when the tax is imposed, then the quantity supplied, q^T, satisfies both

$$q^T = q^S(p^T - 2) \quad \text{and} \quad q^T = q^D(p^T),$$

that is

$$q^T = \frac{2}{3}(p^T - 2) - 4 \quad \text{and} \quad q^T = 20 - 2p^T.$$

Eliminating q^T, we get

$$\frac{2}{3}(p^T - 2) - 4 = 20 - 2p^T,$$

giving $p^T = 19/2$ and $q^T = 20 - 2(19/2) = 1$.

Example 1.4 Suppose that the supply and demand functions for a good are

$$q^S(p) = bp - a \quad \text{and} \quad q^D(p) = c - dp,$$

where a, b, c, d are positive constants. Show that the equilibrium price is $p^* = (c + a)/(b + d)$. If an excise tax of T per unit is imposed ($T \neq 0$), find the resulting market price p^T and show that p^T is strictly less than $p^* + T$.

Solution

The equilibrium price p^* (in the absence of any tax) is found by solving the equations $q^* = bp^* - a = c - dp^*$, which give $p^* = (c + a)/(b + d)$.

When an excise tax of T per unit is imposed, the effective price from the supplier's point of view is $p^T - T$. The quantity sold, q^T, satisfies

$$q^T = b(p^T - T) - a \quad \text{and} \quad q^T = c - dp^T,$$

so that

$$p^T = \frac{c + a}{b + d} + \left(\frac{b}{b + d}\right)T = p^* + \left(\frac{b}{b + d}\right)T.$$

Thus the selling price rises by $(b/(b + d))T$ to its new equilibrium value. Since b and d are both positive, the fraction $b/(b + d)$ is strictly less than 1, and hence the increase is strictly less than T.

Note that we have verified mathematically a qualitative observation: *in cases where the supply and demand sets are described by straight lines with upward and downward slopes, respectively, not all of an excise tax is passed on to the consumer.* This is a case of what is often known as the *Tax Theorem*.

Main Topics

- interpretation of demand and supply sets
- demand, supply, inverse demand and inverse supply functions
- equilibrium price and quantity
- equilibrium price and quantity in the presence of excise tax

Key Terms, Notations and Formulae

- demand set, D
- supply set, S
- demand function, $q^D(p)$; inverse demand function, $p^D(q)$
- supply function, $q^S(p)$; inverse supply function, $p^S(q)$
- equilibrium set, $E = S \cap D$; equilibrium quantity, q^* and price, p^*
- excise tax, T; corresponding equilibrium quantity and price, q^T and p^T

Exercises

1.1 Suppose the supply and demand sets for Glenbowley single malt whisky are as follows: S consists of the pairs (q, p) for which $q - 3p = -5$ and D consists of the pairs (q, p) for which $q + 2p = 145$. Here, p is the price per bottle, measured in dollars, and q is the number of thousands of bottles sold. Determine the equilibrium price and quantity.

Suppose now that the government imposes an excise tax of $\$T$ per bottle. What will be the new selling price and quantity sold?

1.2 The inverse supply and demand functions for a good are given by

$$p^S(q) = 3q + 2 \quad \text{and} \quad p^D(q) = 6 - q,$$

respectively. Find the equilibrium price and quantity.

Suppose that the government decides to impose an excise tax of T on each unit of this good. What price will the consumers end up paying for each unit of the good and how much will be sold?

Find a formula for the amount of money the government obtains from taxing this good.

1.3 The demand and supply functions for a good are, respectively,

$$q^D(p) = 240 - 4p \quad \text{and} \quad q^S(p) = 8p.$$

Find the equilibrium price and quantity.

A *percentage* sales tax of $100r\%$ is introduced (where $0 < r < 1$). Find the new equilibrium price and quantity.

Problems

1.1 If $x + y = 3$ and $x - 2y = -3$, what are x and y?

1.2 Solve the following simultaneous equations:

$$\begin{aligned} 2x + y &= 9, \\ x - 3y &= 1. \end{aligned}$$

1.3 Suppose the market for a commodity is governed by supply and demand sets defined as follows. The supply set S is the set of pairs (q,p) for which $q - 6p = -12$ and the demand set D is the set of pairs (q,p) for which $q + 2p = 40$.

Sketch S and D and determine the equilibrium set $E = S \cap D$, the supply and demand functions q^S and q^D and the inverse supply and demand functions p^S and p^D.

1.4 Suppose that the government decides to impose an excise tax of T on each unit of the commodity discussed in Problem 1.3. What price will the consumers end up paying for each unit of the commodity?

1.5 Find a formula for the amount of money the government obtains from taxing the commodity in the manner described in Problem 1.4. Determine this amount explicitly when $T = 0.5$.

1.6 The supply and demand functions for a commodity are

$$q^S(p) = 12p - 4 \quad \text{and} \quad q^D(p) = 8 - 4p.$$

If an excise tax of T is imposed, what are the selling price and quantity sold, in equilibrium?

2 Mathematical Terms and Notations

2.1 Sets

Mathematics has its own terminology and notation, and therein lies much of its power. Because the notation is both clear and concise, it enables us to carry out calculations and make logical deductions which would be almost impossible without some form of shorthand.

The most basic notion is that of a *set*. This is the mathematical term for a collection of objects defined in a precise way, so that any given object is either in the set or not in the set. We usually denote sets by large letters, X, Y, S, D, and so on. The objects belonging to a set are enclosed in curly brackets, that is $\{\quad\}$. For example, we might define

$$X = \{2, 3, 5, 7, 8, 9\} \quad \text{and} \quad Y = \{1, 4, 5, 7, 9\}.$$

Here, the *members* or *elements* of X and Y are the numbers listed within the brackets. When x is an object in a set S, we write $x \in S$ and say 'x belongs to S' or 'x is a member of S'. Another way of specifying a set is by means of a property which its members must possess; for example

$$Z = \{n \mid n \text{ is a positive whole number less than } 5\}$$

is read as *Z is the set of n such that n is a positive whole number less than* 5. So this particular set could also be written as $\{1, 2, 3, 4\}$. (Some texts use a colon ':' in place of the symbol '$|$'.) The set which has no members is called the *empty set* and is denoted by $\varnothing$.

We say that the set U is a *subset* of the set V and write $U \subseteq V$ if every member of U is a member of V. In symbols, $U \subseteq V$ if and only if $x \in U$ implies $x \in V$. Given two sets A and B, we define the *union* $A \cup B$ to be the set whose members belong to A or B (or both A and B):

$$A \cup B = \{x \mid x \in A \text{ or } x \in B\}.$$

Similarly, we define the *intersection* $A \cap B$ to be the set whose members belong to both A and B:

$$A \cap B = \{x \mid x \in A \text{ and } x \in B\}.$$

Example Let X, Y, Z denote the sets defined above, that is

$$X = \{2,3,5,7,8,9\}, \quad Y = \{1,4,5,7,9\} \quad \text{and} \quad Z = \{1,2,3,4\}.$$

Then we have, for example

$$X \cup Y = \{1,2,3,4,5,7,8,9\}, \quad X \cap Y = \{5,7,9\},$$

$$Y \cup Z = \{1,2,3,4,5,7,9\}, \quad X \cap (Y \cap Z) = (X \cap Y) \cap Z = \varnothing.$$

We also have relationships such as $9 \in X \cap Y$ and $Z \subseteq X \cup Y$. ■

The most important sets considered in mathematics are sets of numbers. There are many extremely interesting questions about numbers, their meaning and their definition, but this is not the place to discuss them. For our purposes we simply have to accept that there is a set $\mathbb{R}$ of *real numbers* which has the properties we associate intuitively with the points on a line. A real number can be described by a decimal representation such as $5834.6234963\ldots$, which may or may not terminate. (The special typeface is used so that we can refer to $\mathbb{R}$ without further explanation: $\mathbb{R}$ always denotes the real numbers, but R can denote any set we please.)

There are several special sets of real numbers (that is, subsets of $\mathbb{R}$) for which it is convenient to have a fixed notation. The set of *non-negative* real numbers $\{x \mid x \geq 0\}$ is denoted by $\mathbb{R}_+$. The set $\{\ldots, -3, -2, -1, 0, 1, 2, 3, \ldots\}$ of *integers* is denoted by $\mathbb{Z}$. The positive integers are also known as *natural numbers*: $\mathbb{N} = \{1, 2, 3, \ldots\}$.

We use the notation $\mathbb{R}^2$ for the set of *ordered pairs* (x, y) of real numbers. Thus $\mathbb{R}^2$ is the set usually depicted as the set of points in a plane, x and y being the coordinates of a point with respect to a pair of axes. The subset $\mathbb{R}^2_+$ consisting of those points for which x and y are both non-negative is sometimes known as the *first quadrant*. This set is particularly relevant in economics because it represents realistic values of the coordinates; for example, in a 'supply and demand' diagram, neither q (quantity) nor p (price) can be negative.

2.2 Functions

Given two arbitrary sets A and B, a *function* from A to B is a rule which assigns one member of B to each member of A. For example, if A and B are both the set $\mathbb{R}$, the rule which says 'multiply by 2' is a function. Normally we express this function by a formula: if we call the function f, we can write the rule which defines f as $f(x) = 2x$. It is worth noting that when we define f in this way the x is a 'dummy variable'; the function could equally well be defined by writing $f(y) = 2y$, or $f(t) = 2t$, or even $f(blob) = 2 \times blob$. The point is that f is the function which doubles any value.

It is often helpful to think of a function as a 'black box' which converts an input into an output (Figure 2.1). If the name of the function is f, then $f(x)$ is the output corresponding to a given input x. The box may represent any rule, such as 'multiply

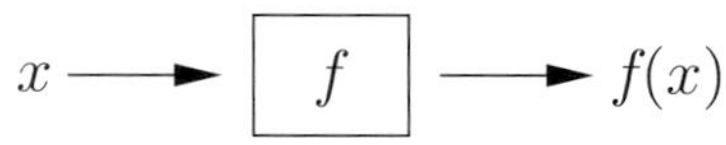

Figure 2.1 Diagrammatic representation of a function.

by 2', or 'add 23', or 'square', provided only that for each input there is a unique, well-defined output.

The diagram is useful because it stresses that a function is a one-way relationship, signified by the direction of the arrows. In general it may not be possible to reverse the arrows: that is, the output may not determine the input. For example, if the function f is the 'square' function, so that $f(x) = x^2$, and the 'output' of the box is 4, we do not know whether the input was 2 or -2, either of which would give the same value. If f is a function for which it *is* possible to reverse the arrows, then the resulting 'reverse' function is called the *inverse* function for f and is denoted by f^{-1}. So if $y = f(x)$, then $x = f^{-1}(y)$ (Figure 2.2).

Figure 2.2 A function and its inverse.

Example Suppose f is given by the formula

$$f(x) = 2x + 7,$$

so that when the input is x the output is $y = 2x + 7$. To obtain a formula for the inverse function, we have to find out what value of x will give a particular y, and we can do this simply by reorganising the equation $y = 2x + 7$ into the form $x = (y - 7)/2$. This uniquely determines x in terms of y, so here we have the inverse function f^{-1}, given by the formula

$$f^{-1}(y) = (y - 7)/2.$$

■

Note that if f and g are functions such that g is the inverse of f, then f is the inverse of g. For $y = f(x)$ if and only if $x = g(y)$, by the definition of $g = f^{-1}$. Here x and y are any values, and their names are irrelevant. So we may interchange the names, giving $y = g(x)$ if and only if $x = f(y)$, and this is just the condition that $f = g^{-1}$.

We can now see that the inverse supply and demand functions discussed in Chapter 1 are particular cases of this general notion: p^S is the inverse function for q^S and p^D is the inverse function for q^D.

2.3 Composite Functions

If f has an inverse function f^{-1}, one can think of f^{-1} as the 'undoing' of f, in the sense that if we take x and form $y = f(x)$ and then apply f^{-1} to y, we obtain x again. For example, consider again the previous example. Here $f(x) = 2x + 7$ and $f^{-1}(y) = (y - 7)/2$. For any x

$$f^{-1}(f(x)) = f^{-1}(2x+7) = \frac{(2x+7)-7}{2} = \frac{2x}{2} = x.$$

Thus, for any $x, f^{-1}(f(x)) = x$.

The observation just made concerns following the action of f by that of f^{-1}. The notion of applying one function directly to the output of another is an important one and can be made precise, as follows. If we are given two functions r and s, then we can apply them consecutively to obtain what is known as the *composite* function, given by the rule $k(x) = s(r(x))$. (See Figure 2.3.) The composite function k is denoted $k = sr$ and is often described in words as 'r followed by s' or as 's after r'. *Note the order: r is applied first, then s, but in the usual notation it comes out as sr.*

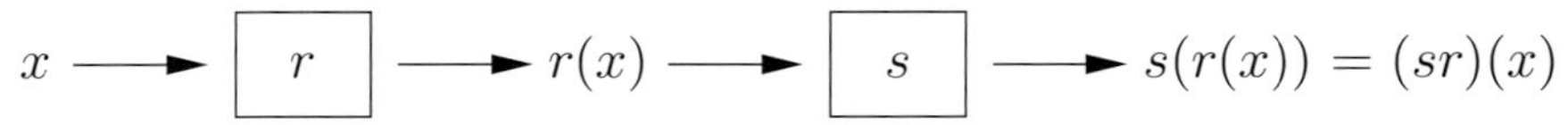

Figure 2.3 A composite function.

Example Suppose that $r(x) = x^2 + 2$ and $s(x) = x^3$. Then the composite function $k = sr$ is given by

$$k(x) = s\,(r(x)) = s(x^2+2) = \left(x^2+2\right)^3.$$

■

A composite function is sometimes called a 'function of a function'. For example, the function k described above can be thought of as the function 'cube' of the function 'square and add 2'. Often it is helpful to split up a given function in this way. Thus, given

$$f(x) = \sqrt{x^4 + x^2 + 5},$$

we can express f as the composite sr of simpler functions $s(x) = \sqrt{x}$ and $r(x) = x^4 + x^2 + 5$.

It is possible to form the composite of more than two functions, by repeating the definition given above. Given three functions, f, g, h, the composite function fgh is defined to be $f(gh)$. In other words, we first form h followed by g, and then form the composite of this function gh followed by f.

(The reader might ask why we don't define fgh as $(fg)h$. In fact, this will always give exactly the same function because here only the order of the operations is important. Technically, we say that composition of functions is an *associative* operation.)

Example Suppose that $f(x) = 1/x$, $g(x) = x^{3/2}$ and $h(x) = x^2 + 2x + 3$. Then

$$(gh)(x) = g(x^2+2x+3) = \left(x^2+2x+3\right)^{3/2}$$

and

$$(fgh)(x) = f\left((x^2+2x+3)^{3/2}\right) = \frac{1}{(x^2+2x+3)^{3/2}}.$$

■

Finally, we note that the function i such that for all x, $i(x) = x$ is known as the *identity function*. This trivial but important function may be compared with the number 0 in arithmetic: it does nothing, but our calculations would be more difficult without it. Thus, the observation that the inverse function f^{-1} is the 'undoing' of f may formally be expressed by the equation $f^{-1}f = i$. Similarly, the composite function ff^{-1} is also equal to i.

2.4 Graphs and Equations

Most of the mathematical techniques which are taught at an elementary level can be described in the 'sets and functions' terminology. For example, we might be asked to draw the 'graph' of $y = x^2 - 5x + 6$. This really means that we should illustrate the set of points

$$G = \{(x, y) \mid y = x^2 - 5x + 6\}$$

represented in the usual way as a subset of $\mathbb{R}^2$ (see Figure 2.4a).

A common basic problem is to 'solve' an equation or a system of equations. Perhaps the simplest case is the single *linear* equation

$$Ax + B = 0,$$

which we learn to solve in elementary algebra: $x = -(B/A)$, provided $A \neq 0$. In other words, the set $\{x \mid Ax + B = 0\}$ contains exactly one member $-(B/A)$. [*Exercise*: what happens if A is zero?]

Another, slightly harder, problem of elementary algebra is to find the set of solutions of a *quadratic* equation

$$ax^2 + bx + c = 0,$$

where we may as well assume that $a \neq 0$, because if $a = 0$ the equation reduces to a linear one. In some cases the quadratic expression can be factorised; for example, $x^2 - 5x + 6 = (x - 2)(x - 3)$, and so we can conclude that the set $\{x \mid x^2 - 5x + 6 = 0\}$ has two members, 2 and 3. Although factorisation may be difficult, there is a general technique for determining the solutions to a quadratic equation. Since $a \neq 0$, we may divide the equation $ax^2 + bx + c = 0$ by a to obtain

$$x^2 + \frac{b}{a}x + \frac{c}{a} = 0,$$

which will have the same solutions (if any) as the original equation. Using a technique known as 'completing the square', this equation may be rewritten as

$$\left(x + \frac{b}{2a}\right)^2 + \frac{c}{a} - \frac{b^2}{4a^2} = 0.$$

This follows from the identity

$$\left(x + \frac{b}{2a}\right)^2 = x^2 + 2x\left(\frac{b}{2a}\right) + \left(\frac{b}{2a}\right)^2 = x^2 + \frac{b}{a}x + \frac{b^2}{4a^2}.$$

The equation we now have to solve is

$$\left(x + \frac{b}{2a}\right)^2 = \frac{b^2}{4a^2} - \frac{c}{a} = \frac{b^2 - 4ac}{4a^2},$$

and from this we can obtain the complete picture. First, if $b^2 - 4ac < 0$ there are no solutions, or (as we might say) the solution set is empty. (Readers familiar with complex numbers will realise that by 'solution' we mean a real solution; that is, a solution which belongs to $\mathbb{R}$.) If $b^2 - 4ac = 0$, then the equation is

$$\left(x + \frac{b}{2a}\right)^2 = 0,$$

which has exactly one solution, namely, $x = -b/(2a)$. If $b^2 - 4ac > 0$, then the equation is

$$\left(x + \frac{b}{2a}\right)^2 = \frac{b^2 - 4ac}{4a^2},$$

so that there are two possibilities:

$$x + \frac{b}{2a} = \frac{\sqrt{b^2 - 4ac}}{2a} \quad \text{or} \quad x + \frac{b}{2a} = -\frac{\sqrt{b^2 - 4ac}}{2a}.$$

Thus, in this case, there are two solutions, x_1 and x_2, given by

$$x_1 = \frac{-b + \sqrt{b^2 - 4ac}}{2a} \quad \text{and} \quad x_2 = \frac{-b - \sqrt{b^2 - 4ac}}{2a}.$$

We often say that the solutions are

$$\frac{-b \pm \sqrt{b^2 - 4ac}}{2a},$$

where the symbol '$\pm$', meaning 'plus or minus', indicates that there are two choices: one where we choose the $+$ sign and another where we choose the $-$ sign.

To summarise, suppose we have the quadratic equation $ax^2 + bx + c = 0$, where $a \neq 0$. Then:

- if $b^2 - 4ac < 0$, the equation has no solutions;
- if $b^2 - 4ac = 0$, the equation has exactly one solution, $x = -b/(2a)$;
- if $b^2 - 4ac > 0$, the equation has two solutions:

$$x_1, x_2 = \frac{-b \pm \sqrt{b^2 - 4ac}}{2a}.$$

Often, the word 'root' is used to describe a solution to a quadratic equation and we may speak, for example, of a quadratic equation having two roots. The quantity $b^2 - 4ac$, which we now see is crucial, is known as the 'discriminant' of the quadratic equation.

Another kind of problem arises when we have several equations which we have to solve 'simultaneously'. This means that we must find the intersection of the solution sets of the individual equations. We have already met this situation in the determination of market equilibrium, where the solution of a pair of simultaneous linear equations

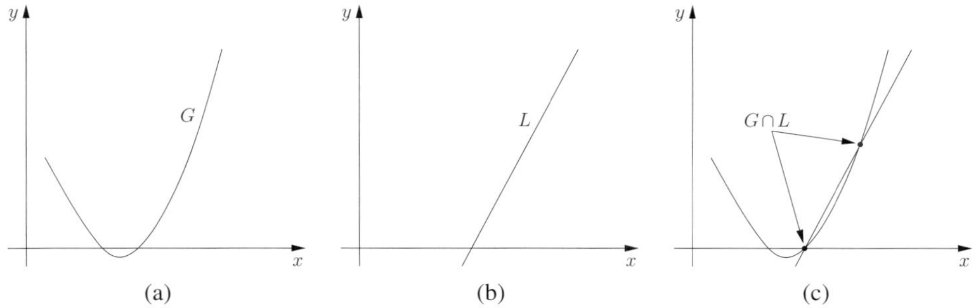

Figure 2.4 The sets G, L and $G \cap L$.

corresponds to finding the point of intersection of the supply set and the demand set. Here is an example in which the equations are not linear.

Example Suppose we wish to determine the set $G \cap L$, where

$$G = \{(x, y) \mid y = x^2 - 5x + 6\} \quad \text{and} \quad L = \{(x, y) \mid y = 2x - 6\}.$$

In other words, we require the points of intersection of the graph G (Figure 2.4a) with the line $y = 2x - 6$ (Figure 2.4b) or, what is the same thing, the solution of the simultaneous equations

$$y = x^2 - 5x + 6, \quad y = 2x - 6.$$

The x-coordinates of the points in $G \cap L$ can be found by eliminating y. This gives the equation $x^2 - 5x + 6 = 2x - 6$, that is $x^2 - 7x + 12 = 0$. Here the quadratic expression factorises as

$$x^2 - 7x + 12 = (x - 4)(x - 3)$$

and the solutions are $x_1 = 4$ and $x_2 = 3$. The corresponding y-coordinates are obtained by substituting in either of the original equations, giving $y_1 = 2$ and $y_2 = 0$. So $G \cap L$ consists of the two points $(4, 2)$ and $(3, 0)$, as shown in Figure 2.4c. ■

Worked Examples

Example 2.1 Suppose that the supply and demand sets for a good are given by

$$S = \{(q, p) \mid q - p = -7\} \quad \text{and} \quad D = \{(q, p) \mid q + 3p = 10\}.$$

Determine the supply function, q^S, the demand function, q^D, the inverse supply function, p^S, and the inverse demand function, p^D. Verify that, for any p and q, $(p^S q^S)(p) = p$ and $(p^D q^D)(p) = p$.

Solution

From the definition of the sets S and D, we get

$$q^S(p) = p - 7 \quad \text{and} \quad p^S(q) = q + 7,$$
$$q^D(p) = 10 - 3p \quad \text{and} \quad p^D(q) = \frac{1}{3}(10 - q).$$

In general, if f^{-1} is the inverse function for f then for all x, $(f^{-1}f)(x) = x$. We shall verify this explicitly for the cases in which f is q^S and q^D. We have

$$(p^S q^S)(p) = p^S\left(q^S(p)\right) = p^S(p - 7) = (p - 7) + 7 = p$$

and

$$(p^D q^D)(p) = p^D(10 - 3p) = \frac{1}{3}(10 - (10 - 3p)) = \frac{1}{3}(3p) = p,$$

as required.

Example 2.2 Suppose that the three functions f, g, h are given by

$$f(x) = x^3, \quad g(x) = 2x + 3 \quad \text{and} \quad h(x) = \frac{1}{x^2 + 1}.$$

Find formulae for the composite functions fg, gf, hfg.

Solution
We have

$$(fg)(x) = f(g(x)) = f(2x + 3) = (2x + 3)^3$$

and

$$(gf)(x) = g(f(x)) = g(x^3) = 2x^3 + 3.$$

Further, $(hfg)(x) = (h(fg))(x)$ and

$$(h(fg))(x) = h((fg)(x)) = h\left((2x + 3)^3\right) = \frac{1}{\left((2x + 3)^3\right)^2 + 1} = \frac{1}{(2x + 3)^6 + 1}.$$

Example 2.3 If f and g are the functions defined in Example 2.2, explain why f and g have inverse functions, and find formulae for them. Show also that fg has an inverse function and that $(fg)^{-1} = g^{-1}f^{-1}$.

Solution
According to the definition, $y = f(x)$ means that $y = x^3$. For each y, this has exactly one solution for x, namely $x = y^{1/3}$, the cube root of y. Hence f has an inverse function, and $f^{-1}(y) = y^{1/3}$. For g, we have $y = g(x)$ if and only if $y = 2x + 3$, which has the unique solution $x = (y - 3)/2$, so g has an inverse function, g^{-1}, given by $g^{-1}(y) = (y - 3)/2$.

Finally, $y = (fg)(x)$ means that $y = (2x + 3)^3$. So $(2x + 3) = y^{1/3}$ and $x = (y^{1/3} - 3)/2$. Therefore, the composite fg has an inverse function $(fg)^{-1}(y) = (y^{1/3} - 3)/2$. Also, the composite function $g^{-1}f^{-1}$ is given by the formula

$$(g^{-1}f^{-1})(y) = g^{-1}\left(f^{-1}(y)\right) = g^{-1}(y^{1/3}) = (y^{1/3} - 3)/2.$$

So we have, for all y, $(fg)^{-1}(y) = (g^{-1}f^{-1})(y)$, and therefore

$$(fg)^{-1} = g^{-1}f^{-1}.$$

[*Exercise*: explain why this is a general rule.]

Example 2.4 Let $Q = \{(x, y) \mid y = x^2 + 3x + 4\}$ and $L = \{(x, y) \mid y = x + 1\}$. Show that $Q \cap L = \varnothing$, and explain your answer graphically.

Solution

The intersection $Q \cap L$ is the set of $(x, y) \in \mathbb{R}^2$ which belong both to Q and to L. So, if (x, y) is in $Q \cap L$, then

$$y = x^2 + 3x + 4 \quad \text{and} \quad y = x + 1.$$

Eliminating y, we get $x^2 + 3x + 4 = x + 1$, which reduces to $x^2 + 2x + 3 = 0$. This is a quadratic equation in which '$b^2 - 4ac$' is equal to -8, which is negative, and so there is no solution. Since there are no possible values of x, there can be no points (x, y) in $Q \cap L$. Graphically, this means that the line L does not meet the curve Q.

Example 2.5 Assume that the catfood market is described by the supply and demand sets

$$S = \{(q, p) \mid 5p - q^2 - 2q = 27\} \quad \text{and} \quad D = \{(q, p) \mid p + q^2 + 2q = 15\}.$$

Write down p^S and p^D and sketch S and D. Determine the equilibrium set $E = S \cap D$ and comment on any interesting features.

Solution

Writing p in terms of q in the usual way, we find

$$p = p^S(q) = \frac{q^2 + 2q + 27}{5} \quad \text{and} \quad p^D(q) = -q^2 - 2q + 15.$$

The sets S and D are the graphs of p^S and p^D, respectively (remember, as always in economics, that the q-axis is the horizontal one). The graph of p^S does not cross the q-axis, since the equation $q^2 + 2q + 27 = 0$ has no solutions ('$b^2 - 4ac$' is $2^2 - 4 \times 1 \times 27$, which is negative). By plotting a few points we quickly arrive at the graph shown in Figure 2.5a. This is known as a *parabola*.

The graph of $p^D(q)$ is also a parabola, but in this case it is 'upside-down' (as is always the case when the coefficient of q^2 is negative). To find where it crosses the q-axis, we solve the equation

$$-q^2 - 2q + 15 = 0.$$

This factorises as $-(q - 3)(q + 5) = 0$, so that the solutions are $q = -5, 3$. The set D is sketched in Figure 2.5b.

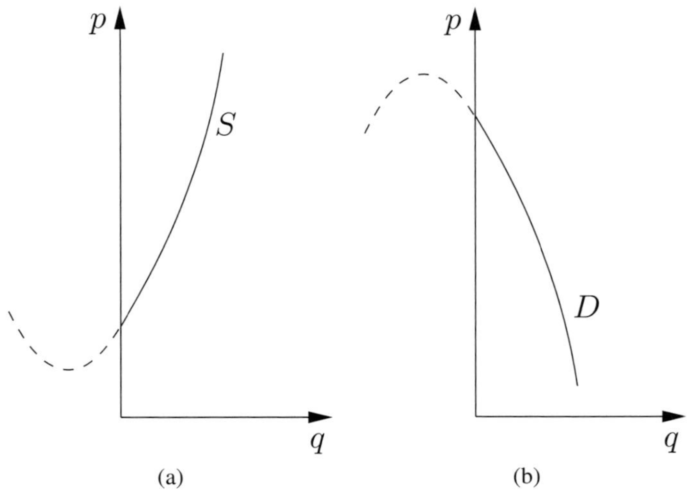

Figure 2.5 The sets S and D for the catfood market.

It appears from the graphs that the set $E = S \cap D$ consists of two points. To be sure, we must verify the result algebraically. We have $(q,p) \in E = S \cap D$ if and only if

$$\frac{1}{5}\left(q^2 + 2q + 27\right) = 15 - q^2 - 2q.$$

Rearranging,

$$q^2 + 2q + 27 - 75 + 5q^2 + 10q = 0, \text{ that is } 6q^2 + 12q - 48 = 0.$$

Dividing by 6, we get $q^2 + 2q - 8 = 0$, which factorises as $(q+4)(q-2) = 0$. The solutions are $q = -4$ and $q = 2$ and the corresponding values of p, given by $p = 15 - q^2 - 2q$, are both equal to 7. So we have $E = \{(-4,7),(2,7)\}$.

The first point has no economic significance because if q represents a quantity of a real commodity, then $q = -4$ is meaningless. The 'economic equilibrium' is therefore given by $q = 2$ and $p = 7$.

Main Topics

- sets and the different ways of describing them
- subsets, unions, intersections
- real numbers and integers and important subsets of these
- functions, inverse functions and composite functions
- graphs and the solutions of equations

Key Terms, Notations and Formulae

- two ways of describing sets: $X = \{\dots\}$ or $X = \{x \mid \dots\}$
- subset, $\subseteq$; union, $\cup$; intersection, $\cap$
- real numbers, $\mathbb{R}$; integers, $\mathbb{Z}$; natural numbers, $\mathbb{N}$; non-negative reals, $\mathbb{R}_+$
- ordered pairs of real numbers, $\mathbb{R}^2$; first quadrant, $\mathbb{R}^2_+$
- inverse function, f^{-1}; composite function, $k = sr$, $k(x) = s(r(x))$
- if $ax^2 + bx + c = 0$ and $b^2 - 4ac \geq 0$, then

$$x = \frac{-b \pm \sqrt{b^2 - 4ac}}{2a}$$

Exercises

2.1 Suppose $A = \{1, 2, 3, 5\}$, $B = \{2, 4, 5, 7, 8\}$ and $C = \{1, 7, 8, 11\}$. Find $A \cap B$ and $(A \cap B) \cup C$. Find also $A \cup C$, $B \cup C$ and $(A \cup C) \cap (B \cup C)$.

2.2 If $f(x) = 3x + 2$, find a formula for $f^{-1}(x)$.

2.3 If $f(x) = \sqrt{x}$ and $g(x) = x^2 + 1$, find a formula for the composition fg. Find also a formula for the composition gf.

2.4 Express the function

$$h(x) = \frac{1}{1 + x^2}$$

as a composition $h = fg$.

2.5 Sketch the curve $y = x^2 + 4x + 3$.

Problems

2.1 For each of the following functions f sketch the graph $y = f(x)$ in $\mathbb{R}^2_+$. Decide whether the inverse function f^{-1} exists, and if it does exist write down the formula for $f^{-1}(y)$.

$$f(x) = 5x, \quad f(x) = x^2 - 4x + 8, \quad f(x) = x^5.$$

2.2 Each of the following definitions specifies a subset K of $\mathbb{R}^2$. (In this question x is the name of the coordinate measured on the horizontal axis and y is the name of the coordinate measured on the vertical axis.)

$$\{(x, y) \mid 3x + 4y = 12\}, \quad \{(x, y) \mid 3x + 4y \leq 12\}, \quad \{(x, y) \mid x^2 + y^2 = 4\},$$

$$\{(x, y) \mid x^2 + y^2 \leq 4\}, \quad \{(x, y) \mid x^2 = 4y\}, \quad \{(x, y) \mid y^2 = 4x\}.$$

In each case sketch the set K and indicate on your sketch the set $K \cap \mathbb{R}^2_+$.

2.3 Suppose that the supply and demand sets for a particular market are

$$S = \{(q, p) \mid 3p - q = 5\} \quad \text{and} \quad D = \{(q, p) \mid 3p + q^2 + 2q = 9\}.$$

Sketch S and D and determine the equilibrium set $E = S \cap D$. Comment briefly on the interpretation of the results.

2.4 The functions f, g, h are given by $f(x) = x^2 + 1$, $g(x) = 1/x^2$ and $h(x) = \sqrt{x}$. Find formulae for the compositions fg, gf, hf, fh, hfg.

2.5 Suppose that the supply and demand sets for a good are given by

$$S = \{(q,p) \mid q - 3p = -1\} \quad \text{and} \quad D = \{(q,p) \mid q + p = 2\}.$$

Determine the supply function, q^S, the demand function, q^D, the inverse supply function, p^S, and the inverse demand function, p^D. Verify that for any p and q, $(p^S q^S)(p) = p$ and $(p^D q^D)(p) = p$.

2.6 For which values of α has the equation

$$x^2 + \alpha x + 1 = 0$$

no solutions, exactly one solution, or two solutions? Determine the solutions in the second and third cases.

3 Sequences, Recurrences, Limits

3.1 Sequences

In this chapter we shall think of a *sequence* of numbers $y_0, y_1, y_2, \ldots$ as a description of how a variable quantity y evolves with respect to time. The general term y_t represents the value of y at the end of the tth time period; for example, y_t might represent the level of unemployment, or the exchange rate for dollars and sterling, at the end of year t.

Often it is possible to specify an economic process by giving an equation which expresses y_t in terms of the previous values y_{t-1}, and so on. For example, if I possess no ties when I am born, and my only supplier of ties is my great-aunt who gives me two for every birthday, then my stock of ties (assuming no wastage!) is described by the equations

$$y_0 = 0, \quad y_t = y_{t-1} + 2 \ \ (t = 1, 2, 3, \ldots).$$

In this example it is very easy to find a *solution* – that is, an explicit formula for y_t. After t years I shall have $2t$ ties, and so $y_t = 2t$.

Generally, an equation which defines y_t in terms of y_{t-1}, y_{t-2}, and so on is known as a *recurrence equation*. You may also see the name 'difference equation', especially in economics books.

3.2 The First-Order Recurrence

A *first-order recurrence* is one in which y_t depends on y_{t-1} but no other previous values. When the relationship has the form

$$y_t = ay_{t-1} + b,$$

where a and b are given constants, we say that the recurrence is *linear*, and that it has *constant coefficients*.

Example Suppose we are given that the values of y_t for $t \geq 1$ satisfy the equation $y_t = 2y_{t-1} - 5$. This is a linear first-order recurrence, with constant coefficients. If we are also given the value of y_0, say $y_0 = 4$, then the entire sequence of values can be obtained by repeated substitution, as follows:

$$\begin{aligned} y_1 &= 2y_0 - 5 = 2 \times 4 - 5 = 3, \\ y_2 &= 2y_1 - 5 = 2 \times 3 - 5 = 1, \\ y_3 &= 2y_2 - 5 = 2 \times 1 - 5 = -3, \end{aligned}$$

and so on. ■

A systematic approach to the solution of first-order linear recurrences is to begin by finding a rather trivial kind of solution, one in which the value of y_t does not depend on t. That is, we try to find a constant y^* such that if $y_{t-1} = y^*$ then the recurrence equation ensures that y_t has the same value, y^*. This simply means that

$$y^* = ay^* + b,$$

which we can rearrange as follows:

$$(1 - a)y^* = b, \quad \text{that is} \quad y^* = b/(1 - a).$$

Thus, if every term y_t has this particular constant value y^*, the equation is satisfied. Because this solution does not vary with t, it is called a *time-independent* solution. Note that the algebra goes wrong if $a = 1$, because then $b/(1 - a)$ is not defined. We shall deal with this case in Section 3.4.

Example (continued) What is the time-independent solution of the recurrence equation $y_t = 2y_{t-1} - 5$?

Here we have $a = 2$ and $b = -5$, so the time-independent solution is

$$y^* = \frac{-5}{1 - 2} = 5.$$

In other words, if every y_t is equal to 5, the equation is satisfied. ■

In practice, we are usually faced with the problem of solving a linear recurrence with a given *initial condition*, which specifies the value of y_0. This makes sense, because if we know y_0 then we can work out y_1 using the equation $y_1 = ay_0 + b$, and once we know y_1 we can work out y_2 using the equation $y_2 = ay_1 + b$, and so on. However, this process will only produce the time-independent solution $y_t = y^* = b/(1 - a)$ if the given initial condition is $y_0 = y^*$. If we are given any other value of y_0, we shall get a different solution.

In order to find the most general form of solution, we use the trick of writing y_t as the sum of the constant y^* and another quantity z_t. Then substituting $y_t = y^* + z_t$ in the equation $y_t = ay_{t-1} + b$, we get

$$y^* + z_t = a(y^* + z_{t-1}) + b.$$

Since $y^* = ay^* + b$ this reduces to the simpler equation

$$z_t = az_{t-1}.$$

Our trick has reduced the equation to one whose solution we can spot straight away. At each step z_t is multiplied by the constant value a, so

$$z_t = az_{t-1} = a^2 z_{t-2} = a^3 z_{t-3} = \cdots = a^t z_0.$$

Here the initial value z_0 is given by $y_0 = y^* + z_0$, so $z_0 = y_0 - y^*$ and

$$y_t = y^* + z_t = y^* + (y_0 - y^*)a^t.$$

Observe that this formula gives a solution for any specified value of y_0. We call $y_t = y^* + (y_0 - y^*)a^t$ the *general solution* of our original equation.

Example (continued) What is the general solution of the recurrence equation $y_t = 2y_{t-1} - 5$?

We have already calculated that $y^* = 5$ in this case, so the general solution is

$$y_t = 5 + 2^t(y_0 - 5).$$

In particular, when $y_0 = 4$ the solution is $y_t = 5 - 2^t$, which agrees with the calculations made earlier. ■

Example When the new Republic of Pushovia was formed, there were initially 32 000 tonnes of grain in the state granary. Each year, half of the existing stock of grain was consumed and another 8000 tonnes of grain were produced. How many tonnes of grain did the state granary contain after 12 years?

We could solve this problem by calculating y_1, y_2, y_3, and so on up to y_{12}, but a better method is to find a general formula for y_t, the amount of grain in the granary after t years. This approach has the advantage that if we then asked for the contents of the granary after, say, 20 years, we need not work out y_{13}, y_{14}, and so on. We need only substitute $t = 20$ in the general formula for y_t. The initial condition is $y_0 = 32\,000$ and the recurrence equation for y_t when $t \geq 1$ is

$$y_t = 0.5y_{t-1} + 8000.$$

In this case $a = 0.5$ and $b = 8000$, so $y^* = 8000/(1 - 0.5) = 16\,000$. The general solution is

$$y_t = 16\,000 + (32\,000 - 16\,000)(0.5)^t = 16\,000(1 + (0.5)^t).$$

The behaviour of the sequence y_t depends on what happens to the term $(0.5)^t$, which is another way of writing $(1/2^t)$. The first few values of this term are

$$\frac{1}{2}, \frac{1}{4}, \frac{1}{8}, \frac{1}{16}, \frac{1}{32}, \ldots.$$

It is clear that these values approach zero as t increases. In particular, the value of $(0.5)^{12}$ is very close to zero, so the granary contains about 16 000 tonnes of grain after 12 years. (Actually, $(0.5)^{12}$ is 0.000244140625, so the granary contains exactly 16003.90625 tonnes of grain after 12 years.) ■

3.3 Limits

In the preceding example the value of y_t approaches y^* as t increases, because $(0.5)^t$ approaches zero. In the general case we have

$$y_t = y^* + (y_0 - y^*)a^t,$$

and the behaviour as t increases depends on what happens to a^t. For example, if $a = 3$, then the sequence of values of a^t is

$$3, 9, 27, 81, 243, 729, 2187, \ldots,$$

which clearly does not approach zero! Similarly, if $a = -2$, we get

$$-2, 4, -8, 16, -32, 64, -128, 256, \ldots,$$

which also does not approach zero.

In mathematics, the theory of *limits* deals with the kind of questions we encounter here. For our purposes it will suffice to rely on simple intuitive notions, but for more advanced work a proper understanding of the theory is needed. When a is greater than 1, as t increases, a^t will eventually become greater than any given number, and we say that a^t *tends to infinity as t tends to infinity*. We write this in symbols as

$$a^t \to \infty \text{ as } t \to \infty \qquad \text{or} \qquad \lim_{t\to\infty} a^t = \infty.$$

On the other hand, when $a < 1$ and $a > -1$, we have

$$a^t \to 0 \text{ as } t \to \infty \qquad \text{or} \qquad \lim_{t\to\infty} a^t = 0.$$

We notice that while a^t gets closer and closer to zero for all values of a in the range $-1 < a < 1$, its behaviour depends to some extent on whether a is positive or negative. When a is negative, the terms are alternately positive and negative, and we say that the approach to zero is *oscillatory*. For example, when $a = -0.2$, the sequence a^t is

$$-0.2, 0.04, -0.008, 0.0016, -0.00032, 0.000064, -0.0000128, 0.00000256, \ldots.$$

When a is less than -1, the sequence is again oscillatory, but it does not approach any limit, the terms being alternately large-positive and large-negative, as in the case $a = -2$ discussed above. In this case, we say that a^t *oscillates increasingly*.

Our original problem was to describe the behaviour of the general solution for y_t which, according to the formula, will depend on the behaviour of a^t. For example, if $a^t \to 0$, then the formula tells us that $y_t \to y^*$. In general, since the solution is $y_t = y^* + (y_0 - y^*)a^t$, its behaviour will also depend on the value of $y_0 - y^*$ (and, in particular, whether it is positive or negative). For instance, consider the case in which $a > 1$. If $y_0 - y^* > 0$, then y_t will tend to infinity, but if $y_0 - y^* < 0$, then y_t will tend to minus infinity. We can tabulate the results as follows. (See Figure 3.1, which illustrates the behaviours for various ranges of a, for the case in which $y_0 > y^*$.)

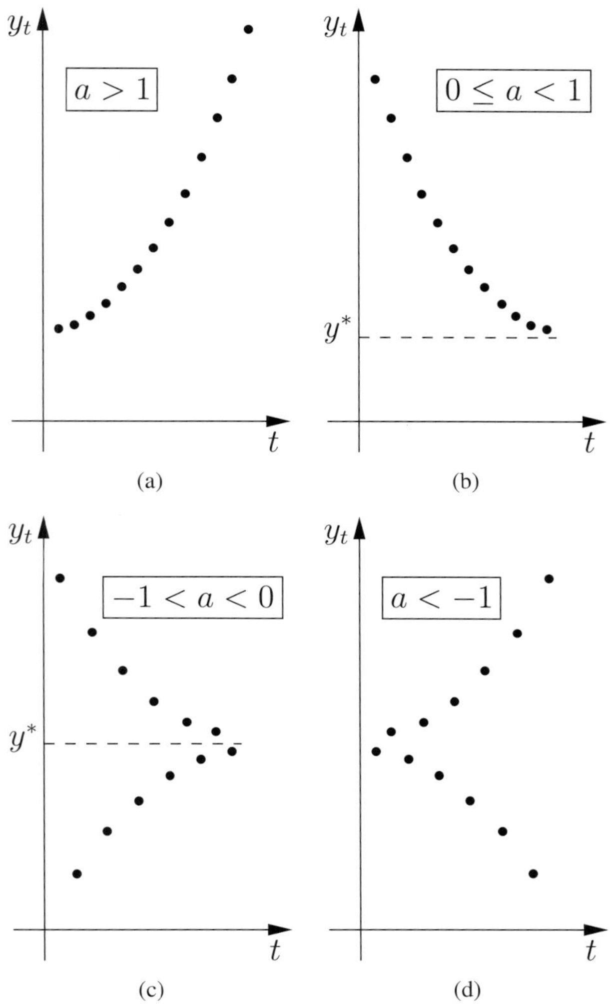

Figure 3.1 The behaviour of y_t (in the cases in which $y_0 > y^*$).

Value of a	*Behaviour of a^t*	*Behaviour of y_t*
$a > 1$	$a^t \to \infty$	$y_t \to \infty$ or $-\infty$
$1 > a \geq 0$	$a^t \to 0$ (decreasing)	$y_t \to y^*$
$0 > a > -1$	$a^t \to 0$ (oscillating)	$y_t \to y^*$
$-1 > a$	oscillates increasingly	oscillates increasingly

3.4 Special Cases

You will notice that the cases $a = 1$ and $a = -1$ do not appear in the table. The case $a = 1$ was specifically omitted at the start, and our method of dealing with first-order recurrence equations breaks down completely, because the expression $b/(1-a)$ for the time-independent solution is meaningless when $a = 1$. Fortunately this case can be dealt with directly. Note that when $a = 1$ the first-order linear recurrence is simply

$$y_t = y_{t-1} + b.$$

This says simply that each term is obtained by adding b to the previous one: $y_1 = y_0 + b$, $y_2 = y_1 + b = y_0 + 2b$, $y_3 = y_2 + b = y_0 + 3b$, and so on. Clearly, the general solution is $y_t = y_0 + tb$. So here $y_t \to \infty$ or $y_t \to -\infty$, depending on whether $b > 0$ or $b < 0$. (If $b = 0$, $y_t = y_0$ for all t.)

In the remaining case, $a = -1$, the sequence of values is y_0, $y_1 = -y_0 + b$, $y_2 = -y_1 + b = y_0, \dots$. Here the values are y_0 and $-y_0 + b$ alternately, and the behaviour is oscillatory. But since the oscillations have constant magnitude, the sequence does not tend to a limit.

Worked Examples

Example 3.1 Find the solution of the recurrence equation

$$y_t = 5y_{t-1} + 6,$$

given that $y_0 = 5/2$.

Solution

If we take $a = 5$ and $b = 6$ in the standard form $y_t = ay_{t-1} + b$ of the first-order recurrence, we have exactly the equation given. The first thing to do is to find the time-independent solution. By the formula, this is

$$y^* = \frac{b}{1-a} = \frac{6}{1-5} = -\frac{3}{2}.$$

We can now write down the general solution and insert the given value of y_0:

$$y_t = y^* + (y_0 - y^*)a^t = -\frac{3}{2} + 4(5^t).$$

It is not absolutely necessary to memorise the formulae, because the method is basically common sense. You need only remember what 'time-independent' means: y^* is the solution obtained by putting $y_t = y_{t-1} = y^*$, that is $y^* = 5y^* + 6$. This gives $y^* = -3/2$. Then we use the trick of putting $y_t = y^* + z_t$, giving

$$\left(-\frac{3}{2} + z_t\right) = 5\left(-\frac{3}{2} + z_{t-1}\right) + 6.$$

This reduces to $z_t = 5z_{t-1}$, which has the obvious solution $z_t = 5^t z_0$. Since $y_0 = 5/2$ and $y^* = -3/2$, it follows that $z_0 = 4$, and we get the result as above.

Example 3.2 Describe in words the behaviour of the following sequences as $t \to \infty$, and explain briefly your reasoning:

$$\text{(a)} \quad \frac{1}{4^t}, \qquad \text{(b)} \quad (1.001)^t.$$

Solution

(a) This sequence is decreasing and tends to zero as t tends to infinity, because $(1/4^t) = (1/4)^t$ and $0 < 1/4 < 1$.

(b) This sequence is increasing and tends to infinity as t tends to infinity, because $1.001 > 1$.

Example 3.3 Find the solution of the recurrence equation

$$3y_t = 2y_{t-1} + 10$$

when $y_0 = 25$, and describe its behaviour as $t \to \infty$. Calculate the least t for which y_t differs from the time-independent solution by less than 0.5.

Solution

The first thing to check is whether the method developed in this chapter can be applied directly. Observe that the given equation is *not* in the standard form $y_t = ay_{t-1} + b$, because there is a factor of 3 multiplying the y_t term. However, dividing both sides of the equation by 3 will not change the solution. We get

$$y_t = \frac{2}{3}y_{t-1} + \frac{10}{3},$$

which *is* in the standard form, with $a = 2/3$ and $b = 10/3$. We now proceed as before, calculating the time-independent solution as

$$y^* = \frac{b}{1-a} = \frac{10/3}{1-2/3} = 10.$$

We can now write down the general solution and insert the given value of y_0:

$$y_t = y^* + (y_0 - y^*)a^t = 10 + 15\left(\frac{2}{3}\right)^t.$$

Since $2/3$ is positive and less than 1, $(2/3)^t$ decreases and tends to zero as $t \to \infty$. Hence $y_t \to y^* = 10$ as $t \to \infty$.

To make y_t differ by less than 0.5 from y^*, we need $y_t - 10 < 0.5$, that is

$$15\left(\frac{2}{3}\right)^t < 0.5 \quad \text{or} \quad \left(\frac{2}{3}\right)^t < \frac{1}{30}.$$

A few trials on a calculator will reveal that the least such t is 9 (see also Section 7.4).

Example 3.4 The new Euro-commissioner for agriculture has inherited a surplus ‘grain mountain’ of 30 000 tonnes, held in a warehouse near Strasbourg. Each year 5% of the grain in the warehouse at the start of the year is eaten by Euro-mice. The commissioner is obliged to add N tonnes to the mountain each year, where N is to be fixed in advance by negotiation between governments. The commissioner is intent on setting a value of N for which the mountain will decrease in size, so that she can claim a great victory. Advise her. If $N = 1300$, how long will it take the grain mountain to shrink to 27 500 tonnes?

Solution

We first find a recurrence equation for y_t, the amount of grain (in tonnes) in the warehouse at the end of year t. We are told that 5% of the grain is eaten by Euro-mice each year, so that 95% of the grain at the beginning of any given year remains at the end. Since there are y_{t-1} tonnes of grain in the warehouse at the end of year $t-1$ (and hence at the beginning of year t), the amount of grain which was in store at the start of year t and which survives until the end of that year is 95% of y_{t-1}, or $0.95y_{t-1}$. During year t, N tonnes of new grain are added, and so the size of the grain mountain at the end of t years is given by

$$y_t = 0.95y_{t-1} + N.$$

We are given that there are initially 30 000 tonnes, so $y_0 = 30\,000$.

Thus we have a recurrence of the standard form, in which $a = 0.95$ and $b = N$ (assumed to be a fixed constant). Proceeding in the usual way, the time-independent solution is $y^* = b/(1-a) = N/(1-0.95) = 20N$, and the solution is

$$y_t = y^* + (y_0 - y^*)a^t = 20N + (30\,000 - 20N)(0.95)^t.$$

The term $20N$ is constant and does not vary with time. Thus, y_t will decrease if and only if $(30\,000 - 20N)(0.95)^t$ decreases with t. Since $(0.95)^t$ decreases, approaching zero, this will be true precisely when the constant factor $(30\,000 - 20N)$ is *positive*. Thus we require $30\,000 - 20N > 0$, or $N < 1500$. Since N is an integer, this condition becomes $N \leq 1499$. Therefore, if the Euro-commissioner manages to negotiate a value of N no larger than 1499, she will ensure victory.

The final part of the problem concerns what happens if the commissioner manages to negotiate a value of N as low as 1300. We first substitute the value $N = 1300$ into the general expression for y_t to obtain the solution in this case. This gives

$$y_t = 20(1300) + (30\,000 - 20(1300))(0.95)^t = 26\,000 + 4000(0.95)^t.$$

It follows that the condition $y_t \leq 27\,500$ is

$$26\,000 + 4000(0.95)^t \leq 27\,500,$$

which reduces to $(0.95)^t \leq 0.375$. Trial with a calculator (or the method to be described in Section 7.4) yields $t \geq 20$. Thus the target of 27 500 will be achieved in the 20th year.

Example 3.5 A closed economy produces an income Y_t in year t, of which a part C_t is consumed and the remainder I_t is invested; thus

$$Y_t = C_t + I_t.$$

It is believed that consumption C_t in year t is one half of the current year's income; that is, $C_t = \frac{1}{2}Y_t$. It is also believed that next year's income is proportional to the current investment; that is, $Y_{t+1} = kI_t$, where k is a constant.

Show that investment I_t satisfies a first-order recurrence equation, and find the condition that investment (and income) rises from year to year.

Solution

This problem is one where a seemingly complex set of relationships becomes very simple when expressed mathematically. We are given three facts relating Y_t, C_t, I_t:

$$Y_t = C_t + I_t, \quad C_t = \frac{1}{2}Y_t, \quad Y_{t+1} = kI_t.$$

Eliminating C_t from the first two equations, we get

$$Y_t = \frac{1}{2}Y_t + I_t, \quad \text{that is} \quad I_t = \frac{1}{2}Y_t.$$

Replacing t by $t-1$ in the third equation gives $Y_t = kI_{t-1}$. Thus

$$I_t = \frac{1}{2}Y_t = \frac{1}{2}(kI_{t-1}) = \frac{k}{2}I_{t-1}.$$

The equation $I_t = (k/2)I_{t-1}$ is plainly a first-order recurrence equation. The solution (which is easy enough to see directly here, or which follows by taking $b = 0$ and $a = k/2$ in the standard form) is

$$I_t = I_0(k/2)^t.$$

If this is to increase with t, then we must have $k/2 > 1$. Thus the condition for investment to increase is that $k > 2$.

Example 3.6 A market is modelled by the demand and supply functions

$$q^D(p) = 4 - p \quad \text{and} \quad q^S(p) = p.$$

Determine the equilibrium price and quantity.

Suppose that for some external reason the market is disturbed and the actual price is reduced to $p_0 = 3/4$, so that demand exceeds supply. Assume that over a given time period (say a month), the resulting change in price is proportional to the excess; that is

$$p_t - p_{t-1} = c(q^D(p_{t-1}) - q^S(p_{t-1})),$$

where c is a positive constant. Solve this equation and show that the price approaches the equilibrium value if and only if $c < 1$.

Describe carefully what happens when (i) $\frac{1}{2} < c < 1$ and (ii) $c = \frac{1}{2}$.

Solution

The equilibrium price p^* and equilibrium quantity q^* are obtained by solving the equations

$$q^* = 4 - p^*, \quad q^* = p^*.$$

We have $4 - p^* = p^*$, so $2p^* = 4$ and $p^* = 2$, giving $q^* = 2$.

Consider the equation $p_t - p_{t-1} = c(q^D(p_{t-1}) - q^S(p_{t-1}))$. Using the expressions for q^D and q^S, this becomes

$$p_t - p_{t-1} = c\left((4 - p_{t-1}) - p_{t-1}\right) = c\left(4 - 2p_{t-1}\right),$$

so that

$$p_t = (1 - 2c)p_{t-1} + 4c.$$

This is a linear first-order recurrence in standard form, $y_t = ay_{t-1} + b$, with p_t in place of y_t and $a = 1 - 2c, b = 4c$. The time-independent solution is

$$p^* = \frac{b}{1-a} = \frac{4c}{1-(1-2c)} = 2,$$

and the general solution when $p_0 = 3/4$ is

$$p_t = 2 + (p_0 - 2)(1 - 2c)^t = 2 - \frac{5}{4}(1 - 2c)^t.$$

The behaviour of p_t is determined by the behaviour of $(1 - 2c)^t$, everything else in this expression being independent of t. Clearly, p_t will tend towards the equilibrium price of 2 if and only if $(1 - 2c)^t \to 0$ as $t \to \infty$, which is the case precisely when $-1 < 1 - 2c < 1$. Remembering that c is given to be positive, the condition $1 - 2c < 1$ is automatically satisfied, and $-1 < 1 - 2c$ implies that $2c < 2$, or $c < 1$.

When $c > 1/2$ the term $1 - 2c$ is negative, and the price will oscillate: it will be alternately greater than 2 and less than 2. If also $c < 1$ then $1 - 2c > -1$ and the oscillations will decrease in magnitude and the price will tend to 2. When $c = 1/2$ we have $1 - 2c = 0$ and the equation for p_t becomes $p_t = 2$, so that the price remains at the equilibrium value for all $t \geq 1$.

Main Topics

- deriving and solving first-order recurrence equations
- the time-independent solution
- the limiting behaviour of a^t
- the limiting behaviour of the solution to a recurrence

Key Terms, Notations and Formulae

- standard first-order linear recurrence, $y_t = ay_{t-1} + b$
- time-independent solution, $y^* = b/(1 - a)$, $(a \neq 1)$
- initial conditions
- general solution, $y_t = y^* + (y_0 - y^*)a^t$
- notation for limits, e.g., $(0.5)^t \to 0$ as $t \to \infty$, or $\lim_{t\to\infty} (0.5)^t = 0$

Exercises

3.1 A sequence of numbers $x_0, x_1, x_2, \ldots$ is such that $x_0 = 4$ and, for each $n \geq 1$,

$$x_n = \frac{1}{3}x_{n-1} + 2.$$

Find an explicit expression for x_n in terms of n. Hence find the behaviour of x_n as $n \to \infty$.

3.2 The number of fish in a lake is currently 10 000. Each year, 5% of the fish who were alive at the start of the year die and 1000 new fish are born. Find an expression, in as simple a form as possible, for the number of fish in the lake N years from now. Describe what happens to the number of fish in the lake in the long run.

Problems

3.1 Find the solution to the following recurrence equation:

$$y_0 = 4, \quad y_t = \frac{1}{2}y_{t-1} + 5 \quad (t = 1, 2, 3, \ldots).$$

3.2 Describe in words the behaviour of the following sequences as $t \to \infty$:

$$\text{(a)} \quad 5 + (0.4)^t, \quad \text{(b)} \quad (-0.999)^t.$$

3.3 Find the solution of the recurrence $y_t = 4y_{t-1} + 3 \quad (t = 1, 2, 3, \ldots)$ given that $y_0 = 7$. Describe its behaviour as $t \to \infty$.

3.4 Find the time-independent solution of the recurrence equation

$$4y_t = y_{t-1} + 9 \quad (t = 1, 2, 3, \ldots).$$

Find the solution when $y_0 = 6$, and describe its behaviour as $t \to \infty$.

3.5 A company currently employs 4000 staff, working a total of 8 million working hours per year. It plans to raise its total employment (in working hours) by E thousand each year, where E is some constant. The company estimates that, each year, due to workers retiring, leaving or reducing their workloads, the total number of working hours employed by the current workforce decreases by W%. Let y_t be the total time worked by the company's employees, in thousands of hours, at the end of t years. Explain why

$$y_t = \left(1 - \frac{W}{100}\right) y_{t-1} + E,$$

for $t \geq 1$. The company plans to maintain at least its current employment in the future. Find an expression for y_t and deduce that E should satisfy $E \geq 80W$.

3.6 A market is modelled by the demand and supply functions

$$q^D(p) = 3 - p \quad \text{and} \quad q^S(p) = p.$$

What is the equilibrium price? The market is initially not in equilibrium and at any time $t \geq 1$ the current price p_t is related to the price in the previous period p_{t-1} by the equation

$$p_t = (1+r)p_{t-1} + k(q^D(p_{t-1}) - q^S(p_{t-1})),$$

where r, k are positive constants with $k > r$. (r may be thought of as a measure of the rate of inflation.) At time $t = 0$ the price is $p_0 = 1$. Solve this equation and show that over time, the price tends to a limiting value if and only if $k < 1 + r/2$.

Show that this limiting value is greater than $3/2$ and less than 3, and so is not equal to the equilibrium price.

4 The Elements of Finance

4.1 Interest and Capital Growth

In this chapter we shall look at one of the basic models used in financial economics, and work out its properties using the general theory of the first-order linear recurrence.

Suppose there is a fixed annual interest rate r available to investors. We shall express r as a number, rather than a percentage, so that what is commonly given as a rate of 8% per annum corresponds to $r = 0.08$. In this case, if we invest \$100 then after 1 year we shall have \$100 + \$8 = \$108. More generally, if we invest P then after 1 year we shall have $P + rP = (1 + r)P$. (From now on we omit the currency units, because they are irrelevant to the calculations.)

We can use this simple model to describe *capital growth*. Suppose we invest P for N years at a constant annual interest rate r. If we let y_t be the capital at the end of the tth year, we have $y_0 = P$ and the recurrence

$$y_t = (1 + r)y_{t-1} \quad (t = 1, 2, 3, \ldots).$$

This is in the standard form discussed in the last chapter, with $a = (1 + r)$ and $b = 0$. The solution is fairly obvious (even without any theory): it is $y_t = (1 + r)^t P$. In particular, the capital C after N years will be y_N, that is

$$C = (1 + r)^N P.$$

It is instructive to think of the relationship between P and C in the language of mathematical functions, as in Figure 4.1. The input is the *principal* P, the function is 'investment for N years at interest rate r' and the output is the capital C as in the formula above. In other words, C is determined as a function of P by the formula $C(P) = (1 + r)^N P$.

Now suppose we reverse the question. If we want to ensure that the final capital is C, what should be the principal P? Clearly, this is found by simply rearranging the equation so that we get P in terms of C, that is

$$P(C) = \frac{C}{(1 + r)^N}.$$

In mathematical language, we have found the inverse of the capital growth function (see Figure 4.1 again). In economics and finance this is known as calculating the *present value* of a capital sum C, due in N years given a fixed interest rate r. It is very important

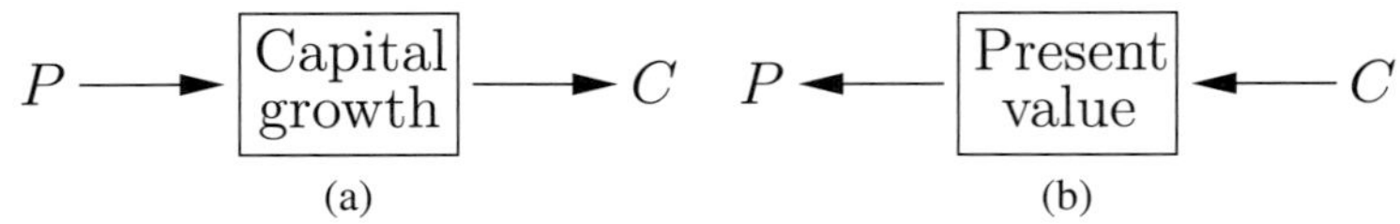

Figure 4.1 Relationships between principal and capital.

because it enables us to compare the guaranteed return from a fixed-rate investment with the return from other ways of using the principal.

Example If I am offered a gift of either \$6000 now or \$10 000 in 7 years time, which should I accept, given the fixed interest rate of 8%?

Here we have to calculate $P(C)$ when $C = 10\,000$, with the parameters $r = 0.08$ and $N = 7$. The formula gives

$$P = \frac{10\,000}{(1.08)^7},$$

and using the approximation $(1.08)^7 = 1.71382427$ we obtain the present value of \$5834.90. So I should accept the \$6000 now. If this is not clear, think of it in the following way. The fact that the present value of \$10 000, due 7 years from now under the prevailing interest rate, is less than \$6000 is equivalent to the statement that \$6000 invested *now* would generate more than \$10 000 after 7 years. In other words, if we are prepared to wait 7 years, then we are equally prepared to take the sum on offer now and place it in a bank account at the given interest rate, leaving it untouched for 7 years. The amount then generated is $6000(1.08)^7 = 10\,282.95$, which is more than 10 000. (Of course, you may decide not to follow this course of action if you have debts, or if you really want that new house, or car, or whatever, *now*! In other words, this analysis makes sense if you are planning to invest the money. If, instead, you are planning to spend it, you should take into account the usefulness of owning the items you purchase.) ■

4.2 Income Generation

As an alternative to capital growth, people often invest their money to provide a regular income, often known as an *annuity*. Suppose we invest P, and withdraw an amount I at the end of each year for N years, at which time the capital is used up. What income can be generated from the principal P, or (mathematically speaking) what is I as a function of P?

Here the recurrence equation is

$$y_t = (1 + r)y_{t-1} - I, \quad \text{where} \quad y_0 = P.$$

This is another case of the first-order linear recurrence, in standard form with $a = (1 + r)$ and $b = -I$. The time-independent solution is therefore $y^* = I/r$. The general solution is $y_t = y^* + (y_0 - y^*)a^t$, and since $y_0 = P$ we obtain

$$y_t = \frac{I}{r} + \left(P - \frac{I}{r}\right)(1+r)^t.$$

In order to determine I as a function of P, we must use the condition that nothing is left after N years, that is $y_N = 0$. This condition is

$$\frac{I}{r} + \left(P - \frac{I}{r}\right)(1+r)^N = 0,$$

and rearranging we get

$$\frac{I}{r}\left((1+r)^N - 1\right) = P(1+r)^N,$$

so that

$$I(P) = \left(\frac{r(1+r)^N}{(1+r)^N - 1}\right)P.$$

Here too it is natural to consider the inverse function. The question here is: what principal P is required to provide an annual income I for the next N years, or what is the inverse function $P(I)$? Rearranging the equation gives the result

$$P(I) = \frac{I}{r}\left(1 - \frac{1}{(1+r)^N}\right).$$

This tells us the present value of an annuity generating I each year guaranteed for the next N years.

Example What is the present value of an annuity generating \$10 000 a year for the next 7 years, given the fixed interest rate of 8%?

The formula gives

$$P = \frac{10\,000}{0.08}\left(1 - \frac{1}{(1.08)^7}\right).$$

Using the approximation $(1.08)^7 = 1.71382427$, we obtain the answer

$$P = \frac{10\,000}{0.08}\left(1 - \frac{1}{1.71382427}\right) = 52\,063.70.$$

Equivalently, \$52 063.70 invested at 8% will provide \$10 000 per year for the next 7 years. ■

4.3 The Interval of Compounding

In the foregoing discussion we have taken r to be the annual rate of interest, and assumed that the interest is added as a single lump sum at the end of each year. It is instructive to ask what happens if interest is added more frequently. Suppose, for example, that 4% interest is added twice yearly, once at the middle of the year and once at the end. (We say that the 'equivalent annual rate' is 8%, although, as we shall see, such an arrangement does not have the same effect as a single payment of 8%.) If the principal is \$100, the capital after 1 year will be

$$100(1 + 0.04)^2 = 108.16,$$

which is slightly more than the \$108 which results from the single annual addition. If the interest is added quarterly, the capital after 1 year will be

$$100(1 + 0.02)^4 = 108.24$$

approximately. In general, when the year is divided into m equal periods, the equivalent rate is r/m over each period, and the capital after 1 year is

$$100\left(1 + \frac{r}{m}\right)^m.$$

Our numerical experiments indicate that this quantity increases steadily as the number m of compounding intervals increases. [*Exercise*: work out the answer if the interest is added daily, that is $m = 365$.] An obvious question, which we shall answer in Chapter 7, is what happens as $m \to \infty$?

Worked Examples

Example 4.1 Write down explicit formulae for:

(a) the final value $C(P)$ of an amount P invested at 5% annual interest for 10 years, if there is no annual withdrawal;

(b) the constant annual income $I(P)$ generated by an amount P invested at 5% annual interest over 10 years, if there is no final capital.

[Use the approximation $(1.05)^{10} = 1.629$.]

Write down the inverse functions $P(C)$ and $P(I)$ and give their interpretation in terms of present value.

Solution

For (a), we use the fact that a principal invested at constant annual rate r for t years will grow to $P(1 + r)^t$. Here, the interest rate is 5%, so $r = 0.05$. The answer is therefore $C(P) = P(1 + 0.05)^{10}$, or $C(P) = 1.629P$ approximately.

For part (b), we use the formula given in Section 4.2, with $r = 0.05$ and $N = 10$:

$$I(P) = \left(\frac{r(1 + r)^N}{(1 + r)^N - 1}\right) P = \left(\frac{0.05(1.05)^{10}}{(1.05)^{10} - 1}\right) P$$

$$= \left(\frac{(0.05)(1.629)}{0.629}\right) P = 0.1295P.$$

To determine the inverse function $P(C)$, note that $C = 1.629P$ is equivalent to $P = C/1.629 = 0.6139C$, so that $P(C) = 0.6139C$. This is the amount of money which will provide the capital sum C after 10 years at a constant interest rate of 5%;

in other words, the present value of C due 10 years from now, given a fixed 5% interest rate.

To determine the inverse function $P(I)$, we note that $I = 0.1295P$ is equivalent to $P = I/0.1295 = 7.722I$. Thus, $P(I) = 7.722I$. This is the principal which will generate a yearly income I for the next 10 years; that is, the present value of an annuity of I for the next 10 years, given a fixed 5% interest rate.

Example 4.2 Suppose that you have won a competition in a national newspaper and you can choose either to receive a lump sum of \$100 000 now, or a payment of \$20 000 at the end of each year for the next 7 years. Which prize should you choose, assuming that the highest interest rate you can obtain is a constant 7% over the 7-year period?

Solution

We calculate the present value of the stream of income (that is, the annuity) and compare this with \$100 000. Using the formula with $r = 0.07, N = 7, I = 20\,000$, we see that the present value of 20 000 at the end of each year for the next 7 years is

$$P = \frac{I}{r}\left(1 - \frac{1}{(1+r)^N}\right) = \frac{20\,000}{0.07}\left(1 - \frac{1}{(1.07)^7}\right) = 107\,785.79.$$

Since this is greater than 100 000, if you act rationally you should accept the annuity rather than the cash sum. (Of course, as mentioned in the example of Section 4.1, this analysis only makes sense if you would be investing the lump sum, were you to accept it.)

Example 4.3 An amount of \$1000 is invested and attracts interest at a rate equivalent to 10% per annum. Find the total after 1 year if the interest is compounded (a) annually, (b) quarterly, (c) monthly, (d) daily. (Assume the year is not a leap year.)

Solution

We use the fact that if the interest is paid in m equally spaced instalments, then the total after 1 year is $1000\left(1 + \frac{r}{m}\right)^m$, where $r = 0.1$ and $m = 1, 4, 12, 365$ in the four cases. Therefore the answers are as follows:

(a) $1000\,(1 + 0.1) = 1100.$

(b) $1000\left(1 + \dfrac{0.1}{4}\right)^4 = 1103.81.$

(c) $1000\left(1 + \dfrac{0.1}{12}\right)^{12} = 1104.71.$

(d) $1000\left(1 + \dfrac{0.1}{365}\right)^{365} = 1105.16.$

Example 4.4 Show that the present value of an annuity of I for N years, given the fixed interest rate r, is

$$P = \frac{I}{1+r} + \frac{I}{(1+r)^2} + \frac{I}{(1+r)^3} + \cdots + \frac{I}{(1+r)^N}.$$

Use the formula $(1-x)(x+x^2+x^3+\cdots+x^N) = x - x^{N+1}$ to show that this yields the same expression for $P(I)$ as that given in Section 4.2.

Solution

To help explain the general approach, imagine that you were offered \$1000, 1 year from now, *together with* \$1000, 2 years from now, and that the interest rate was constant at 8%. How would you determine the present value of this annuity? One way is to observe that the present value of the first \$1000 is $1000/1.08$ (since it is due 1 year from now), the present value of the second \$1000 (due 2 years from now) is $1000/(1.08)^2$, and so the present value of the annuity must be the sum of these two present values; that is $1000/1.08 + 1000/(1.08)^2$.

Consider now the general case, where the annual income is I and the interest rate is r. The present value of the first payment is $I/(1+r)$, the present value of the second payment is $I/(1+r)^2$ and, generally, the present value of the ith payment is $I/(1+r)^i$. It follows that the present value of the annuity, which is the sum of the present values for $i = 1, 2, \ldots, N$, is as stated above.

Using the formula with $x = 1/(1+r)$, we have

$$\frac{1}{1+r} + \frac{1}{(1+r)^2} + \frac{1}{(1+r)^3} + \cdots + \frac{1}{(1+r)^N} = \frac{\frac{1}{1+r} - \frac{1}{(1+r)^{N+1}}}{1 - \frac{1}{1+r}}$$

$$= \frac{1}{r}\left(1 - \frac{1}{(1+r)^N}\right).$$

Multiplying by I, we obtain the required expression for $P(I)$.

Main Topics

- interest, capital growth and the meaning of present value
- recurrence equations in annuity problems
- the interval of compounding

Key Terms, Notations and Formulae

- principal, P; capital, C
- with annual compounding, $C = (1+r)^N P$
- present value with annual compounding, $P(C) = \dfrac{C}{(1+r)^N}$
- annuity generates income, $I(P) = \left(\dfrac{r(1+r)^N}{(1+r)^N - 1}\right) P$
- present value of annuity, $P(I) = \dfrac{I}{r}\left(1 - \dfrac{1}{(1+r)^N}\right)$

Exercises

4.1 Sarah saves money in a bank account paying interest at a fixed rate of 4%, where the interest is paid once per year, at the end of the year. She deposits an amount A at the beginning of each of the next N years. Find an expression, in terms of A and N, and in as simple a form as possible, for the final amount saved (the amount just after the last deposit).

4.2 Martin takes out a loan of \$10 000 at an interest rate of 2% per month. He has to make monthly repayments of \$$D$ at the beginning of each subsequent month for the next 25 months in order to repay the loan.

Explain why he owes $10\,000(1.02) - D$ after his first repayment.

How much does he owe after n repayments? (Simplify your answer as far as possible.)

Given that 1.02^{25} is approximately 1.64, show that D is approximately 512.50.

4.3 An ex-student donates a sum of money to his former school in order to provide an annual prize of \$$P$, to be awarded N times, starting 1 year from the date when he makes the donation. The school plans to invest the money in a bank account paying interest annually at a rate of 5%. Find an expression, in as simple a form as possible, for the minimum donation that the ex-student will have to make to the school. Justify your answer fully, showing all steps in your reasoning and calculations.

Problems

4.1 Suppose that a savings account pays interest annually at a rate of 5%. An investor deposits an amount \$$P$ which is large enough to ensure that, each year for the next 15 years, she can withdraw \$1000 from the account at the end of the year, maintaining a non-negative balance. Let y_t be the amount of money in the account after t years, so that $y_0 = P$. Explain why

$$y_t = 1.05\, y_{t-1} - 1000,$$

for $t = 1, 2, 3, \ldots$. Find an expression for y_t and hence show that

$$P \geq 20\,000 \left(1 - \frac{1}{(1.05)^{15}}\right).$$

4.2 Imagine you have \$200 000 to invest, at a constant rate of 9%, and that you want to withdraw a fixed amount I at the end of each year for the next 20 years. What is the maximum possible value of I for which this is possible?

Answer the same question if the money is withdrawn at the *beginning* of each of the next 20 years.

4.3 How much should you invest now in a bank account where the interest rate is a constant 7%, in order to be able to withdraw \$1000 at the end of each of the next 30 years?

4.4 Suppose you have won a competition and that you are given the choice between \$180 000 now or \$10 000 at the start of each year, for the rest of your life. Assume that the bank has a constant interest rate of 6% and that you currently have no debts (so that your decision is a purely rational one, based on 'present value'). Which option should you choose if you think you will live (a) until 65, (b) until 100, (c) forever? (Ignore (a) if you are over 65 and ignore (a) and (b) if you are over 100!)

4.5 An amount of \$2000 is invested and attracts interest at a rate equivalent to 8% per annum. Find the total after 1 year if the interest is compounded (a) annually, (b) quarterly, (c) daily. (Assume the year is not a leap year.)

5 The Cobweb Model

5.1 How Stable Is Market Equilibrium?

In this chapter we shall discuss the stability of the market for a single good. The questions we address concern the way in which the market operates when there is a time lag in the actions of the suppliers, and the way in which the market reacts when external factors affect the supply.

We consider an agricultural product for which there is a yearly 'crop', and for which the supply and demand sets take the typical form shown in Figure 5.1. If there are no disturbances, the equilibrium price p^* and quantity q^* will be as shown in the diagram. Suppose that one year, for some external reason such as drought, there is a shortage, so that the quantity falls and the price rises to p_0. During the winter the farmers plan their production for the next year on the basis of this higher price, and so an increased quantity appears on the market in the next year; specifically, $q_1 = q^S(p_0)$. Because the quantity is greater, the price consumers pay falls, to the value $p_1 = p^D(q_1)$. Overall, the effect of the disturbance on the price is that it goes from p_0, which is greater than p^*, to p_1, which is less than p^*.

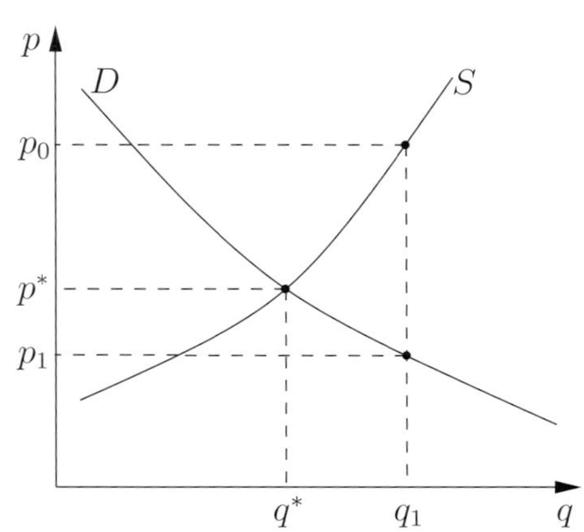

Figure 5.1 The effect of a disturbance, over a 1-year cycle.

But we are not finished. The process is repeated again in the following year: this time the lower price p_1 leads to a decrease in production q_2 and that in turn means a higher price p_2. The next year a similar process takes place, and so on. When the

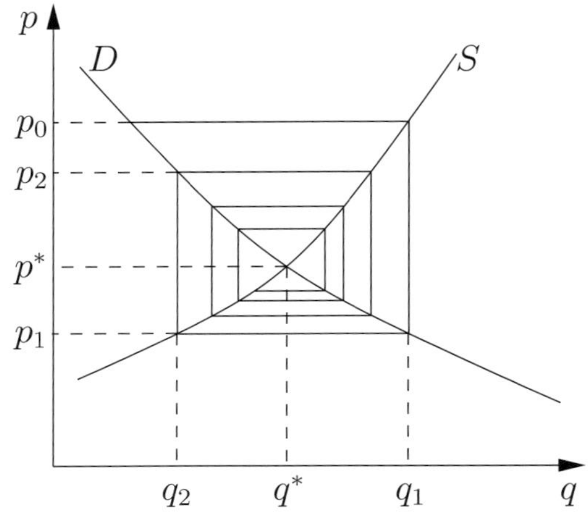

Figure 5.2 A cobweb.

sequences $p_0, p_1, p_2, \ldots$ and $q_1, q_2, \ldots$ are plotted on the supply and demand diagram, we get a picture like Figure 5.2. This is the reason for the name 'cobweb model'.

5.2 An Example

It appears from the diagram that the sequence p_t, $t = 0, 1, 2, \ldots$ oscillates, the terms being alternately greater and less than the equilibrium price p^*, and the values approach p^* as the years go by. In general, the sequence of prices is determined by the initial price p_0, the supply function q^S and the inverse demand function p^D. Generalising the argument above, we see that p_{t-1} determines q_t, which in turn determines p_t, according to the rules

$$q_t = q^S(p_{t-1}) \quad \text{and} \quad p_t = p^D(q_t).$$

In any given case we can use these equations to obtain a recurrence equation for p_t.

Example Suppose that the demand and supply sets are

$$D = \{(q, p) \mid q + p = 24\} \quad \text{and} \quad S = \{(q, p) \mid 2q + 18 = p\}.$$

Then the equilibrium quantity and price are $q^* = 2$ and $p^* = 22$. The relevant functions are

$$q^S(p) = 0.5p - 9 \quad \text{and} \quad p^D(q) = 24 - q,$$

so the equations linking p_{t-1}, q_t and p_t are thus

$$q_t = 0.5p_{t-1} - 9 \quad \text{and} \quad p_t = 24 - q_t.$$

Eliminating q_t, we obtain a first-order linear recurrence for p_t:

$$p_t = 33 - 0.5p_{t-1}.$$

This is in the standard form $y_t = ay_{t-1} + b$, with p_t replacing y_t and $a = -0.5, b = 33$. The time-independent solution is $b/(1-a) = 33/(3/2) = 22$, and the explicit solution in terms of p_0 is

$$p_t = 22 + (p_0 - 22)(-0.5)^t.$$

Not surprisingly, the time-independent solution is the equilibrium price $p^* = 22$, and in this case the sequence approaches p^* in an oscillatory way, as suggested by the discussion in Section 5.1. For example, if we start with a price higher than the equilibrium price, such as $p_0 = 23$, then we have

$$p_1 = 21.5, \quad p_2 = 22.25, \quad p_3 = 21.875, \quad \text{and so on.}$$

■

5.3 The General Linear Case

We now look at more general features of the model. We know that p_t is determined by q_t, which is determined by p_{t-1}, according to the supply and demand equations given above. In fact, we can bypass q_t by putting the two equations together; this gives p_t as a 'function of a function' of p_{t-1}:

$$p_t = p^D(q^S(p_{t-1})).$$

However, all this is a bit abstract, and so we shall concentrate on the case where S and D are straight lines. In general, we may take

$$S = \{(q,p) \mid q = bp - a\} \quad \text{and} \quad D = \{(q,p) \mid q = c - dp\}.$$

The reason for writing the equation in this form is that when a, b, c, d are positive, the sets S and D take the 'economically respectable' form shown in Figure 5.3.

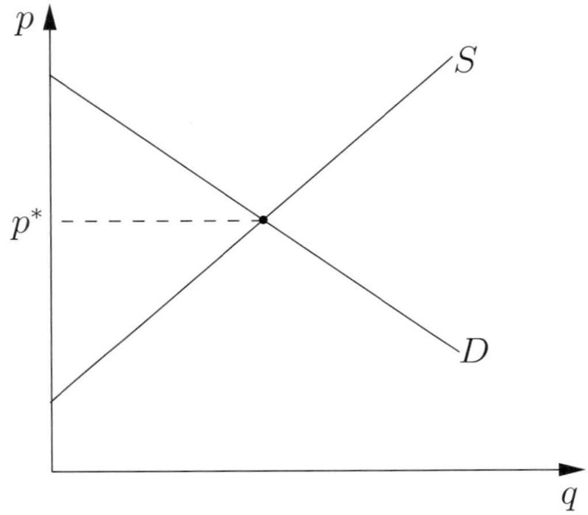

Figure 5.3 The linear case.

We shall need the explicit formulae for the functions q^S and p^D, which are

$$q^S(p) = bp - a \quad \text{and} \quad p^D(q) = \frac{c-q}{d}.$$

Following the method used in the example, we obtain

$$p_t = \frac{c - q_t}{d} = \frac{c - (bp_{t-1} - a)}{d},$$

which simplifies to

$$p_t = \left(-\frac{b}{d}\right) p_{t-1} + \left(\frac{c+a}{d}\right).$$

This is just another linear first-order recurrence. To solve it, we first find the time-independent solution

$$p^* = \frac{c+a}{b+d}.$$

We observe that this is the equilibrium value of the price, in other words, the value of p where the lines $q = bp - a$ and $q = c - dp$ intersect! This is not really surprising, since, if the process begins at the equilibrium price $p_0 = p^*$, we would expect all subsequent values of p_t to remain at p^* also.

What happens if $p_0 \neq p^*$? In this case we need to look at the general solution to the recurrence which, according to the general theory, is

$$p_t = p^* + (p_0 - p^*)\left(-\frac{b}{d}\right)^t.$$

At this point it may help to look at another example.

Example Suppose the supply and demand sets are given by the equations $q = p - 2$ and $q = 7 - 2p$, respectively. Then we have $b = 1, d = 2$ in the above notation, and the recurrence is

$$p_t = -0.5p_{t-1} + 4.5.$$

The equilibrium price p^* is $(7 + 2)/(1 + 2) = 3$, and the general solution is

$$p_t = 3 + (p_0 - 3)(-0.5)^t.$$

Suppose that we start with a high price, such as $p_0 = 3.5$. Then we find that $p_1 = 2.75$, $p_2 = 3.125$, $p_3 = 2.9375$, and so on. The prices oscillate around the equilibrium value $p^* = 3$, and because $(-0.5)^t$ approaches zero as $t \to \infty$, the magnitude of the oscillations decreases and $p_t \to p^*$ as $t \to \infty$. We say that the market equilibrium is *stable* in this case. This means that, if the process begins at a non-equilibrium price, then it will nevertheless approach the equilibrium price in the course of time. ■

In the example, the reason that $p_t \to p^*$ as $t \to \infty$ is that -0.5 is strictly between -1 and 1, and so $(-0.5)^t \to 0$ as $t \to \infty$. In the general case, -0.5 is replaced by the appropriate value of $-b/d$, and so it is this which determines whether or not the equilibrium is stable.

Recall that b and d are both positive, so the criterion is simply whether or not b is less than d. If it is, then $-b/d$ lies between -1 and 0, and $(-b/d)^t \to 0$. On the other hand, if b is greater than d, then $-b/d < -1$ and $(-b/d)^t$ oscillates with increasing magnitude. In this case the equilibrium is *unstable*: if the initial price is different from p^* then, however small the difference, the price will not approach p^*.

5.4 Economic Interpretation

We can get still more insight into the economic interpretation of the cobweb model by noticing that b and d measure the slopes of the S and D sets, respectively. So the stability condition $b < d$ simply means that the S line is 'steeper' than the D line. In plain language, if a change in the quantity affects the suppliers' price more than the consumers' price, then the equilibrium will be stable. (*Warning:* in books which take the p-axis to be the horizontal one, the final conclusion is the same, but the argument about the steepness of the lines is reversed.)

Of course, we might wish to look at the question of stability when the S and D sets are not straight lines. Clearly that will require a mathematical technique which enables us to define the slope of non-linear functions. This technique, the differential calculus, will be introduced in the next chapter.

Worked Examples

Example 5.1 Consider a market in which the supply and demand sets are

$$S = \{(q,p) \mid q = 3p - 7\} \quad \text{and} \quad D = \{(q,p) \mid q = 38 - 12p\}.$$

Write down the recurrence equations which determine the sequence p_t of prices, assuming that the suppliers operate according to the cobweb model. Find the explicit solution given that $p_0 = 4$, and describe in words how the sequence p_t behaves. Write down a formula for q_t, the quantity on the market in year t.

Solution

Here, $q^S(p) = 3p - 7$ and $p^D(q) = \frac{1}{12}(38 - q)$. The equations determining p_t are

$$q_t = 3p_{t-1} - 7 \quad \text{and} \quad p_t = \frac{38 - q_t}{12},$$

which, on eliminating q_t, lead to the single recurrence

$$p_t = \frac{38 - (3p_{t-1} - 7)}{12} = 3.75 - 0.25p_{t-1}.$$

The time-independent solution to this equation is

$$p^* = \frac{3.75}{1.25} = 3$$

and the explicit solution is

$$p_t = p^* + (p_0 - p^*)(-0.25)^t = 3 + (-0.25)^t .$$

The behaviour of p_t is determined by the behaviour of the term $(-0.25)^t$. This oscillates with decreasing magnitude and approaches zero as t tends to infinity; thus, p_t oscillates and tends to $p^* = 3$ as t tends to infinity.

To find q_t, we use the fact that

$$q_t = 3p_{t-1} - 7 = 3\left(3 + (-0.25)^{t-1}\right) - 7 = 2 + 3(-0.25)^{t-1} .$$

Example 5.2 Suppose that the supply and demand functions for a good are

$$q^S(p) = 3p - 21\,500 \quad \text{and} \quad q^D(p) = 8500 - p.$$

Assuming that the suppliers operate according to the cobweb model, find a recurrence equation for the sequence p_t of prices. Find the explicit solution given that $p_0 = 7499$, and describe in words how the sequence p_t behaves.

Solution

Here, $q^S(p) = 3p - 21\,500$ and $p^D(q) = 8500 - q$. The sequence p_t of prices is determined by the recurrence equations

$$q_t = 3p_{t-1} - 21\,500 \quad \text{and} \quad p_t = 8500 - q_t,$$

where q_t is the sequence of quantities. Eliminating q_t, we obtain the single recurrence

$$p_t = 8500 - (3p_{t-1} - 21\,500) = 30\,000 - 3p_{t-1}.$$

The time-independent solution to this equation is $p^* = 30\,000/4 = 7500$, and the explicit solution is

$$p_t = p^* + (p_0 - p^*)(-3)^t = 7500 - (-3)^t .$$

The behaviour of p_t is determined by the behaviour of the term $(-3)^t$. This oscillates increasingly as t tends to infinity; thus, p_t oscillates increasingly and the equilibrium is unstable. It is questionable whether such behaviour is economically realistic. Indeed, if t is taken large enough, the model even predicts that the price p_t will be negative. One can imagine that, in reality, the behaviour described by an unstable cobweb process is only transitory and that the supplier and consumers realise what is happening and cease to act in the manner described by the model.

Example 5.3 Without solving any equations, determine whether the cobweb model predicts stable equilibrium for the market with

$$q^S(p) = 5p - 10 \quad \text{and} \quad q^D(p) = 6 - 2p.$$

Solution

Note that here the supply and demand functions are of the form $q^S(p) = bp - a$ and $q^D(p) = c - dp$, where $a = 10, b = 5, c = 6, d = 2$. Since $b > d$, the theory tells us that the equilibrium will be unstable.

Main Topics

- the cobweb model in general terms
- how to derive a recurrence equation when demand and supply are linear
- solving this recurrence to find the sequences of prices and quantities
- analysing the stability of the cobweb model

Key Terms, Notations and Formulae

- cobweb model, $q_t = q^S(p_{t-1})$ and $p_t = p^D(q_t)$
- stable and unstable cobwebs
- when $q^S(p) = bp - a$ and $q^D(p) = c - dp$, stability if $b < d$

Exercises

5.1 The supply and demand functions for a good are

$$q^S(p) = 0.05p - 4 \qquad \text{and} \qquad q^D(p) = 20 - 0.15p.$$

Find the equilibrium price. What is the inverse demand function $p^D(q)$? Suppose that the sequence of prices p_t is determined by $p_t = p^D(q_t)$ and $q_t = q^S(p_{t-1})$ as in the cobweb model. Find an expression for p_t.

5.2 The supply and demand functions for a good are

$$q^S(p) = p - 4 \qquad \text{and} \qquad q^D(p) = 20 - p.$$

Find the equilibrium price. Suppose the suppliers operate according to the cobweb model and that the initial price is 10. Find an expression for p_t and describe how it behaves as t tends to infinity.

Problems

5.1 Suppose that the supply and demand sets for a certain good are

$$S = \{(q, p) \mid 2p - 3q = 12\} \quad \text{and} \quad D = \{(q, p) \mid 2p + q = 20\},$$

and suppliers operate according to the cobweb model, so that if p_t and q_t are (respectively) the price and quantity in year t, then $p_t = p^D(q_t)$ and $q_t = q^S(p_{t-1})$. Suppose also that the initial price is $p_0 = 10$. Find an expression for p_t. How does p_t behave as t tends to infinity? How does q_t behave as t tends to infinity?

5.2 The supply and demand functions for a good are

$$q^S(p) = 15p - 41 \quad \text{and} \quad q^D(p) = 40 - 12p.$$

Suppose the suppliers operate according to the cobweb model and that the initial price is 2.5. Find explicit formulae for p_t and q_t, the price and quantity in year t.

5.3 Determine whether the cobweb model predicts stable or unstable equilibrium for the market with

$$q^S(p) = 2p - 3 \quad \text{and} \quad q^D(p) = 18 - p.$$

6 Introduction to Calculus

6.1 The Rate of Change of a Function

There are many problems in economics which require us to take account of how a function changes with respect to its input. For example, how does a small change in price affect the quantity of a good which consumers will buy? Again, we know that governments are very concerned about 'economic growth', which is measured by calculating how certain indicators are changing with respect to time. In order to explain precisely what all this means, we shall have to make some clear definitions.

Mathematicians use the picture in Figure 6.1 in order to explain the rate of change of a function $f : \mathbb{R} \to \mathbb{R}$. The idea is to compare the value of the function at x with its value at $x+h$, where h is a small quantity. The change in the value of f is $f(x+h)-f(x)$, and this, when divided by the change h in the 'input', measures the average rate of change. The quantity $(f(x+h)-f(x))/h$ is represented diagrammatically by the gradient of the chord joining the points $(x,f(x))$ and $(x+h,f(x+h))$.

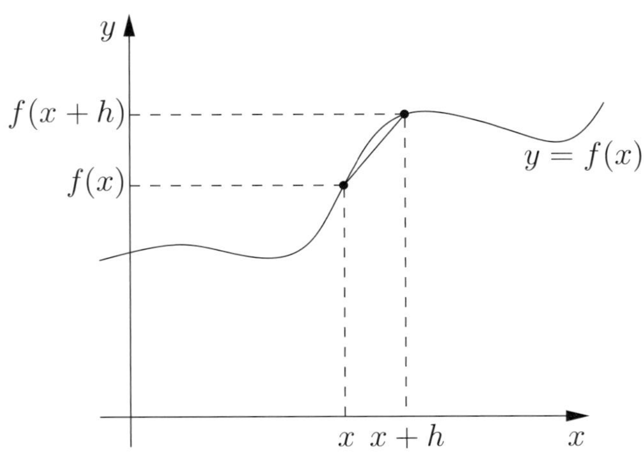

Figure 6.1 Definition of the derivative.

In the diagram, as h approaches zero the chord approaches the tangent at $(x,f(x))$, and its gradient approaches a value which we may take to be the 'instantaneous' rate of change of f at x. This leads us to *define* the gradient of f at x to be the limiting value of the quantity

$$\frac{f(x+h)-f(x)}{h},$$

as h approaches zero, provided this limiting value exists. The notation used for this limiting value is

$$f'(x) = \lim_{h \to 0} \frac{f(x+h)-f(x)}{h}.$$

Note that for each value x, $f'(x)$ is a number, the rate of change of f at the given value. The function f' which tells us the rate of change of f is called the *derivative* of f. The process of finding the derivative is often known as *differentiation*.

Example Suppose $f(x) = x^3$. In order to work out the derivative, we calculate as follows:

$$\frac{f(x+h)-f(x)}{h} = \frac{(x+h)^3 - x^3}{h} = \frac{(x^3 + 3x^2h + 3xh^2 + h^3) - x^3}{h}$$
$$= 3x^2 + h(3x+h).$$

The first term is independent of h and the second term approaches zero as h approaches zero, so the derivative is the function given by $f'(x) = 3x^2$. This tells us the slope of the tangent to the graph for any value of x. For example, at $x = 2$, which corresponds to the point $(2, 8)$ on the graph, the slope is $3 \times 2^2 = 12$. ■

Another way of looking at the definition of f' is to think of it as an *approximation*, which tells us how a small change in the input x affects the output $f(x)$. If we denote a small change in x by Δx (instead of h), then the resulting change in $f(x)$ is

$$\Delta f = f(x + \Delta x) - f(x).$$

Since $f'(x)$ is the limit of $\Delta f / \Delta x$ as Δx approaches zero, for small values of Δx we have

$$f'(x) \simeq \frac{\Delta f}{\Delta x}, \quad \text{or} \quad \Delta f \simeq f'(x)\Delta x,$$

where the symbol '$\simeq$' means 'is approximately equal to'. This is the origin of the much-used d-notation in which we write

$$\frac{df}{dx} \quad \text{instead of} \quad f'(x).$$

Example What is the approximate change in the function $f(x) = x^3$ when x changes from 4 to 4.01? Using the fact that the derivative is $3x^2$, we have

$$\Delta f \simeq f'(4)\Delta x = 3 \times 4^2 \times 0.01 = 0.48.$$

■

6.2 Rules for Finding the Derivative

The ideas outlined in the previous section have been around for about 300 years now, and in that time mathematicians have worked out many useful rules for finding the derivative. This means that we do not have to go back to the definition every time a derivative has to be calculated. The basic rules are given below. You may have been taught these rules in the d-notation, in which df/dx is used to denote $f'(x)$. In that case, you should translate each rule given here into a more familiar one.

The first rule tells us the derivative of a power of x.

- If $p(x) = x^k$, then $p'(x) = kx^{k-1}$.

For example, in the previous section we showed by an explicit calculation that the derivative of x^3 is $3x^2$, which is the case $k = 3$ of the rule. It should be noted that the rule holds for all values of k, not just positive integers.

Example Suppose the demand set for tins of caviar (Figure 6.2) is

$$D = \{(q,p) \mid p^3 q = 8000\},$$

where q is the number of tins (in thousands per week) and p is the price per tin in dollars. (The authors are unfamiliar with this commodity and apologise if the assumptions are unreasonable!) If p is increased from \$20 to \$21, what will be the expected fall in sales, approximately? If, on the other hand, production were to be increased from 1000 tins per week to 1100, what would be the expected fall in price?

The demand function and its derivative are

$$q^D(p) = 8000p^{-3}, \quad q^{D'}(p) = 8000 \times (-3)p^{-4} = -24\,000/p^4.$$

Therefore, when $p = 20$ and $\Delta p = 1$ we have

$$\Delta q \simeq (-24\,000/p^4)\Delta p = -24\,000/20^4 = -0.15.$$

Remembering that the units are thousands of tins, it follows that 150 fewer tins will be sold per week.

For the second question we have to consider p as a function of q, and so we need the inverse demand function and its derivative:

$$p^D(q) = 20q^{-1/3}, \quad p^{D'}(q) = 20 \times (-1/3)q^{-4/3}.$$

So when $q = 1$ and $\Delta q = 0.1$, we have

$$\Delta p \simeq (-20/3)q^{-4/3}\Delta q = (-20/3) \times 0.1 = -2/3.$$

The conclusion is that the price falls by about 67 cents. ■

The next three rules tell us how to find the derivative of the sum, product and quotient of given functions.

- *The sum rule:* If $h(x) = f(x) + g(x)$ then $h'(x) = f'(x) + g'(x)$.
- *The product rule:* If $h(x) = f(x)g(x)$ then $h'(x) = f'(x)g(x) + f(x)g'(x)$.

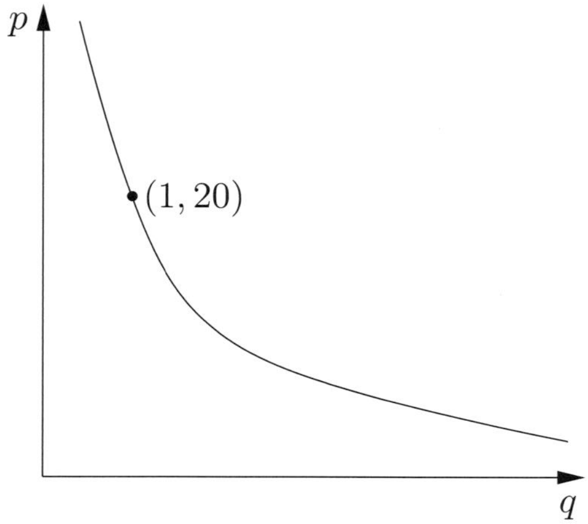

Figure 6.2 The demand set $p^3q = 8000$.

- *The quotient rule:* If $h(x) = f(x)/g(x)$ and $g(x) \neq 0$ then

$$h'(x) = \frac{f'(x)g(x) - f(x)g'(x)}{g(x)^2}.$$

Example In order to calculate the derivative of the function k defined by the formula $k(x) = (x^3 + 1)(x^2 + 5x)$, we observe that $k(x)$ is the product of two functions, each of which is a sum of powers of x. Using the appropriate rules we have

$$k'(x) = (3x^2)(x^2 + 5x) + (x^3 + 1)(2x + 5) = 5x^4 + 20x^3 + 2x + 5.$$

Of course, an alternative method would be to multiply out the original expression and then find the derivative of each term. [*Exercise*: check that this gives the same result.] ■

6.3 Marginal Cost as a Derivative

Suppose that a firm which makes electric light bulbs knows that in order to produce q light bulbs it will have to pay out $C(q)$ dollars in wages, materials, overheads, and so on. We say that C is the firm's *cost function*.

In this case a change in production from q to $q+1$ light bulbs is relatively small, and may be described as 'marginal'. The corresponding increase in cost is $C(q+1) - C(q)$, and this can be thought of as the 'marginal cost' of making one more light bulb, when the level of production is q. In the Δ notation of Section 6.1, we see that the marginal cost is the change ΔC corresponding to a change $\Delta q = 1$. In general, the relationship between Δq and ΔC is given by the approximation

$$\Delta C \simeq C'(q)\Delta q,$$

and so, when $\Delta q = 1$, the resulting ΔC is approximately $C'(q)$. For this reason it makes sense to *define* the *marginal cost function* to be the derivative of the cost function C. (Indeed, in the traditional language of economics, the derivative of any function F is referred to as the marginal of F, and is often denoted by MF.) What we have shown is that if the units are small, as for example in the production of light bulbs, $C'(q)$ represents the cost of producing one more unit when q units are being produced. The concept of marginal cost is equally useful when the units are larger, provided we remember the basic idea.

Example Suppose that the costs of a firm making bicycles are \$50 000 per week in overheads and \$25 for every bicycle made. Then its cost function in dollars is

$$C(q) = 50\,000 + 25q, \quad \text{so that} \quad C'(q) = 25.$$

Here the marginal cost is 25 dollars, independent of the level of production q.

Slightly more realistically, suppose that in order to produce a substantially larger weekly output of bicycles it would be necessary for the firm to incur extra costs, possibly because its increased consumption of a raw material would drive the price of that raw material upwards. We can account for this by introducing an additional cost term, say $0.001q^2$, which is trivial when q is small but which is more significant as q increases. In this case we have

$$C(q) = 50\,000 + 25q + 0.001q^2, \quad \text{so that} \quad C'(q) = 25 + 0.002q.$$

Thus the marginal cost is \$25.2 if the output is 100 bicycles per week, but it would rise to \$45 if the output were 10 000 bicycles per week. ■

6.4 The Derivative of a Composite Function

In addition to the rules given in Section 6.2, there is one other very useful rule for differentiation – the *composite function rule*, also known as the *function of a function rule* or *chain rule*.

Suppose that k is a composite function sr, that is $k(x) = s(r(x))$. When x is altered by a small amount Δx there is a corresponding change Δr in $r(x)$, and this in turn produces a change Δs in $s(r(x))$. Since $k(x) = s(r(x))$, Δs is also the change in $k(x)$ resulting from Δx, that is $\Delta k = \Delta s$. It follows that

$$\frac{\Delta k}{\Delta x} = \frac{\Delta s}{\Delta x} = \frac{\Delta s}{\Delta r}\frac{\Delta r}{\Delta x}.$$

Replacing the Δ-quotients by the corresponding derivatives, we obtain the rule for differentiating a composite function:

$$\frac{dk}{dx} = \frac{ds}{dr}\frac{dr}{dx}.$$

This is easy to remember, but in some ways the d-notation is confusing and obscure; in particular, it does not indicate at what values the derivatives are to be evaluated. In 'functional notation', everything is explicit: the derivative $k' = (sr)'$ is given by

- $(sr)'(x) = s'(r(x))\, r'(x)$.

Example What is the derivative of $k(x) = (5x^2 + 3x + 7)^2$ when $x = 1$? Here $k = sr$, where

$$r(x) = 5x^2 + 3x + 7$$

and $s(x) = x^2$. We have

$$r'(x) = 10x + 3 \quad \text{and} \quad s'(x) = 2x.$$

So $r'(1) = 13$ and $s'(r(1)) = s'(15) = 30$, giving $k'(1) = 13 \times 30 = 390$.

In general, the derivative $k' = (sr)'$ is given by

$$k'(x) = s'(r(x))r'(x) = 2(r(x))r'(x) = 2(5x^2 + 3x + 7)(10x + 3).$$

Note that substituting $x = 1$ in this expression gives $k'(1) = 390$, as before. ■

6.5 The Derivative of an Inverse Function

The rule for finding the derivative of a composite function can also be used to find the derivative of an inverse function. Recalling the definitions given in Section 2.2, we know that if $g = f^{-1}$ then the composite gf is such that $g(f(x)) = x$. In other words, $gf = i$, the identity function. So the derivative of the composite gf is the derivative of i, and since $i(x) = x$ this is identically equal to one (by the power rule for x^1, if you wish). Using the 'function of a function' rule, we get

$$1 = (gf)'(x) = g'(f(x))f'(x).$$

We can rearrange this so that it gives g':

$$g'(f(x)) = \frac{1}{f'(x)}, \quad \text{when} \quad g = f^{-1}.$$

Equivalently, a more succinct rule is

- $g'(y) = \dfrac{1}{f'(x)}, \quad \text{when} \quad y = f(x) \quad \text{and} \quad x = g(y).$

Example The supply set $S = \{(q, p) \mid 2p - 3q = 12\}$ was considered in Example 1.2. We found that the supply function and its inverse are

$$q^S(p) = \frac{2}{3}p - 4 \quad \text{and} \quad p^S(q) = 6 + \frac{3}{2}q.$$

Clearly, the derivatives of q^S and p^S are $2/3$ and $3/2$, which is as predicted by the general theory, since $3/2 = \frac{1}{2/3}$. ■

The inverse function rule is rather more memorable when expressed in terms of small changes Δx and Δy. If $y = f(x)$ and $x = g(y)$, we know that

$$\Delta y \simeq f'(x)\Delta x \quad \text{and} \quad \Delta x \simeq g'(y)\Delta y.$$

It follows that the rule $g'(y) = 1/f'(x)$ is equivalent to the statement that

$$\frac{\Delta x}{\Delta y} = 1\Big/\frac{\Delta y}{\Delta x},$$

which is a simple fact of elementary algebra.

Example In Section 6.2 we considered the demand set consisting of pairs (q,p) for which $p^3 q = 8000$. Here the demand function is $q^D(p) = 8000p^{-3}$. Given that the derivative of q^D is $-24\,000p^{-4}$, we can obtain the derivative of the inverse function p^D from the general rule as follows:

$$p^{D'}(q) = \frac{1}{q^{D'}(p)} = \frac{1}{-24\,000p^{-4}} = \frac{-p^4}{24\,000} = \frac{-(20q^{-1/3})^4}{24\,000} = \frac{-20q^{-4/3}}{3},$$

which is the same result as we obtained by differentiating p^D directly. We also found that, when $q = 1$ and $p = 20$,

$$\frac{\Delta q}{\Delta p} \simeq \frac{-0.15}{1} \quad \text{and} \quad \frac{\Delta p}{\Delta q} \simeq \frac{-2/3}{0.1}.$$

This checks with the fact that $\Delta q/\Delta p$ is the reciprocal of $\Delta p/\Delta q$. ■

Worked Examples

Example 6.1 Calculate the derivative of the function

$$h(x) = \frac{x^3 + 1}{x^2 + 1}.$$

Solution

We observe that $h(x) = f(x)/g(x)$, where $f(x) = x^3 + 1$ and $g(x) = x^2 + 1$, and we shall therefore use the rule for differentiating a quotient. The derivatives of f and g are

$$f'(x) = 3x^2 \quad \text{and} \quad g'(x) = 2x,$$

and so

$$h'(x) = \frac{f'(x)g(x) - f(x)g'(x)}{g(x)^2} = \frac{3x^2(x^2+1) - (x^3+1)(2x)}{(x^2+1)^2} = \frac{x^4 + 3x^2 - 2x}{(x^2+1)^2}.$$

Example 6.2 Use the derivative to find the approximate change in the function $f(x) = x^4$ when x changes from 3 to 3.005. Compare this with the actual change.

Solution

The derivative of f is $f'(x) = 4x^3$ and we therefore have the approximation

$$\Delta f \simeq f'(3)\Delta x = 4(3)^3 \times 0.005 = 0.540.$$

The actual value of the change is

$$\Delta f = f(3.005) - f(3) = (3.005)^4 - 3^4 = 0.54135,$$

so that our approximation is correct to two decimal places.

Example 6.3 Suppose that the demand set for copies of a *Mathematics for Economics* book is

$$D = \{(q,p) \mid p^2 q = 6000\},$$

where q is the number of copies (in thousands) and p is the price in dollars. If the price is increased from \$20 to \$21, what is the approximate fall in expected sales?

Solution

The demand function and its derivative are

$$q^D(p) = 6000p^{-2}, \quad q^{D'}(p) = 6000 \times (-2)p^{-3} = \frac{-12\,000}{p^3}.$$

Therefore, when $p = 20$ and $\Delta p = 1$ we have

$$\Delta q \simeq (-12\,000/p^3)\Delta p = -12\,000/20^3 = -1.5.$$

So the price rise will result in approximately 1500 fewer copies being sold.

Example 6.4 A firm has cost function $C(q) = 1500 + 15q - 3q^2 + q^3$. Show that its marginal cost is always positive.

Solution

The marginal cost is

$$C'(q) = 15 - 6q + 3q^2 = 3(q^2 - 2q + 5).$$

Completing the square (see Section 2.4) gives

$$q^2 - 2q + 5 = (q-1)^2 + 4,$$

and so

$$C'(q) = 3(q-1)^2 + 12.$$

Thus, as $(q-1)^2$ is never less than zero, we see that $C'(q) \geq 12$ for all q.

Example 6.5 Calculate the derivative of the function

$$f(x) = \sqrt{x^2 + 1}.$$

Solution

Note that $f(x) = (x^2 + 1)^{1/2}$. Thus, $f = sr$ where $s(x) = x^{1/2}$ and $r(x) = x^2 + 1$. Using the composite function rule and the facts that $s'(x) = \frac{1}{2}x^{-1/2}$ and $r'(x) = 2x$, we have

$$f'(x) = s'(r(x))\, r'(x) = \frac{1}{2}\left(x^2 + 1\right)^{-1/2}(2x) = \frac{x}{\sqrt{x^2 + 1}}.$$

Example 6.6 What is the derivative of the inverse of the function $f(x) = x^3 + 1$?

Solution

If $y = x^3 + 1$ then $x = (y - 1)^{1/3}$. In other words, the inverse function $g = f^{-1}$ is given by $g(y) = (y - 1)^{1/3}$. According to the analysis above, the derivative of g is

$$g'(y) = \frac{1}{f'(x)}.$$

Now, $f'(x) = 3x^2$ and so

$$g'(y) = \frac{1}{3x^2} = \frac{1}{3\left((y-1)^{1/3}\right)^2} = \frac{1}{3(y-1)^{2/3}} = \frac{1}{3}(y-1)^{-2/3}.$$

Note that the same result is obtained by differentiating the expression for g directly.

Main Topics

- the definition of the derivative as the rate of change
- approximation using the derivative
- the sum, product and quotient rules
- marginal cost (and marginals in general)
- the composite function rule
- the derivative of an inverse function

Key Terms, Notations and Formulae

- derivative, $f'(x) = \dfrac{df}{dx} = \lim_{h \to 0} \dfrac{f(x+h) - f(x)}{h}$
- approximation, $\Delta f = f(x + \Delta x) - f(x) \simeq f'(x)\Delta x$
- if $p(x) = x^k$, then $p'(x) = kx^{k-1}$
- sum rule: if $h(x) = f(x) + g(x)$, then $h'(x) = f'(x) + g'(x)$
- product rule: if $h(x) = f(x)g(x)$, then $h'(x) = f'(x)g(x) + f(x)g'(x)$
- quotient rule: if $h(x) = f(x)/g(x)$, then $h'(x) = \dfrac{f'(x)g(x) - f(x)g'(x)}{g(x)^2}$
- cost function, marginal cost, marginal cost function
- composite function rule: if $k = sr$, then $k'(x) = s'(r(x))r'(x)$

- if $g = f^{-1}$ then $g'(y) = \dfrac{1}{f'(x)}$, where $x = g(y) = f^{-1}(y)$

Exercises

6.1 Find the derivatives of

$$(3x+7)^{15} \quad \text{and} \quad \sqrt{x^4+1}.$$

6.2 Use differentiation to find the approximate change in $\sqrt{x}$ as x increases from 100 to 101. Show, more generally, that when n is large, the change in $\sqrt{x}$ as x increases from n to $n+1$ is approximately $1/(2\sqrt{n})$.

Problems

6.1 Work out the derivatives of the following functions, using the standard rules:

$$f(x) = (2x^2+5x+4)^3; \quad g(x) = \frac{x^2}{x-3}.$$

Explain why the rules do not apply to g when $x = 3$.

6.2 The function f is given by $f(x) = x^5 + 3$.
Calculate the derivative of f^{-1}:
(a) by using the rule for inverse functions, and
(b) by differentiating the explicit expression for f^{-1}.

6.3 Find the derivative of the function

$$h(x) = (x^3+8)^{1/3}.$$

For what value of x is $h'(x)$ not defined? Find also the derivative of

$$f(x) = \sqrt{x + (x^3+8)^{1/3}}.$$

6.4 A manufacturer's cost function is

$$C(q) = 1000 + 20q + q\sqrt{1+q}.$$

Find the marginal cost function.

6.5 Suppose that the demand function for a good is

$$q^D(p) = \frac{80\,000}{p^2+1},$$

where q is the quantity and p is the price in dollars. If the price is decreased from \$9 to \$8.50, what is the approximate increase in the quantity sold?

7 Some Special Functions

7.1 Powers

We start this chapter with a discussion of what it means to raise a number to a power. This is a topic which, to a certain extent, has been taken for granted in earlier chapters. However, a more careful discussion of powers will reveal some important questions that cannot be answered without some further thought.

Let a be any positive number. When n is a positive integer, the nth power of a, a^n, is simply the product of n copies of a, that is

$$a^n = \underbrace{a \times a \times a \times \cdots \times a}_{n \text{ times}}.$$

As a consequence of this definition, we can easily verify the *power rules*

$$a^x a^y = a^{x+y} \quad \text{and} \quad (a^x)^y = a^{xy},$$

whenever x and y are positive integers. For example,

$$a^3 \times a^2 = (a \times a \times a) \times (a \times a) = a \times a \times a \times a \times a = a^5 = a^{3+2}.$$

In order to define a^x when x is not a positive integer, we must be guided by the power rules. First, we clearly need to define $a^0 = 1$, since then we have $a^{m+0} = a^m \times a^0 = a^m \times 1 = a^m$.

Next, what should we mean by a^{-n}, when $-n$ is a negative integer? We cannot multiply a by itself a negative number of times, but the rules tell us what a^{-n} must be. Since we must have

$$a^n a^{-n} = a^{n-n} = a^0 = 1,$$

it follows that $a^{-n} = 1/a^n$. In other words, we *define* a^{-n} to be $1/a^n$. For instance, 2^{-5} is $1/2^5 = 1/32$.

Note that the power rules hold when x and y are any integers, positive, negative or zero. For example, when m is a positive integer and $-n$ is a negative integer, with $m > n$, we have

$$\begin{aligned}a^m a^{-n} &= a^m\left(\frac{1}{a^n}\right)\\ &= \underbrace{a \times a \times \cdots \times a}_{m\text{ times}} \times \frac{1}{\underbrace{a \times \cdots \times a}_{n\text{ times}}}\\ &= \underbrace{a \times a \times \cdots \times a}_{m-n\text{ times}}\\ &= a^{m-n} = a^{m+(-n)}.\end{aligned}$$

Continuing in the same vein, if n is a positive integer, what should $a^{1/n}$ mean? The second power rule suggests that

$$(a^{1/n})^n = a^{(1/n)\times n} = a^1 = a.$$

In words, raising $a^{1/n}$ to the nth power gives a: that is $a^{1/n}$ is the 'nth root of a'.

Moving on again, it is now easy to assign a meaning to $a^{m/n}$, where m and n are integers and n is positive. We simply define

$$a^{m/n} = \left(a^{1/n}\right)^m.$$

Numbers which can be written in the form m/n are called *rational* numbers; they are the numbers that can be expressed as *ratios* of integers. So we have now defined a^x whenever x is a rational number, and the power rules hold.

Example Suppose we want to calculate $8^{2/3}$. By pressing a few buttons on a calculator, the answer appears: 4. In order to understand what this means, and why the answer is an integer, we can apply the above definitions:

$$8^{2/3} = \left(8^{1/3}\right)^2 = (2)^2 = 4.$$

■

Notice that we have *not* yet defined what a^x should mean when x is not a rational number. For example, what does $2^{\sqrt{2}}$ mean? The number $\sqrt{2}$ is not a rational number m/n (a fundamental fact that has been known for over 2000 years), and so we cannot obtain $2^{\sqrt{2}}$ by taking the mth power of the nth root of 2. Before we answer this question, we shall make an important diversion.

7.2 The Exponential Function and Its Properties

In Section 4.3 we observed that the return obtained from investing \$100 for 1 year when the interest is compounded in m equal periods is

$$100\left(1 + \frac{r}{m}\right)^m,$$

where r is the equivalent annual rate. As the number m increases, the return increases, but it appears to increase fairly slowly, and in fact it turns out that it does *not* tend to infinity, but rather it approaches a finite limit. Of course, the limit depends on the value of r.

Following on from this line of argument we define, for any real number x,

$$\exp(x) = \lim_{m\to\infty} \left(1 + \frac{x}{m}\right)^m.$$

The function exp is known as the *exponential function*. It can be proved that the definition works (that is, the limit exists and is finite) for all $x \in \mathbb{R}$.

The exponential function has many remarkable properties, all of which can be derived from the definition. For our purposes the detailed mathematical arguments are not needed, and we shall merely use the definition to motivate the discussion in general terms. The most important property concerns the relationship between the general value $\exp(x)$ and the particular value $\exp(1)$. The latter is denoted by the standard symbol e, and we have

$$e = \exp(1) = \lim_{n\to\infty} \left(1 + \frac{1}{n}\right)^n = 2.71828\ldots.$$

Given a rational number p/q, where p and q are positive integers, we have explained in Section 7.1 what is meant by $e^{p/q}$. Using the definition of e, we have

$$e^{p/q} = \left(\lim_{n\to\infty} \left(1 + \frac{1}{n}\right)^n\right)^{p/q} = \lim_{n\to\infty} \left(1 + \frac{1}{n}\right)^{np/q}.$$

Now put $m = np/q$. Then $1/n = (p/q)/m$ and, since $m \to \infty$ as $n \to \infty$,

$$e^{p/q} = \lim_{m\to\infty} \left(1 + \frac{p/q}{m}\right)^m = \exp(p/q).$$

This argument shows that $\exp(p/q)$ is just 'the (p/q)th power of the number e'. For example

$$\exp(3/2) = e^{3/2} = \sqrt{(2.71828\ldots)^3} = \sqrt{20.08553\ldots} = 4.48169\ldots.$$

If x is not a rational number, $\exp(x)$ is still defined (by the limit formula), and so it makes sense to *define* e^x to be $\exp(x)$. In fact, as we shall see in Section 7.4, this is a crucial step towards defining a^x for any value of a.

Henceforth we shall use the notations $\exp(x)$ and e^x interchangeably. Many important properties of the exponential function follow from the fact that the two expressions have the same value. For example

$$\exp(0) = 1 \quad \text{and} \quad \exp(x) \to \infty \text{ as } x \to \infty.$$

Furthermore, since $e^{-x} = 1/e^x$, we see that $\exp(x) \to 0$ as $x \to -\infty$. The graph of the exponential function is shown in Figure 7.1.

There are two other useful rules. The power rule, $e^x e^y = e^{x+y}$, translates into the property

$$\exp(x)\exp(y) = \exp(x + y).$$

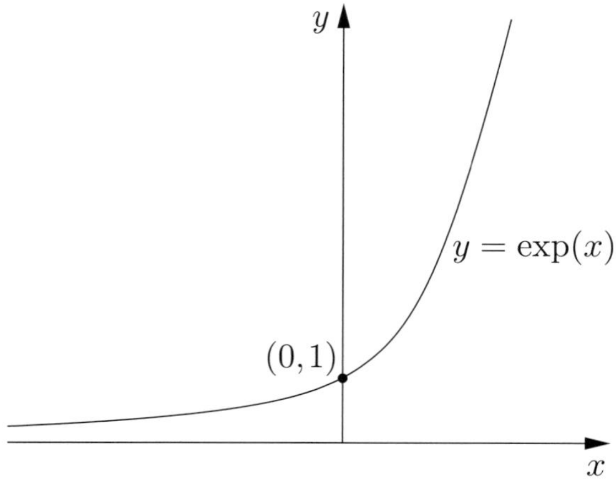

Figure 7.1 The graph of the exponential function.

Finally, a very important property is the remarkable fact that the derivative of the exponential function is the same function:

- *the derivative of* $\exp(x)$ *is* $\exp(x)$.

This can be deduced from the properties we already know, but the proof is not particularly illuminating, and we shall omit it.

7.3 Continuous Compounding of Interest

We can now use the exponential function to work out explicitly the relationship between capital and principal when the year is divided into m equal intervals, and m tends to infinity. This is usually referred to as *continuous compounding* of interest. If the equivalent annual rate is r, the capital resulting from an amount P invested for 1 year is

$$P \times \left(\lim_{m\to\infty} \left(1 + \frac{r}{m}\right)^m \right).$$

The limit is $\exp(r)$, by the definition of the exponential function, and so the capital after 1 year is $P\exp(r)$. Suppose now that P is invested for t years, where t need not be an integer. The resulting capital will be

$$C = P(\exp(r))^t = P\exp(rt).$$

We can also write this as $C = Pe^{rt}$. In order to find the present value of an amount C due t years from now, we simply rearrange the equation. Since $1/e^{rt} = e^{-rt}$, this gives $P = Ce^{-rt}$, or

$$P = C\exp(-rt).$$

Example Suppose that \$2000 is invested in an account where interest is compounded continuously at a constant annual equivalent rate of 7%. How much money is in the account after (a) 4 years; (b) 6.5 years?

The amount in the account after t years is $2000\exp(rt)$, where $r = 0.07$. Note that in this formula, t need not be an integer. Taking $t = 4$, we have that, after 4 years, the amount is $2000e^{0.07\times 4} = 2646.26$. After 6.5 years, the amount is $2000e^{0.07\times 6.5} = 3152.35$. ■

7.4 The Logarithm Function

The graph of $\exp(x)$ (Figure 7.1) shows that for each positive value of y, there is exactly one number x such that $\exp(x) = y$. Thus, if we restrict attention to positive values of y, the exponential function has an inverse, known as the *logarithm* function. In symbols,

$$\text{if} \quad y = \exp(x), \quad \text{then} \quad x = \ln(y).$$

This relationship is illustrated in Figure 7.2.

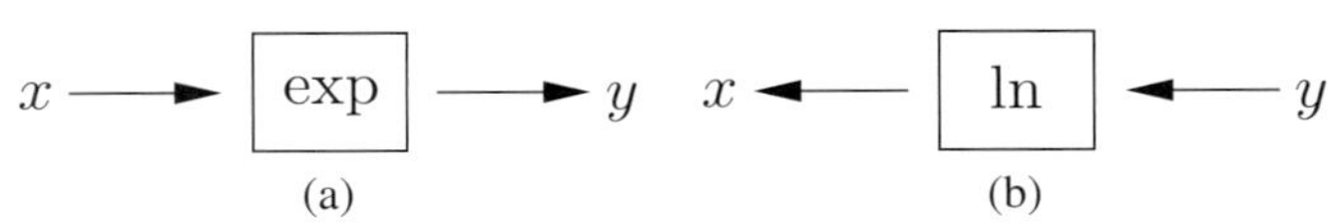

Figure 7.2 The relationship between exp and ln.

Often we write $\ln y$ rather than $\ln(y)$. To be precise, $\ln y$ is the *natural* logarithm of y. The natural logarithm is related to another function, the 'log to the base 10' function, denoted by $\log_{10}$ or simply log, which you may have met. The relationship is very simple (see Problem 7.8):

$$\log_{10}(y) = 0.4343\ldots \times \ln(y).$$

As with $\exp(x)$, most pocket calculators will compute values of $\ln(x)$. It is important to stress again that $\ln(x)$ is only defined when x is positive, and your calculator *should* display an error message if you ask it to find the logarithm of a negative number. (Also, your calculator might well have a 'log' key and a 'ln' key. Be careful to distinguish between them.)

The graph of the function $\ln(x)$, for $x > 0$, is sketched in Figure 7.3.

General properties of the logarithm function can be deduced from the corresponding properties of the exponential function. For example, from the rule $\exp(x)\exp(y) = \exp(x + y)$ it is easy to deduce (see Example 7.5) that

$$\ln(ab) = \ln a + \ln b.$$

Before the invention of the pocket calculator this rule was used extensively to perform complicated calculations. Using printed tables of the logarithm function, multiplication of two numbers could be reduced to addition of their logarithms.

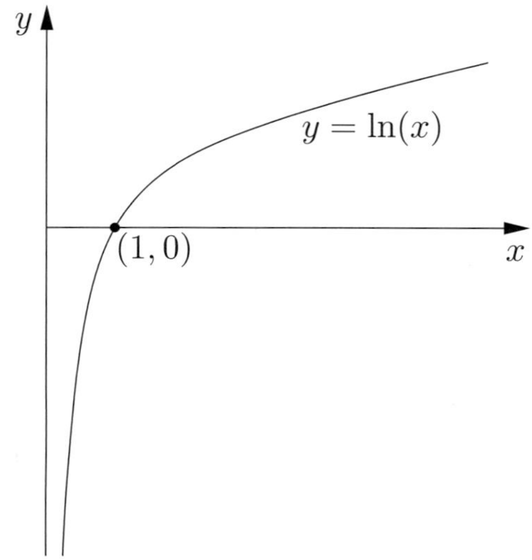

Figure 7.3 The graph of the logarithm function.

For our purposes, there are several useful properties of the logarithm. First, since exp and ln are inverse functions, we have $a = \exp(\ln a)$ for all positive numbers a. Taking the xth power and using what we know about exp, we have

$$a^x = (\exp(\ln a))^x = (e^{\ln a})^x = e^{x \ln a} = \exp(x \ln a).$$

This shows that any power of any positive number can be written in terms of the exp and ln functions. In Section 7.1 we were only able to define a^x when x is a rational number. We now have a satisfactory *definition* of what a^x means, in general, even when x is not rational: for any $a > 0$ and any number x, we *define* $a^x = \exp(x \ln a)$.

Example We now know how to define $2^{\sqrt{2}}$:

$$2^{\sqrt{2}} = \exp(\sqrt{2} \ln 2).$$

Everything on the right-hand side of this equation makes sense. Although it looks somewhat complicated, and perhaps artificial, there really is no simpler definition: we cannot say 'multiply 2 by itself $\sqrt{2}$ times'. The proper definition also tells us how to find the derivative of a^x, something we could not do without it (see Example 7.3). ■

Another useful fact is obtained by 'taking logs' of the equation

$$a^x = \exp(x \ln a),$$

which gives

$$\ln(a^x) = x \ln(a).$$

Example The last part of Example 3.3 amounted to finding the smallest positive integer t such that

$$\left(\frac{2}{3}\right)^t < \frac{1}{30}.$$

Applying the logarithm function to both sides, we require the least positive integer t such that $\ln(2/3)^t < \ln(1/30)$, which can be rewritten as

$$t \ln(2/3) < \ln(1/30), \quad \text{that is} \quad (-0.405)t < -3.401.$$

This is equivalent to $t > 8.398$, so the answer (which must be an integer) is 9. ■

Example In Chapter 11 we shall discuss how a firm's output q depends on its inputs, specifically its capital k and labour l. Often it is convenient to assume that the relationship takes the *Cobb–Douglas* form, $q = k^\alpha l^\beta$, where α and β are constants characteristic of a particular firm. Although this looks complicated, it can be simplified by 'taking logarithms'. Thus

$$\ln q = \ln(k^\alpha l^\beta) = \ln(k^\alpha) + \ln(l^\beta) = \alpha \ln k + \beta \ln l.$$

So, if we define new variables $Q = \ln q$, $K = \ln k$ and $L = \ln l$, the relationship $q = k^\alpha l^\beta$ takes a more familiar form, the linear form $Q = \alpha K + \beta L$. ■

To find the derivative of the logarithm function, let $f(x) = \exp x$, so that $g(y) = \ln y$ is the inverse for f. (We have to remember yet again that y must be a positive number.) The formula for the derivative of an inverse function (Section 6.5) says that

$$g'(y) = 1/f'(x), \quad \text{when} \quad y = f(x) \text{ and } x = g(y).$$

Here we have $f(x) = \exp x$ and $f'(x) = \exp x$, therefore

$$g'(y) = \frac{1}{\exp x} = \frac{1}{y}.$$

In words:

- *the derivative of* $\ln y$ *is* $1/y$.

Example What is the derivative of $g(x) = \ln \ln x$? Using the composite function rule, and the rule for the derivative of ln, we get

$$g'(x) = \left(\frac{1}{\ln x}\right)\left(\frac{1}{x}\right) = \frac{1}{x \ln x}.$$

■

7.5 Trigonometrical Functions

We assume that the reader has studied trigonometrical functions before. It will be recalled that the three main trigonometrical functions are the sine function $\sin x$, the cosine function $\cos x$ and the tangent function $\tan x$, which is defined as $\tan x = \sin x / \cos x$.

In mathematics we usually assume that the angle x is measured in radians, rather than in degrees. 180 degrees equals π radians, so to convert an angle of α degrees to radians we multiply α by $\pi/180$.

The rules for differentiating trigonometrical functions are based on the following two results:

- If $f(x) = \sin x$, then $f'(x) = \cos x$.
- If $g(x) = \cos x$, then $g'(x) = -\sin x$.

There is a useful way of helping to remember these results. We know that one of the derivatives has a minus sign. To remember which, simply say 'minus sign' to yourself; this sounds the same as 'minus sine'!

Example We shall find the derivative of the function $f(x) = \tan(x^2 + 1)$. Note that f is the composition sr, where $s(x) = \tan x$ and $r(x) = x^2 + 1$. The derivative of $r(x)$ is $r'(x) = 2x$. Since

$$\tan x = \frac{\sin x}{\cos x},$$

its derivative is, by the quotient rule,

$$\frac{(\cos x)(\sin x) - (\sin x)(-\sin x)}{(\cos x)^2} = \frac{(\cos x)^2 + (\sin x)^2}{(\cos x)^2}.$$

Since $(\sin x)^2 + (\cos x)^2 = 1$ for all x, the derivative of s is

$$s'(x) = \frac{1}{(\cos x)^2}.$$

Therefore

$$f'(x) = s'(r(x))r'(x) = \left(\frac{1}{(\cos(x^2+1))^2}\right)(2x) = \frac{2x}{(\cos(x^2+1))^2}.$$

■

Worked Examples

Example 7.1 What is the present value of \$1000 due 4 years from now when the annual interest rate is 5% and (a) interest is paid once each year, at the end of the year; (b) interest is continuously compounded?

Solution

Part (a) is just like the examples we considered in Chapter 4. The answer is that, in dollars, $P = 1000/(1 + 0.05)^4 = 822.70$. For part (b), we use the fact that the present value of an amount C due t years from now when interest is continuously compounded at an annual equivalent rate of r is $P(C) = C/\exp(rt) = C\exp(-rt)$

$= Ce^{-rt}$. Thus, the answer to part (b) is $1000e^{-4(0.05)} = 818.73$. Observe that this is *less* than the answer to part (a). A little thought shows why this is to be expected. The present value of \$1000 due 4 years from now is the principal which, if invested now, will grow to \$1000 in 4 years. Since money grows *faster* under the continuous compounding described in (b) than under the annual payment regime described in (a), the principal required under continuous compounding is *lower* than that required under the annual interest scheme.

Example 7.2 Calculate the derivative of the function

$$g(x) = \ln\left(x + \sqrt{x^2+1}\right).$$

Solution

We note first that since $x + \sqrt{x^2+1}$ is positive for all values of x, we can take its logarithm, so the definition of g is valid for all x. (Recall that we cannot take the logarithm of a negative number.) We may calculate $g'(x)$ using the composite function rule, as follows: $g = uv$ where $u(x) = \ln x$ and $v(x) = x + \sqrt{x^2+1}$. Hence

$$\begin{aligned} g'(x) &= u'(v(x))v'(x) \\ &= \frac{1}{x+\sqrt{x^2+1}}\left(1 + \frac{x}{\sqrt{x^2+1}}\right) = \frac{1}{x+\sqrt{x^2+1}}\,\frac{\sqrt{x^2+1}+x}{\sqrt{x^2+1}}, \end{aligned}$$

which simplifies to

$$g'(x) = \frac{1}{\sqrt{x^2+1}}.$$

(In calculating $v'(x)$, we use the result of Example 6.5, which is itself obtained using the composite function rule. Therefore, the differentiation of this example uses the composite function rule twice.)

Example 7.3 Using the fact that $a^x = \exp(x \ln a)$, find the derivative of a^x.

Solution

If $f(x) = a^x$, the function f is the composition sr, where $s(x) = \exp x$ and $r(x) = x \ln a$. Using the rule for differentiating a composition, and noting that $s'(x) = \exp x$ and $r'(x) = \ln a$, we see that

$$f'(x) = s'(r(x))r'(x) = \exp(x \ln a)(\ln a).$$

Thus, $f'(x) = (\ln a)a^x$. (This result shows that e is the only number a such that the derivative of a^x is itself; for this is true if and only if $\ln a = 1$, which means $a = e$.)

Example 7.4 What are the derivatives of (a) $e^{x^2} \sin x$ and (b) $\sin(\cos x)$?

Solution

(a) Note first that, by the composite function rule, the derivative of e^{x^2} is $e^{x^2}(2x) = 2xe^{x^2}$. The derivative of $\sin x$ is $\cos x$. Therefore, the derivative of $e^{x^2} \sin x$ is $2xe^{x^2} \sin x + e^{x^2} \cos x$, using the product rule.

(b) Here we use the rule for differentiating a composite function $f = sc$, where $s(x) = \sin x$ and $c(x) = \cos x$. Thus, since $s'(x) = \cos x$ and $c'(x) = -\sin x$, we have

$$f'(x) = s'(c(x))c'(x) = \cos(\cos x)(-\sin x) = -\sin x \cos(\cos x).$$

Example 7.5 Use the fact that the logarithm function is the inverse of the exponential function, together with the equation $\exp(x + y) = \exp(x)\exp(y)$, to show that

$$\ln(ab) = \ln a + \ln b.$$

Solution

Take $x = \ln a$ and $y = \ln b$. Then $a = \exp(x)$ and $b = \exp(y)$, so we have

$$ab = \exp(x)\exp(y) = \exp(x + y).$$

By definition of the logarithm function, $\exp(x + y) = ab$ is equivalent to $x + y = \ln(ab)$. It follows that

$$\ln(ab) = x + y = \ln a + \ln b,$$

as required.

Main Topics

- integer and rational powers of positive numbers
- the exponential function and its properties, including its derivative
- capital growth and present value under continuous compounding
- the logarithm function and its properties, including its derivative
- irrational powers of positive numbers
- differentiating trigonometrical functions

Key Terms, Notations and Formulae

- for $a > 0$, $m \in \mathbb{Z}$, $n \in \mathbb{N}$, $a^{m/n} = \left(a^{1/n}\right)^m$
- exponential function, $\exp(x) = \lim_{m\to\infty}\left(1 + \frac{x}{m}\right)^m = e^x$
- $\exp(x + y) = \exp(x)\exp(y)$
- if $f(x) = \exp(x)$, then $f'(x) = \exp(x)$; if $g(x) = \exp(kx)$, then $g'(x) = k\exp(kx)$
- under continuous compounding, $C = P\exp(rt)$, $P = C\exp(-rt)$
- logarithm function, $\ln = \exp^{-1}$
- $\ln(ab) = \ln a + \ln b$; $\ln(a^x) = x\ln a$
- general definition of a^x: $a^x = \exp(x\ln a)$
- if $f(x) = \ln x$, then $f'(x) = 1/x$
- sine, cosine: if $f(x) = \sin x$, then $f'(x) = \cos x$; if $g(x) = \cos x$, then $g'(x) = -\sin x$

Exercises

7.1 Two investments are made where the principal of one is 80% of the other. If the smaller and larger principals are invested in accounts where interest is continuously compounded at constant annual equivalent rates of 5% and 2%, respectively, how long will it take for the two accounts to have the same balance?

7.2 Suppose $g(x) = x^x$. Determine the derivative g'.

7.3 The function f is defined for $x > 1$ by

$$f(x) = \frac{1}{x} \ln\left(\frac{1}{x}\right) + \left(1 - \frac{1}{x}\right) \ln\left(1 - \frac{1}{x}\right).$$

Show that the derivative of f is

$$f'(x) = \frac{\ln(x-1)}{x^2}.$$

Problems

7.1 Evaluate

$$\text{(a)} \quad (16)^{7/4}; \quad \text{(b)} \quad (125)^{-1/3}; \quad \text{(c)} \quad (100)^{-5/2}.$$

7.2 Suppose you invest \$60 000 in a special savings account where, for the first 10 years, interest of 6% is paid annually at the end of each year and, thereafter, interest is continuously compounded at an annual equivalent rate of 7%. How much money do you have in the account after 15 years if you remove no money from it during that period?

7.3 What is the present value of \$30 000 due 10 years from now when the annual interest rate is 8% and (a) interest is paid once each year, at the end of the year; (b) interest is continuously compounded?

7.4 Work out the derivatives of the following functions, using the standard rules:

$$\text{(a)} \quad f(x) = \ln(2x^2 + 5x + 4); \quad \text{(b)} \quad g(x) = 3x^4 \exp x.$$

In each case state explicitly the set of values of x for which the derivative is defined.

7.5 Let f be the function

$$f(x) = \ln\left(\frac{x^4 + 6x^2 + 9}{x^2 + 1}\right).$$

By observing that

$$f(x) = \ln\left(x^4 + 6x^2 + 9\right) - \ln\left(x^2 + 1\right),$$

determine the derivative $f'(x)$. Show that one obtains the same result by differentiating $f(x)$ directly.

7.6 Calculate the derivatives of the functions $\sin(\ln x)$, $\tan(e^{x^2})$.

7.7 If x and y are related by the equation $x^{1/4}y^{2/3} = 8$, use the logarithm function to define X and Y in terms of x and y so that the relationship between X and Y is linear, that is of the form $aX + bY = c$.

7.8 The function $\log_{10}$ is defined as the inverse of the power function 10^x, in just the same way as ln is defined as the inverse of e^x. Write $x = \ln(y)$ and $z = \log_{10}(y)$, and use the fact that $10^z = \exp(z \ln 10)$ to show that

$$\ln(y) = (\ln 10) \log_{10}(y) = (2.3025\ldots) \log_{10}(y).$$

8 Introduction to Optimisation

8.1 Profit Maximisation

The most frequent use of the derivative occurs in problems where we have to find the 'best' value of some quantity. Such problems belong to the part of mathematics known as *optimisation*.

Suppose that a firm has cost function C, so that it costs $C(q)$ to produce q units of its product. Suppose also that the product can be sold at a price $P(q)$ per unit, depending on the quantity produced. Then the firm's *revenue* from producing q units is

$$R(q) = qP(q),$$

and its *profit* is

$$\Pi(q) = R(q) - C(q) = qP(q) - C(q).$$

Clearly, the 'best' value of q, from the firm's point of view, is that which maximises the profit. For example, suppose that the cost function is $C(q) = 9 + 5q$ and the price function is $P(q) = 6 - 0.01q$, as in Figure 8.1. Then the profit function, also sketched in Figure 8.1, is

$$\Pi(q) = q(6 - 0.01q) - (9 + 5q) = -9 + q - 0.01q^2.$$

Observe that, in this example, the cost function C and the price function P take simple but reasonable forms. The cost is the sum of two components: a *fixed cost* 9 and a *variable cost* $5q$. We may think of the fixed cost as representing overheads, and the variable cost as indicating that the cost of producing each of the q items is 5. The price function tells us that the goods will sell for a price of 'about' 6 per item, although there is a slight fall in price as the number of items increases. Later we shall look at these matters more generally.

From the graph it is clear that in this case there is a value q_m where the profit Π is a maximum. Furthermore, the graph is horizontal at this point. In other words, the gradient of the tangent to the graph at that point is 0. In terms of the derivative, this means that $\Pi'(q_m) = 0$. In general, if we are given a function f, then a value c for which $f(c)$ is a maximum will satisfy $f'(c) = 0$. Although this principle is perhaps the most important application of calculus to practical situations, it has to be used with care. In the rest of this chapter we shall look more closely at the relationship between a function f and its derivative f'.

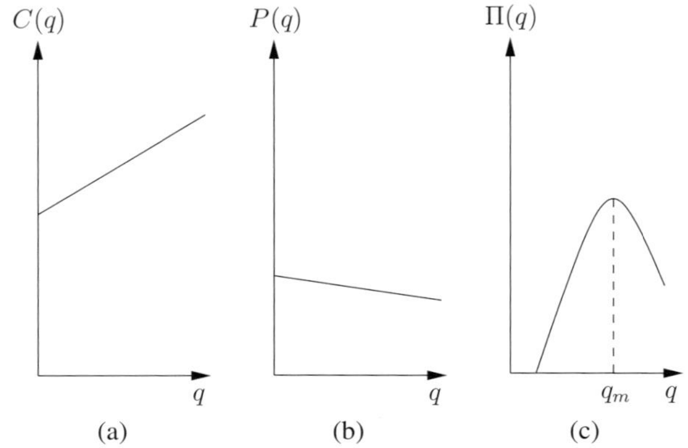

Figure 8.1 Graphs of cost, price and profit functions.

8.2 Critical Points

We defined the derivative $f'(x)$ as a measure of the gradient of f at x. It follows from this that we can tell whether a function is increasing or decreasing at a given point, simply by working out its derivative at that point.

- *If $f'(x) > 0$, then f is increasing at x.*
- *If $f'(x) < 0$, then f is decreasing at x.*

It is also clear that at a point c for which $f'(c) = 0$, the function f is neither increasing nor decreasing: in this case we say that c is a *critical point* (or *stationary point*) of f.

For example, we can find the maximum point q_m of the profit function discussed in Section 8.1, using the fact that it is a critical point. By the elementary rules of differentiation, we have

$$\Pi'(q) = 1 - 0.02q,$$

so the condition $\Pi'(q_m) = 0$ gives $1 - 0.02q_m = 0$, that is $q_m = 50$.

It must be stressed that a function can have more than one kind of critical point. For example, the function whose graph is shown in Figure 8.2 has four critical points: two of them (a and d) are *maximum* points, one of them (b) is a *minimum* point and the other one (c) is neither a maximum nor a minimum, it is an *inflexion* point.

We can decide the nature of a given critical point by considering what happens to f' in its vicinity. Looking at the point a in Figure 8.2, we observe that the gradient (the derivative) is positive for values just less than a, zero at a and negative for values just greater than a. This means that f' is decreasing at a, so the derivative of f' is negative at a. We call the derivative of f' the *second derivative* of f, and denote it by

$$f''(x) \quad \text{or} \quad \frac{d^2f}{dx^2}.$$

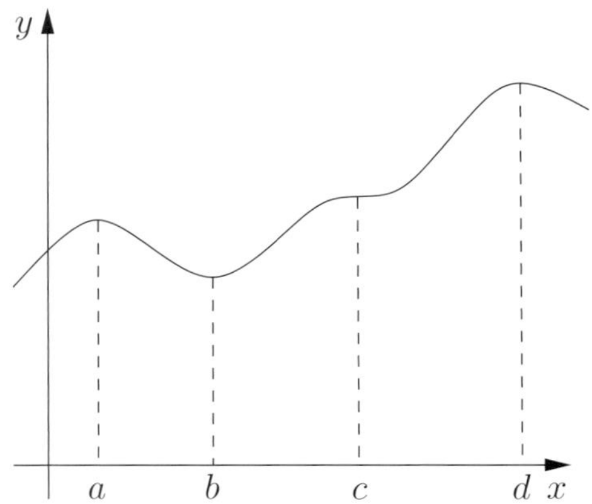

Figure 8.2 A function with four critical points.

Although the graphical argument is not completely general, it can in fact be proved that a sufficient condition for f to have a maximum at a critical point a is $f''(a) < 0$. There is a similar condition for a minimum. Summarising:

- *If $f'(a) = 0$ and $f''(a) < 0$, then the point a is a maximum of f.*
- *If $f'(b) = 0$ and $f''(b) > 0$, then the point b is a minimum of f.*

These observations together form the *second-order conditions* for the nature of a critical point.

In Figure 8.2, the inflexion point c is such that $f''(c) = 0$, and it is clear that, in general, this condition must be satisfied at any inflexion point. However, if f'' is zero at a critical point then we *cannot* conclude that the point is an inflexion point. For example, if $f(x) = x^4$ then $f''(0) = 0$, but f does not have an inflexion point at 0; it has a minimum there (see Example 8.4). The general point is worth repeating:

- *If $f'(c) = 0$ and $f''(c) = 0$, then c may be a maximum, a minimum or an inflexion point.*

However, the second-order conditions are usually adequate for classifying critical points, as the following examples show.

Example The critical points of the function $f(x) = x^3 - 12x^2 + 21x + 100$ are the points where $f'(x) = 0$; that is

$$3x^2 - 24x + 21 = 0 \quad \text{or} \quad 3(x-1)(x-7) = 0.$$

Thus the critical points are at 1 and 7. The second derivative is $f''(x) = 6x - 24$, so

$$f''(1) = -18 < 0 \quad \text{and} \quad f''(7) = 18 > 0.$$

It follows that 1 is a maximum and 7 is a minimum. [*Exercise*: sketch the graph of f.] ■

Example In Sections 1.3 and 1.4 we looked at the market for a good in which the demand and supply functions are

$$q^D(p) = 40 - 5p \text{ and } q^S(p) = \frac{15}{2}p - 10.$$

Suppose the government wishes to raise revenue by imposing an excise tax on this good. Clearly, a small tax will bring in little revenue but, on the other hand, if the tax is too large consumption will fall dramatically and the revenue will also be hit. What is the best policy?

In Section 1.4 we found that the equilibrium price and quantity in the presence of an excise tax T are

$$p^T = 4 + \frac{3}{5}T \text{ and } q^T = 20 - 3T.$$

The revenue $R(T)$ is the product of the quantity sold, q^T, and the excise tax, T. That is

$$R(T) = q^T \times T = (20 - 3T)T = 20T - 3T^2.$$

To find the value of T_m where this is a maximum, we first set $R'(T) = 0$. We have $R'(T) = 20 - 6T$, so that $10/3$ is the only critical point. The second derivative is $R''(T) = -6$, which is negative, so $T_m = 10/3$ is the maximum point. The maximum revenue is $R(T_m) = 100/3$. ■

8.3 Optimisation in an Interval

The maximum and minimum points discussed in the previous section are, strictly speaking, *local* maxima and minima. For example, in Figure 8.2, the point a is a local maximum, because the value at a is greater than at neighbouring points, but it is clearly not the largest value of the function overall. To make this distinction clearer we have to specify the set of x-values under consideration. Usually this is an *interval*, that is the set of x between two values u and v, denoted by

$$[u, v] = \{x \mid u \leq x \leq v\}.$$

In practice, we are usually faced with a *global* optimisation question, in the form: what is the largest (or smallest) value of $f(x)$ when x is in $[u, v]$? For example, there will usually be an upper bound L on the quantity of its product that a firm can make, so the realistic profit-maximisation problem is to find the greatest value of $\Pi(q)$ when q is in $[0, L]$.

Looking again at Figure 8.1 we observe that if, for example, the firm cannot produce more than 40 units of its product, then the local maximum of Π at $q_m = 50$ is irrelevant. The global maximum of Π in the interval $[0, 40]$ is at 40; thus, the maximum profit achievable for q between 0 and 40 is the profit $\Pi(40)$. On the other hand, if the firm can produce up to 80 units, then it should produce only 50, because that gives the largest value of Π in the interval $[0, 80]$. This example illustrates the fact that in order to find the global maximum of a function in a given interval, we need to use a combination of calculus and common sense.

We conclude by summarising the rules we have found for solving the problem of the global maximum. Of course there is a parallel statement concerning the global minimum.

- *Suppose f is a function for which f' and f'' both exist in the interval $[u, v]$. Then f has a global maximum point m in $[u, v]$, and* either *(i) $u < m < v$, $f'(m) = 0$ and $f''(m) \leq 0$,* or *(ii) m is one of u, v.*

Example What are the maximum and minimum values of the function $f(x) = x^3 - 8x^2 + 16x - 1$ in the interval $[0, 2]$?

First we find the critical points.

$$f'(x) = 3x^2 - 16x + 16 = (3x - 4)(x - 4),$$

so $f'(x) = 0$ at the points $4/3$ and 4. In the interval $[0, 2]$ only the critical point $4/3$ is relevant. We have $f''(x) = 6x - 16$, so $f''(4/3) = 24/3 - 16$, which is negative, so $4/3$ is a local maximum. In order to find the global maximum and minimum in $[0, 2]$, we must compare the values at the end-points with the value at the relevant critical point. We have

x:	0	4/3	2
$f(x)$:	-1	229/27	7

Since $229/27$ is greater than 7, the maximum value in the interval $[0, 2]$ is at the critical point $4/3$ and the minimum value in this interval is at the end-point 0. (See Figure 8.3.) ■

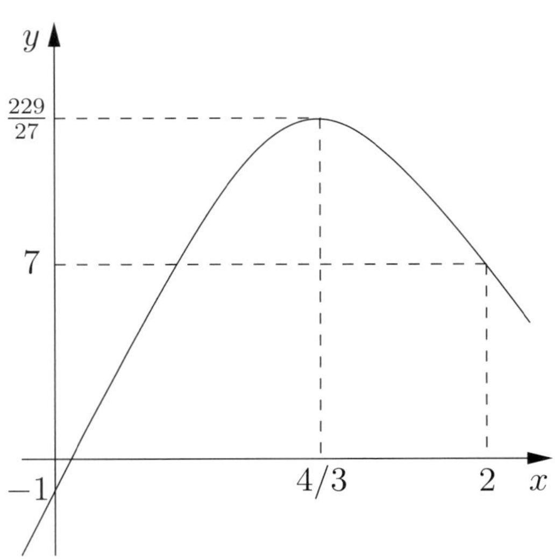

Figure 8.3 The function $x^3 - 8x^2 + 16x - 1$.

8.4 Infinite Intervals

Occasionally, the constraint on the values of x that may be considered is of the form $x \geq u$ for some number u; in other words, there is a lower bound on the allowable x-values, but no upper bound of the form $x \leq v$. We may easily extend the interval notation to such a situation. We say that the set $\{x \mid x \geq u\}$ is the *infinite interval* $[u, \infty)$. Here, ∞ is merely a symbol denoting the fact that the set of x is not bounded above; it should never be thought of as a number itself.

If we have to optimise a function f on an interval of the form $[u, \infty)$, we follow a procedure very similar to that for optimising on a bounded interval $[u, v]$. However, there is one important difference. Although it is true that any 'well-behaved' function f has a largest value in an interval of the form $[u, v]$, it is possible that f does not have a largest value in an interval of the form $[u, \infty)$. For example, the function $f(x) = x^2$ has no largest value on the interval $[0, \infty)$; this follows from the fact that x^2 tends to infinity as $x \to \infty$.

The following result summarises the infinite case.

- *Suppose f is a function for which f' and f'' both exist in the interval $[u, \infty)$. If f has a global maximum point m in $[u, \infty)$,* then either *(i) $m > u$, $f'(m) = 0$ and $f''(m) \leq 0$,* or *(ii) $m = u$. A similar statement holds with 'maximum' replaced by 'minimum'.*

Example Suppose you are given a nineteenth-century painting currently worth \$2000, and you estimate that its value will increase steadily at \$500 per annum, so that the amount realised by selling the painting after t years will be $2000 + 500t$. (As usual, we omit the currency units.) In the context of continuous compounding of interest at 10%, the present value of the amount realised is

$$P(t) = (2000 + 500t)e^{-0.1t}.$$

What is the optimum time to sell?

By routine application of the rules for differentiation, we get

$$P'(t) = 500e^{-0.1t} + (2000 + 500t)(-0.1)e^{-0.1t} = e^{-0.1t}(300 - 50t).$$

Since this is zero when $t = 6$, that is a critical point of P. Differentiating again we get

$$P''(t) = (-0.1)e^{-0.1t}(300 - 50t) + e^{-0.1t}(-50) = e^{-0.1t}(5t - 80).$$

It follows that $P''(6) < 0$, so the critical point $t = 6$ is a local maximum.

The fact that $t = 6$ is indeed the maximum in $[0, \infty)$ can be verified by common-sense arguments. We know that $t = 6$ is the only critical point of $P(t)$, and that it is a local maximum. It follows that $P(t)$ must decrease steadily for $t > 6$, because if at any stage it started to increase again, it would have to pass through a critical point first. A few calculations will establish that the graph of $P(t)$ has the form shown in Figure 8.4; in particular, as $t \to \infty$, $P(t) \to 0$. ■

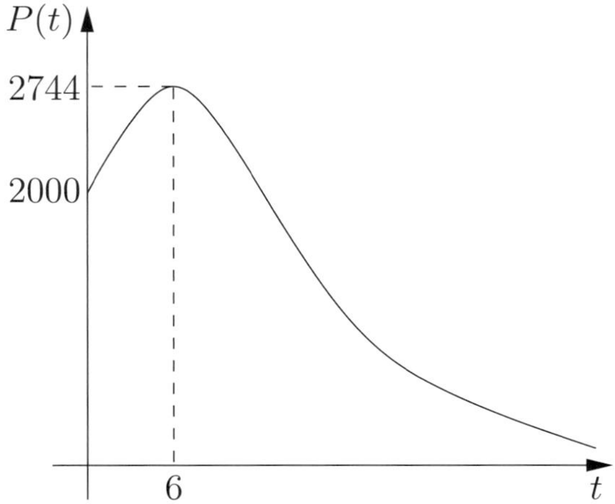

Figure 8.4 The graph of $P(t)$.

Worked Examples

Example 8.1 Find the points where the graph of the function $f(x) = x^3 - 6x^2 + 11x - 6$ crosses the axes, and sketch the graph.

Solution

We first determine where the graph of f crosses the x-axis; that is where $f(x) = 0$. Trying $x = 1$, we see that $f(1) = 1 - 6 + 11 - 6 = 0$, so one of the factors of $x^3 - 6x^2 + 11x - 6$ is $(x - 1)$. We know, therefore, that

$$f(x) = x^3 - 6x^2 + 11x - 6 = (x - 1)(ax^2 + bx + c),$$

for some numbers a, b, c. By comparing the x^3 terms and the constant terms on each side of this equation (in other words, by 'comparing coefficients'), we obtain $a = 1$ and $c = 6$. Thus

$$x^3 - 6x^2 + 11x - 6 = (x - 1)(x^2 + bx + 6) = x^3 + (b - 1)x^2 + (6 - b)x - 6.$$

It is clear that $b = -5$, and we have

$$f(x) = (x - 1)(x^2 - 5x + 6) = (x - 1)(x - 2)(x - 3).$$

Thus, $f(x) = 0$ when $x = 1, 2$ or 3. To find where the graph of f crosses the y-axis, we simply observe that $f(0) = -6$, so it crosses the y-axis at $(0, -6)$. For large values of x (positive or negative), the dominant term in $f(x)$ is x^3; therefore, as $x \to \infty$, $f(x) \to \infty$ and as $x \to -\infty, f(x) \to -\infty$. All this information is sufficient to sketch the graph of f; see Figure 8.5.

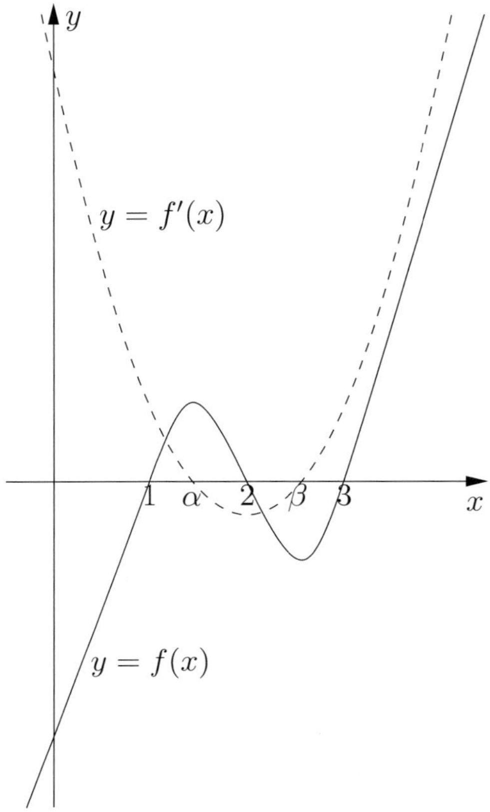

Figure 8.5 The function $x^3 - 6x^2 + 11x - 6$ and its derivative.

Example 8.2 Let f be as in Example 8.1. Sketch the graph of f' and explain how the properties of this graph correspond to the properties of the graph of f.

Solution

The derivative of f is $f'(x) = 3x^2 - 12x + 11$. Setting $f'(x) = 0$ and using the formula for the solutions to a quadratic equation given in Section 2.4, the critical points of f are

$$\alpha = \frac{1}{6}(12 - \sqrt{12}) = 1.423\ldots, \quad \beta = \frac{1}{6}(12 + \sqrt{12}) = 2.577\ldots.$$

The second derivative is $f''(x) = 6x - 12$, and it is easily checked that $f''(\alpha)$ is negative while $f''(\beta)$ is positive. Hence α is a local maximum of f and β is a local minimum of f.

The graph of f' is sketched in Figure 8.5. We have already found that $f'(x) = 0$ when $x = \alpha$ and $x = \beta$, so we have determined where the graph of f' crosses the x-axis. The function f' is a quadratic and the coefficient of x^2 is positive, so the graph has the shape illustrated. It crosses the y-axis when $x = 0$, that is when $y = 11$. The point where f' attains its minimum can be found by setting $f''(x) = 0$: since $f''(x) = 6x - 12$ this gives $x = 2$.

There are several points of comparison between the two graphs. First, we note that the points where f has its maximum and minimum correspond to the points where the graph of f' crosses the x-axis, since, at these points, $f'(x) = 0$. Furthermore, when f is increasing, which is for $x \leq \alpha$ and $x \geq \beta$, f' is positive and the graph of f' lies above the x-axis. When f is decreasing, between α and β, the graph of f' lies below the x-axis.

Example 8.3 In each of the following cases, find: (a) all the points where the derivative of the function is zero; (b) the points in the given interval where the function attains its maximum and minimum values.

(i) $2x^3 - 9x^2 + 12x$ in $[0, 2]$
(ii) $2x^3 - 9x^2 + 12x$ in $[0, 3]$
(iii) $x \sin x + \cos x$ in $[0, 2]$

Solution

(i) We find the critical points:

$$f'(x) = 6x^2 - 18x + 12 = 6(x^2 - 3x + 2) = 6(x - 1)(x - 2),$$

so the critical points are 1 and 2, both of which lie in the interval under consideration. The values of f at these points are $f(1) = 5$ and $f(2) = 4$. The value of f at the (remaining) end-point of the interval is $f(0) = 0$. Hence, the maximum of f in $[0, 2]$ is at the point 1 and the minimum is at 0.

(ii) Here, we consider the same function, but in the interval $[0, 3]$. The value of f at the end-point is $f(3) = 9$; it now follows that the maximum value in $[0, 3]$ is attained at 3, and the minimum at 0.

(iii) Recall that the derivative of $\cos x$ is $-\sin x$ and the derivative of $\sin x$ is $\cos x$. So if $f(x) = x \sin x + \cos x$, we have

$$f'(x) = \sin x - x \cos x - \sin x = x \cos x.$$

Thus, the critical points in $[0, 2]$ are the points where $x \cos x = 0$. These are $x = 0$ and $x = \pi/2$. (As usual, angles are measured in radians.) The corresponding function values are $f(0) = 1$ and $f(\pi/2) = \pi/2 = 1.571$, approximately.

It remains to check the values at the end-points. We have already considered 0; at 2 we have $f(2) = 1.402$, approximately. It follows that the minimum value of f on the interval $[0, 2]$ is at 0 and the maximum is at $\pi/2$.

Example 8.4 Verify that when $g(x) = x^4$, we have $g'(0) = 0$ and $g''(0) = 0$. Explain why the critical point 0 is a minimum point of g.

Solution

We have $g'(x) = 4x^3$ and $g''(x) = 12x^2$, so it is clear that $g'(0) = 0$ and $g''(0) = 0$. In other words, 0 is a critical point, but the second-order conditions fail to classify it.

However, the nature of the critical point can be determined quite simply, by looking at how $g(x) = x^4$ behaves in the neighbourhood of $x = 0$. Since $x^4 = (x^2)^2$, the values are positive, whether x is positive or negative. Hence, 0 must be a minimum of g.

Example 8.5 The supply and demand sets for a good are

$$S = \{(q,p) \mid q = bp - a\}, \quad D = \{(q,p) \mid q = c - dp\},$$

where a, b, c, d are all positive. Suppose the government wishes to raise as much money as possible by imposing an excise tax on the good. What should be the value of the excise tax? What is the resulting government revenue?

Solution

The tax revenue is $R(T) = Tq^T$. In order to calculate q^T, we first note that when the excise tax is imposed, the selling price at equilibrium, p^T, is such that

$$q^T = b(p^T - T) - a = c - dp^T.$$

Solving for p^T, we obtain

$$p^T = \frac{c+a}{b+d} + \frac{bT}{b+d}.$$

(See Example 1.5.) Then

$$q^T = c - dp^T = \frac{bc - ad}{b+d} - \frac{bdT}{b+d},$$

so that

$$R(T) = \left(\frac{bc - ad}{b+d}\right) T - \left(\frac{bd}{b+d}\right) T^2.$$

Setting $R'(T) = 0$, we discover that there is only one critical point:

$$T_m = \frac{bc - ad}{2bd}.$$

The second derivative, $R''(T) = (-2bd)/(b+d)$, is constant and negative (since b and d are positive). Hence $R''(T_m) < 0$ and T_m is a maximum point. Therefore, T_m is the level of excise tax the government should impose. The resulting government revenue is

$$R(T_m) = \frac{(bc - ad)^2}{4bd(b+d)}.$$

Example 8.6 Suppose that you have just inherited an asset whose current market value is \$2000. Assume that the market value will increase steadily at a rate of \$300 per annum, and that the interest on a bank deposit will be compounded continuously at the equivalent annual rate of 6%. Explain why the present value of the amount realised by selling the asset after t years is

$$P(t) = (2000 + 300t)\exp(-0.06t),$$

and determine the optimum time to sell.

Solution

The market value after t years is given by $C(t) = 2000 + 300t$. The present value of $C(t)$ in t years under continuous compounding at an equivalent annual rate of 0.06 is

$$P(t) = C(t)e^{-0.06t} = (2000 + 300t)\exp(-0.06t).$$

In order to maximise this quantity (and hence determine the optimal time to sell), we calculate the derivative of $P(t)$. We have

$$\begin{aligned} P'(t) &= 300\exp(-0.06t) + (2000 + 300t)(-0.06)\exp(-0.06t) \\ &= \exp(-0.06t)\,(180 - 18t)\,. \end{aligned}$$

For a critical point, we must have $P'(t) = 0$, that is $t = 10$. Calculating $P''(t)$ we find that $P''(10)$ is negative, and so this is a local maximum. (Alternatively, we could observe that, for $t < 10$, $P'(t) > 0$ and for $t > 10$, $P'(t) < 0$.) As t tends to infinity, the negative exponential factor $\exp(-0.06t)$ tends to zero, and in fact this dominates the term $(2000 + 300t)$, the net effect being that $P(t) \to 0$. Furthermore, $P(0) = 2000$ and $P(10) > 2000$. It follows that $t = 10$ gives the *global* maximum of $P(t)$ for $t \geq 0$ and hence the asset should be sold after 10 years. (As in the example of Section 8.4, we could instead note that the local maximum at $t = 10$ must be the global maximum since it is the only critical point: $P(t)$ must decrease steadily for $t > 10$, because if it started to increase again, it would have to pass through a critical point first.)

Main Topics

- revenue and profit
- finding critical points
- classifying critical points using second derivative
- optimisation in an interval

Key Terms, Notations and Formulae

- revenue, $R(q) = qP(q)$
- profit function, $\Pi(q) = R(q) - C(q)$
- critical point: $f'(c) = 0$
- maximum, minimum, inflexion
- second derivative of f, $f'' = (f')'$, also denoted $\dfrac{d^2f}{dx^2}$
- second-order conditions
- if $f'(a) = 0$ and $f''(a) < 0$, then the critical point a is a maximum
- if $f'(b) = 0$ and $f''(b) > 0$, then the critical point b is a minimum
- global maximum, global minimum
- interval $[u, v]$; infinite interval $[u, \infty)$
- global maximum of f on $[u, v]$ is at u, v or c, with $u < c < v$, $f'(c) = 0$
- if m is global maximum of f on $[u, \infty)$, then $m = u$ or $m > u$ and $f'(m) = 0$

Exercises

8.1 Find the critical points of the function

$$f(x) = 2x^3 - 9x^2 + 1.$$

Determine whether each is a maximum, minimum or inflexion point. Sketch the graph of f.

8.2 Find the critical points of $f(x) = xe^{-x^2}$. Determine their natures by considering the behaviour of the derivative. Determine their natures, also, by using the second-derivative test.

8.3 The function f is given by

$$f(x) = (x^2 + 10x - 10)e^{-x}.$$

Find the critical points of f and, for each, determine whether it is a maximum, minimum or inflexion point.

Problems

8.1 The function g is given by $g(x) = x^3 - 6x^2 + 12x - 1$. Show that g has only one critical point. Determine whether this point is a maximum, a minimum or an inflexion point.

8.2 Find the maximum and minimum values of the function $x^3 - 8x^2 + 16x - 1$ in the interval $[2, 5]$. (Note that this function was discussed in Section 8.3.)

8.3 Find the maximum and minimum values of the function $f(x) = x^4 - 8x^3 + 16x^2 - 7$ in the interval $[1, 4]$.

8.4 Find the maximum value of the function $f(x) = -2x^3 + 3x^2 + 12x + 9$ in the interval $[0, \infty)$.

8.5 Find the maximum and minimum values of x^2e^{-x} in the interval $[0, 5]$.

8.6 Suppose you have inherited an antique whose current market value is \$500. Let us assume that the market value will increase steadily at a rate of \$100 per annum, and that interest on a bank deposit will be compounded continuously at the equivalent annual rate of 5%. Write down the expression for the present value of the amount realised by selling the antique after t years, and determine the optimum time to sell (assuming you can find someone to buy it then, at the market value).

8.7 Suppose you own a piece of land whose value $V(t)$ after t years is $V(t) = e^{\sqrt{t}}$. Assuming that interest on a bank deposit will be compounded continuously at the equivalent annual rate of 12.5%, write down an expression for the present value of the amount realised by selling the land after t years, and determine the optimum time to sell.

9 The Derivative in Economics I

9.1 Elasticity of Demand

When the market is described by a downward-sloping demand set D, an increase in the selling price of a good will lead to a decrease in the quantity sold. In Chapter 6 we explained how the derivative of the demand function q^D determines the relationship between the changes Δp and Δq: specifically, we showed that $\Delta q \simeq q'\Delta p$. (Here, and in what follows, we avoid cumbersome notation by writing q instead of $q^D(p)$ and q' instead of $q^{D'}(p)$.)

In practice, a more important question is how *revenue* will change if the selling price is increased. If the price p rises, the quantity sold q falls; but the revenue $R = qp$ is the product of these two things, and it may rise or fall. In order to determine which, we need a little more mathematics.

Of course, the answer to this question also depends upon the assumptions which we make about the firm's place in the market. We shall assume that a price rise applies uniformly to the entire supply of the good under consideration, whether or not it is produced by that firm. This would certainly be the case if the firm is a 'monopoly' (see Section 9.3), which produces the entire supply itself. In this case the firm can decide to increase the selling price without having to consider the effect of competition from other suppliers. Given that this is an option, the firm needs to know whether such an increase would increase its revenue.

Using the product rule to differentiate $R = qp$ with respect to p, remembering that q is a function of p, we get

$$R' = q'p + q.$$

Thus the condition that revenue increases, $R' > 0$, is equivalent to $q'p + q > 0$. It is customary to write this inequality in the equivalent form

$$-\frac{q'p}{q} < 1,$$

where the expression $-q'p/q$ is called the *elasticity of demand*, and denoted by $\varepsilon(p)$. Thus the elasticity determines whether revenue increases or decreases as price increases, as follows.

- If $\varepsilon(p) < 1$, a small increase in price results in an increase in revenue. We say the demand is *inelastic*.

- If $\varepsilon(p) > 1$, a small increase in price results in a decrease in revenue. We say the demand is *elastic*.

Example Suppose the demand function is

$$q^D(p) = \frac{K}{p^c} = Kp^{-c},$$

where K and c are positive constants. Then the elasticity of demand is

$$\varepsilon(p) = -\frac{q'p}{q} = -\frac{\left(-cKp^{-c-1}\right)p}{Kp^{-c}} = c.$$

Here the elasticity of demand is *constant*: $\varepsilon(p) = c$ for all prices p. Thus if $c < 1$ the demand is inelastic, and if $c > 1$ it is elastic. ■

There is another way of thinking about elasticity. Suppose that the selling price is changed by a small amount Δp and that, as a result, the quantity sold changes by an amount Δq. Then we know that $\Delta q \simeq q'\Delta p$, and so we can replace q' by $\Delta q/\Delta p$ in the definition of $\varepsilon(p)$:

$$\varepsilon(p) = -\frac{q'p}{q} \simeq -\frac{\Delta q}{\Delta p}\frac{p}{q}.$$

Rearranging the right-hand side we see that $\varepsilon(p)$ is approximately equal to

$$\frac{(-\Delta q/q)}{(\Delta p/p)}.$$

In other words, it is the ratio of the *relative* decrease in quantity to the *relative* increase in price.

This formula shows that demand is inelastic ($\varepsilon < 1$), and revenue rises with price, if $-\Delta q/q < \Delta p/p$, that is if the proportional decrease in quantity sold is less than the proportional increase in price. Putting it another way, the demand is inelastic if a percentage increase in price results in a smaller percentage decrease in quantity sold.

Example Suppose the demand set is $D = \{(q,p) \mid q + 5p = 30\}$. Then the demand function is $q(p) = 30 - 5p$ and the elasticity of demand is

$$\varepsilon(p) = -\frac{q'p}{q} = -\frac{(-5)p}{30-5p} = \frac{p}{6-p}.$$

This is not defined when $p = 6$ (since then the denominator is zero). The values of p which are economically significant are those for which both p and q are non-negative, that is for $0 \leq p \leq 6$, and the elasticity is defined for all such p other than 6. The demand is elastic when $p/(6-p) > 1$; that is when $3 < p < 6$, and it is inelastic when $p/(6-p) < 1$, which is the case for $0 \leq p < 3$. ■

9.2 Profit Maximisation Again

In the previous section we discussed how revenue changes as a function of the price. In this section we shall look at revenue and profit as functions of the quantity; in particular, we shall study the question of how to maximise profit. We have already discussed a very simple and special case of this problem in Section 8.1, but here we shall begin with a more general approach.

The relevant variables are the firm's revenue function $R(q)$, its cost function $C(q)$ and its profit $\Pi(q)$, which are related by the fundamental equation

$$\Pi(q) = \text{profit} = \text{revenue} - \text{cost} = R(q) - C(q).$$

The 'profit maximisation principle' suggests that the firm will seek to produce the quantity q_m for which $\Pi(q_m)$ is a maximum. In practice, the range of possible values of q is an interval $[0, L]$, where L is the upper limit on the amount that can be produced. Although we can be sure that the function Π attains a maximum value somewhere in $[0, L]$, it may be that the maximum occurs at one of the end-points 0 and L. In that case the firm will maximise its profit either by producing nothing, or by producing the most that it can. These are both conceivable situations, but we put them aside for the moment and suppose instead that the maximum profit is achieved by producing an amount strictly between 0 and L. In this case, the profit is maximised at a point q_m where

$$\Pi'(q_m) = 0, \quad \text{that is} \quad R'(q_m) = C'(q_m).$$

Recall (Section 6.3) that the derivative C' of the cost function is usually referred to as the *marginal cost*, representing the cost of producing one more unit. In the same way, we can think of R' as the *marginal revenue*, the revenue which results from producing one more unit. Using this terminology we have a basic economic principle:

- *The optimum production level occurs when marginal revenue equals marginal cost.*

There is a simple way of thinking about this principle. Clearly, if the cost of producing one extra unit is less than the revenue which results, the extra unit should be produced; in other words, the firm should produce more. On the other hand, if the extra cost exceeds the extra revenue, the firm should not produce more. When the extra cost is equal to the extra revenue, the firm is at the optimum production level. For example, a fruit-drop firm should produce the number of fruit-drops for which the cost of producing one more fruit-drop is equal to the revenue from producing one more fruit-drop.

9.3 Competition versus Monopoly

In general, a firm's revenue is the product of the quantity q produced and the price per unit which holds when q is available, which we denote by $P(q)$. In symbols

$$R(q) = qP(q).$$

The form of $P(q)$ depends upon the assumptions we make about the place of the firm in the market. At one extreme is the case of *perfect competition*, where the firm is small and its output does not affect the market price of its good. From the firm's point of view, however much it produces the price remains fixed, so here we have $P(q) = p_0$, a constant independent of q.

At the other extreme is the case of *monopoly*, where the firm supplies the entire quantity of the good under consideration. Here the price is determined by the demand set D; in other words, $P(q)$ is $p^D(q)$, the price the consumers will pay when q is available.

When $P(q)$ is known, the profit function is

$$\Pi(q) = qP(q) - C(q),$$

and the profit maximisation condition $\Pi'(q) = 0$ becomes

$$qP'(q) + P(q) - C'(q) = 0.$$

In any particular case we can use this equation to solve the profit maximisation problem. The case of perfect competition will be studied in Chapter 10, but we shall look first at an example involving a monopoly.

For a monopoly, then, the profit function is

$$\Pi(q) = qp^D(q) - C(q).$$

This has to be maximised for q in an interval $[0, L]$ representing the feasible limits of production. It may be that the maximum is attained at one of the endpoints, and this possibility must not be forgotten. Furthermore, when we use the profit maximisation condition $\Pi'(q_m) = 0$, we must remember that not all critical points are maxima, and so the second-order conditions need to be checked too.

Example The only firm manufacturing a certain kind of machine tool can produce up to 100 per week. The demand set for these items is

$$D = \{(q,p) \mid q + 5p = 850\},$$

where p is measured in suitable units. The cost (in the same units) of producing q items per week is

$$C(q) = 300 - 10q + q^2.$$

How many items should be produced each week in order to maximise profit?

Since the firm is the only one producing the machine tools, it is a monopoly. The inverse demand function is $p^D(q) = 170 - 0.2q$, and it follows that the weekly profit from producing q machine tools is

$$\begin{aligned}\Pi(q) = qp^D(q) - C(q) &= q(170 - 0.2q) - (q^2 - 10q + 300)\\ &= -300 + 180q - 1.2q^2.\end{aligned}$$

The condition $\Pi'(q) = 0$ is $180 - 2.4q = 0$, so there is one critical point $q_m = 75$. The profit $\Pi(75)$ is 6450.

In order to be sure that this is indeed the optimum value, we have to check two things. First, would it not be better to produce either zero or the upper limit of 100 items per week? Since

$$\Pi(0) = -300 \quad \text{and} \quad \Pi(100) = 5700,$$

both of which are less than 6450, we conclude that neither of these possibilities is relevant. Finally, we observe that $\Pi''(75) = -2.4$, which is negative, so $q_m = 75$ is indeed a maximum. (In fact, this conclusion also follows by common-sense arguments here, but in general it must be verified.) ■

Worked Examples

Example 9.1 The demand set for a good is

$$D = \{(q, p) \mid q(1 + p^2) = 100\}.$$

Determine the elasticity of demand $\varepsilon(p)$ as a function of p. For what values of p is the demand inelastic?

Solution

The demand function is $q^D(p) = 100/(1 + p^2)$, which has derivative

$$q' = -200p/(1 + p^2)^2.$$

The elasticity of demand is

$$\varepsilon(p) = -\frac{q'p}{q} = \frac{200p^2/(1 + p^2)^2}{100/(1 + p^2)} = \frac{2p^2}{1 + p^2}.$$

The demand is inelastic where $\varepsilon(p) < 1$. In this case $2p^2/(1 + p^2) < 1$ implies that $p^2 < 1$ and since $p \geq 0$, this means that the demand is inelastic when $p < 1$.

Example 9.2 Calculus Corporation is a monopoly with cost function

$$C(q) = q + 0.02q^2$$

and the upper limit on its production is 200. The demand set for its product is $D = \{(q, p) \mid q + 20p = 300\}$. Work out (a) the inverse demand function; (b) the profit function; (c) the optimal value q_m and the maximum profit; (d) the corresponding price.

Solution

(a) The inverse demand function is

$$p^D(q) = (300 - q)/20 = 15 - 0.05q.$$

(b) The profit function is

$$\Pi(q) = qp^D(q) - C(q) = q(15 - 0.05q) - (q + 0.02q^2) = 14q - 0.07q^2.$$

(c) We have $\Pi'(q) = 14 - 0.14q$, so $q = 100$ is a critical point. The second derivative of Π is $\Pi''(q) = -0.14$, which is negative, so the critical point is a local maximum. The value of the profit there is $\Pi(100) = 1400 - 700 = 700$, whereas $\Pi(0) = 0$ and $\Pi(200) = 0$. Since the maximum profit in the interval $[0, 200]$ must be either at a local maximum or an end-point, it follows that the maximum profit is 700, obtained when $q = 100$.
(d) The price when $q = 100$ is $p^D(100) = 15 - (0.05)(100) = 10$.

Example 9.3 Integration Incorporated is a monopoly with cost function

$$C(q) = 100 + 80q - 50q^2 + 0.5q^3,$$

and the demand set for its product is

$$D = \{(q,p) \mid 2p + q^2 - 20q = 100\}.$$

Sketch the graph of the profit function for $q > 0$. Find the level of production which maximises the firm's profit, if the upper limit on its output is (i) 30, (ii) 50.

Solution

The inverse demand function is easily obtained from the definition of the demand set: $p^D(q) = 50 + 10q - 0.5q^2$. The profit function is therefore

$$\begin{aligned}\Pi(q) = qp^D(q) - C(q) &= q(50 + 10q - 0.5q^2) - (100 + 80q - 50q^2 + 0.5q^3)\\ &= -q^3 + 60q^2 - 30q - 100.\end{aligned}$$

The critical points are found by setting $\Pi'(q) = 0$, that is

$$-3q^2 + 120q - 30 = -3(q^2 - 40q + 10) = 0.$$

Using the formula for the solutions of a quadratic equation, we find critical points α and β given by

$$\alpha = 20 - \sqrt{390} = 0.2516, \quad \beta = 20 + \sqrt{390} = 39.7484.$$

The second derivative of Π is $\Pi''(q) = -6q + 120$. Since $\Pi''(\alpha) > 0$, α is a local minimum and, since $\Pi''(\beta) < 0$, β is a local maximum. The corresponding values of Π are $\Pi(\alpha) = -103.766$ and $\Pi(\beta) = 30703.8$.

We also have

$$\Pi(0) = -100, \quad \Pi(30) = 26000, \quad \Pi(50) = 23400.$$

The graph of Π is shown in Figure 9.1. It follows that the maximum profit in the interval $[0, 30]$ is attained when $q = 30$, whereas the maximum profit in the interval $[0, 50]$ is attained when $q = \beta = 39.75$ approximately.

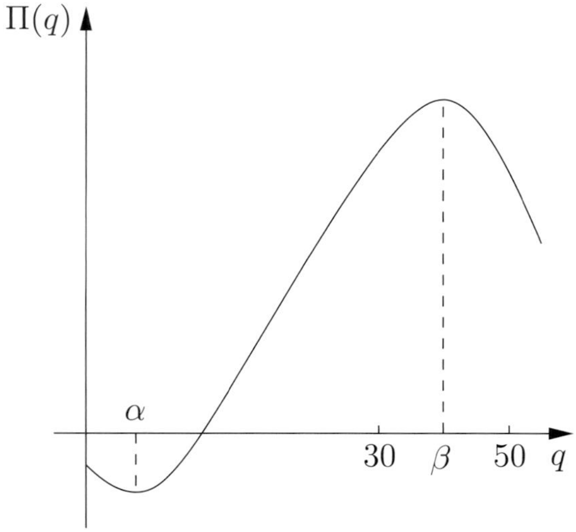

Figure 9.1 The graph of $\Pi(q)$.

Main Topics

- elasticity of demand and its relation to revenue
- profit maximisation in general
- perfect competition and monopolies

Key Terms, Notations and Formulae

- elasticity of demand, $\varepsilon(p) = -\dfrac{q'p}{q}$, where $q = q^D(p)$
- demand is elastic if $\varepsilon(p) > 1$, inelastic if $\varepsilon(p) < 1$
- at optimum production, marginal revenue = marginal cost
- for a monopoly, $\Pi(q) = qp^D(q) - C(q)$

Exercises

9.1 Given that a market's demand equation is

$$q^2(4 + p^3) = 31,$$

find the elasticity of demand and show that it never exceeds $3/2$.

9.2 A firm is a monopoly with cost function

$$C(q) = q + 0.02q^2.$$

The demand equation for its product is $q+20p = 300$. Work out (a) the inverse demand function; (b) the profit function; (c) the optimal value q_m and the maximum profit; (d) the corresponding price.

Problems

9.1 Calculate the elasticity of demand when the demand function is given by

$$q^D(p) = 70 - 4p.$$

For what range of values of p is your expression valid, and for which of these values is the demand inelastic?

9.2 Show that if the demand function for a good is $q^D(p) = a - bp$ for some positive constants a and b, then the demand is inelastic in the range $0 < p < a/2b$.

9.3 Suppose the demand set for a commodity is

$$D = \{(q,p) \mid q^2(2+p^3) = 200\}.$$

Determine the values of p where the demand is elastic.

9.4 Prove that, at the price which maximises revenue, the elasticity of demand equals 1.

9.5 Didlum Corporation is a monopoly with cost function

$$C(q) = q + 0.03q^2,$$

and the demand set for their product is

$$D = \{(q,p) \mid q + 20p = 500\}.$$

Work out (i) the inverse demand function; (ii) the profit function; (iii) the optimal value, q_m, of q, given that the upper limit on its production is 200. Explain why the optimal value is a maximum.

9.6 Idlers Incorporated is a monopoly with cost function

$$C(q) = q^3 - 105q^2 + 140q + 200,$$

the demand set for its product is

$$D = \{(q,p) \mid p + q^2 - 5q = 100\}$$

and the upper limit on its production is 150. Find the level of production, q_m, which maximises the firm's profit and determine the maximum profit. Sketch a graph of the profit function $\Pi(q)$.

10 The Derivative in Economics II

10.1 The Efficient Small Firm

In this chapter we shall use elementary calculus to describe the behaviour of a small firm under perfect competition. We shall take the definition of *small* to be that the price of the firm's product in the market is not affected by its own level of production. In other words, there is a given market price p_0 (sometimes called the *going price*) which is not under the firm's control. Economists say that the firm is a 'price-taker'.

Suppose that the cost function C for the firm is known. Then the firm is said to be *efficient* if it behaves in the following way. When the given market price is p_0, it will produce the amount q_0 for which the profit is a maximum, remembering that in some circumstances it may be best to produce nothing.

Our task is to determine the production level q_0 of an efficient small firm in terms of its cost function C and the market price p_0. Of course, the relationship between q_0, the amount the firm will supply, and p_0, the going price, is just the firm's supply set S. Our task, therefore, is to study the supply set for an efficient small firm.

Common sense suggests that S will take the form indicated in Figure 10.1.

When p_0 is below a certain level p_s, the firm will produce nothing, but when p_0 reaches p_s, the firm will 'start up' with a production level q_s. Production will then increase as the price increases, until p_0 reaches the level p_L at which the firm is at the upper limit of its production, L. When p_0 exceeds p_L, the firm produces L.

We shall show that, under certain assumptions, a supply set S resembling the one illustrated in Figure 10.1 can be derived from the firm's cost function C.

Suppose the firm were to produce a quantity q when the going price is p_0. Then its revenue $R(q)$ would be qp_0 and the cost would be $C(q)$, so its profit would be

$$\Pi(q) = R(q) - C(q) = qp_0 - C(q).$$

Since the firm is efficient, the amount q_0 which it actually produces will be such that q_0 gives the maximum value of this function. As we know, q_0 is found by equating the derivative to 0, that is

$$p_0 - C'(q_0) = 0.$$

So if the firm decides to go into production when the price is p_0, the amount q_0 it will produce is determined by the equation

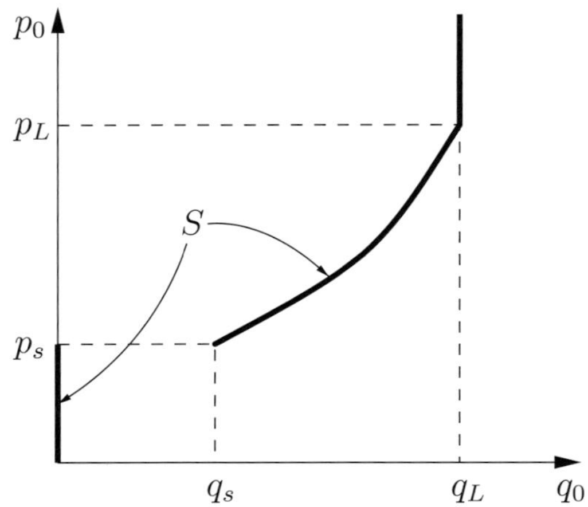

Figure 10.1 The supply set for an efficient small firm.

$$p_0 = C'(q_0).$$

By this simple argument we have solved a large part of the problem of determining the supply set S. The rule which determines p_0 as a function of q_0 is just the inverse supply function p^S. But we have just shown that this function is C', the marginal cost function. This is our first economic principle for efficient small firms:

- *Under perfect competition a firm's inverse supply function is equal to its marginal cost function.*

Referring to Figure 10.1, we have shown that the 'sloping' part of the supply set is just the graph of C'.

10.2 Startup and Breakeven Points

For convenience, we now drop the subscripts on p_0 and q_0. Thus we assume that when the market price p is given, and provided it lies within a certain range, the firm will produce a quantity q satisfying $p = C'(q)$. In this range, the profit from producing q is

$$\Pi(q) = qp - C(q) = qC'(q) - C(q).$$

Example Consider an efficient small firm with cost function

$$C(q) = 800 + 70q - 12q^2 + q^3.$$

We have $C'(q) = 70 - 24q + 3q^2$ so the profit function is

$$\begin{aligned}\Pi(q) &= q(70 - 24q + 3q^2) - (800 + 70q - 12q^2 + q^3)\\ &= -800 - 12q^2 + 2q^3.\end{aligned}$$

■

We must now look at what happens when the market price p is small. In this situation it may be in the firm's best interests to produce nothing. In the example, the profit when zero units are produced is

$$\Pi(0) = -C(0) = -800.$$

In general, $C(0)$, which is known as the *fixed cost*, is positive, so the profit from producing nothing is negative; in plain words, it is a loss. When q is just greater than zero, it is reasonable to expect that the loss will exceed the fixed cost, because the return from producing a small amount does not justify the expense. Figure 10.2 illustrates this behaviour for the profit function $\Pi(q) = -800 - 12q^2 + 2q^3$ discussed above.

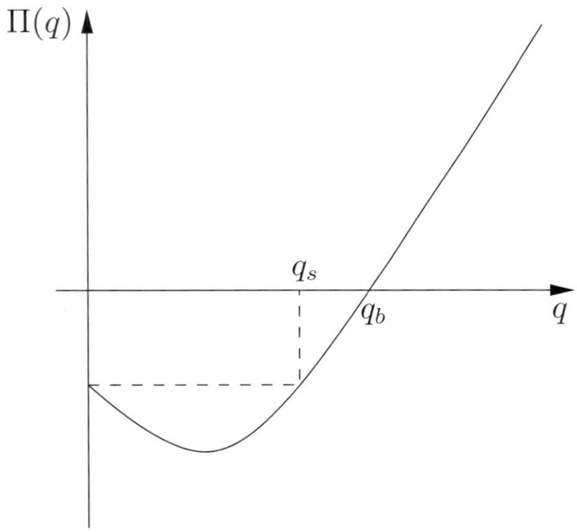

Figure 10.2 The profit function of the firm in the example.

The *startup point* is the production level q_s where the loss becomes equal to the fixed cost, so that it is worthwhile to start production, at least in the short run. Thus q_s can be found by solving the equation $\Pi(q_s) = \Pi(0)$; explicitly

$$q_s C'(q_s) - C(q_s) = -C(0).$$

The corresponding value of the market price, p_s, is determined by the fact that $p_s = C'(q_s)$, and the equation can be rearranged to give this explicitly:

$$p_s = C'(q_s) = \frac{C(q_s) - C(0)}{q_s}.$$

The quantity $C(q) - C(0)$ represents the total cost less the fixed cost, and is known as the *variable cost*. It is sometimes denoted by VC. Similarly, dividing by q we obtain the *average variable cost*, denoted by AVC. Writing the marginal cost $C'(q)$ as MC, the last equation says that, at q_s, $MC = AVC$.

- *At the startup point, marginal cost is equal to average variable cost.*

Finally we turn to the question of real profit; or when does $\Pi(q)$ become positive? In the long run this is the only significant question, because a firm which continually makes a loss will not survive. We define the *breakeven point* q_b by the equation $\Pi(q_b) = 0$. Using the formula for Π again we get

$$q_b C'(q_b) - C(q_b) = 0; \quad \text{that is} \quad C'(q_b) = \frac{C(q_b)}{q_b}.$$

Using the 'marginal' and 'average' terminology, we have our third principle.

- *At the breakeven point, the marginal cost is equal to the average cost.*

All three principles obtained above are illustrated in Figure 10.3. The first principle tells us that the *MC* curve is the graph of the firm's inverse supply function (for $q \geq q_s$). The other two principles tell us that the *MC* curve intersects the *AVC* curve and the *AC* curve at q_s and q_b, respectively. Figure 10.3 also indicates two other results which follow from similar calculations (see Example 10.2):

- *At the startup point, the derivative of the average variable cost is zero.*
- *At the breakeven point, the derivative of the average cost is zero.*

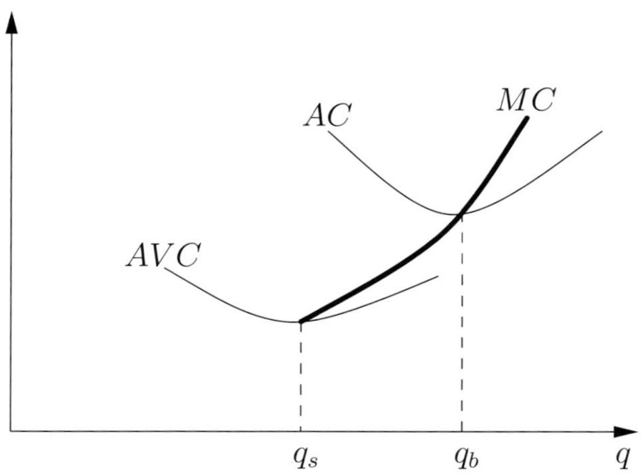

Figure 10.3 Relations between *MC*, *AC* and *AVC*.

Example (continued) Consider again the efficient small firm described in the earlier example. To find the startup point we have to solve the equation $\Pi(q) = \Pi(0)$, which is

$$-800 - 12q^2 + 2q^3 = -800, \quad \text{that is} \quad 2q^2(-6 + q) = 0.$$

The solution is $q_s = 6$, and the going price at which the firm starts production is $p_s = C'(q_s) = 34$.

To find the breakeven point we have to solve the equation $\Pi(q) = 0$:

$$-800 - 12q^2 + 2q^3 = 0, \quad \text{that is} \quad 2(10 - q)(40 + 4q + q^2) = 0.$$

Now the quadratic equation $q^2 + 4q + 40 = 0$ has no solutions, since $4^2 < 4 \times 1 \times 40$, so the breakeven point is $q_b = 10$. So provided the firm's upper limit of production L exceeds 10, the analysis given above holds, and the firm's supply set is like the one illustrated in Figure 10.1.

It is easy to verify all the 'principles' in this case. For example, the marginal cost function is $MC = 70 - 24q + 3q^2$ and the average variable cost function is $AVC = (C(q) - C(0))/q = q^2 - 12q + 70$. By substituting $q = q_s = 6$ we can check that $MC = AVC$ at the startup point. ■

Of course, the example is artificial, because the cost function has been chosen to produce the suggested behaviour. Another way of setting up the cost function will be discussed in Chapter 21.

Worked Examples

Example 10.1 Suppose that Alpern and Co. is an efficient small firm which cannot produce more than 6 units of its product each week. If the cost function is $C(q) = 100 + 20q - 6q^2 + q^3$, determine: (a) their fixed cost; (b) their profit function; (c) their startup point; (d) their breakeven point; (e) their supply set. Sketch the profit function $\Pi(q)$ and the supply set.

Solution

(a) The fixed cost is the cost when nothing is produced, which is $C(0) = 100$.
(b) Suppose the firm is in production; that is $0 < q \leq L$, where $L = 6$ is the given upper limit on capacity. Then $p = C'(q)$ and the profit function is $\Pi(q) = qC'(q) - C(q)$. Hence

$$\Pi(q) = q(20 - 12q + 3q^2) - (100 + 20q - 6q^2 + q^3) = 2q^3 - 6q^2 - 100.$$

(c) The startup point is the value q_s such that $\Pi(q_s) = \Pi(0) = -C(0)$. Solving the equation $2q^3 - 6q^2 - 100 = -100$, or $q^2(q - 3) = 0$, it follows that $q_s = 3$. The corresponding value of p is $p_s = C'(3) = 11$.
(d) The breakeven point q_b satisfies $\Pi(q_b) = 0$. We therefore need to solve the equation $2q^3 - 6q^2 - 100 = 0$. Trial and error reveals the factorisation

$$2q^3 - 6q^2 - 100 = 2(q - 5)(q^2 + 2q + 10).$$

The quadratic equation $q^2 + 2q + 10 = 0$ has no solutions, since $2^2 - 4(1)(10) < 0$, so 5 is the only solution and the breakeven point is $q_b = 5$.
(e) We have found that the startup point is $q_s = 3$, and the corresponding going price is $p_s = 11$. Thus the firm supplies nothing when $p < 11$, and for $p \geq 11$ the amount q supplied is related to p by

$$p = C'(q) = 3q^2 - 12q + 20.$$

But this can hold only when q is in the feasible range $q \leq L$, where $L = 6$. The corresponding price p_L is given by $p_L = C'(6) = 56$. Thus when $p > 56$, the firm will produce its full capacity of 6.

The firm's supply set is the union of three pieces:

- the points $(0, p)$ for $0 < p < 11$;
- the points (q, p) such that $p = 3q^2 - 12q + 20$ for $11 \leq p \leq 56$;
- the points $(6, p)$ for $p > 56$.

The profit function and the supply set are sketched in Figure 10.4.

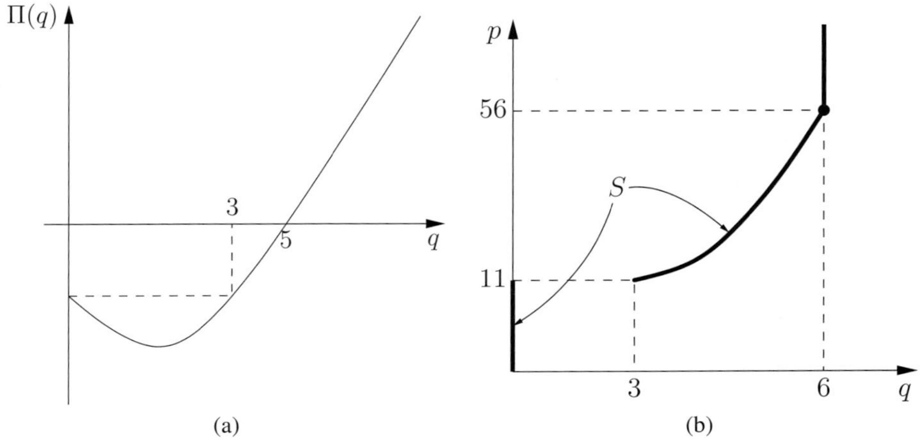

Figure 10.4 The profit function and the supply set (Example 10.1).

Example 10.2 Prove that, at the breakeven point for an efficient small firm, the derivative of average cost is zero.

Solution

If q_b is the breakeven point, then, by definition, $\Pi(q_b) = 0$. But $\Pi(q) = qC'(q) - C(q)$ for $q > 0$, therefore $q_bC'(q_b) - C(q_b) = 0$. Now the average cost is $AC(q) = C(q)/q$, so by the quotient rule

$$AC'(q) = \frac{C'(q)q - C(q)}{q^2}.$$

Then $AC'(q_b) = (C'(q_b)q_b - C(q_b))/q_b^2 = 0/q_b^2 = 0$, as required.

Main Topics

- efficient small firm in a perfectly competitive market
- startup and breakeven points

- relations between marginal cost, average cost, average variable cost
- determining the supply curve of an efficient small firm

Key Terms, Notations and Formulae

- efficient small firm
- for efficient small firm, $p^S(q) = C'(q)$
- profit function, $\Pi(q) = qC'(q) - C(q)$
- fixed costs, $C(0)$
- startup point, q_s: $\Pi(q_s) = \Pi(0)$
- breakeven point, q_b: $\Pi(q_b) = 0$
- MC: marginal cost
- variable cost, $VC = C(q) - C(0)$
- average cost, $AC = \dfrac{C(q)}{q}$
- average variable cost, $AVC = \dfrac{VC}{q} = \dfrac{C(q) - C(0)}{q}$
- at startup, $MC = AVC$
- at breakeven, $MC = AC$

Exercises

10.1 A firm has cost function

$$C(q) = 50\,000 + 25q + 0.001q^2.$$

Find the fixed cost FC and the marginal cost MC. What is the marginal cost when the output is 100? What is the marginal cost when the output is 10 000?

10.2 An efficient small firm with cost function

$$C(q) = q^3 - 3q^2 + 5q + 4$$

has a maximum level of weekly production of 5. Determine: (a) their fixed cost; (b) their profit function; (c) their startup point; (d) their breakeven point, (e) their supply set.

Problems

10.1 Suppose that Quality Widgets Limited is an efficient small firm with cost function

$$C(q) = q^3 - 10q^2 + 100q + 196$$

and suppose also that the maximum level of weekly production is 10. Determine: (a) their fixed cost; (b) their profit function; (c) their startup point; (d) their breakeven point; (e) their supply set.

10.2 In the case of Quality Widgets Limited (Problem 10.1), verify that (a) at startup, the marginal cost equals the average variable cost; (b) at startup, the derivative of the average variable cost is zero; (c) at breakeven, the marginal cost equals the average cost; (d) at breakeven, the derivative of the average cost is zero.

10.3 Prove that, at the startup point for an efficient small firm, the derivative of the average variable cost is zero.

10.4 The theory in this chapter has indicated how, given a cost function C, we may (in some circumstances) obtain a supply set S_C which has the form shown in Figure 10.1. Study the theory carefully and try to formulate reasonable conditions on C which will ensure that S_C does indeed take such a form.

11 Partial Derivatives

11.1 Functions of Several Variables

Recall that a function f may be thought of as a 'black box', which accepts an input x and produces an output $f(x)$. In this chapter we shall look at functions for which the input consists of a pair of numbers (x, y). The theory extends in an obvious way to the general case where the input consists of n numbers $(x_1, x_2, \ldots, x_n)$, but the case $n = 2$ is quite sufficient to illustrate all the important points.

We use the notation $f : \mathbb{R}^2 \to \mathbb{R}$ for a function f which assigns to each (x, y) in $\mathbb{R}^2$ a value $f(x, y)$ in $\mathbb{R}$, and we often refer to f as a 'function of two variables'. For example, the following formulae define functions of two variables:

$$g(x, y) = \exp(x + 4xy^2) - 17, \quad h(x, y) = x^3 + y^3.$$

A typical example from economics is the *production function* for a firm (Figure 11.1). Here we make the reasonable assumption that the quantity q which the firm can produce depends upon the amounts of capital k and labour l available. In other words, we have a rule for finding $q(k, l)$ when k and l are known. Obviously it is useful to know how q changes in response to changes in k and l, and this is the kind of question we shall study in the rest of the chapter.

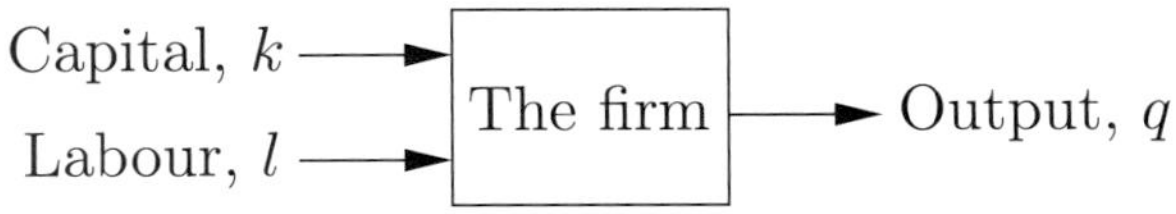

Figure 11.1 The production function $q = q(k, l)$ for a firm.

11.2 Partial Derivatives

Let $f : \mathbb{R}^2 \to \mathbb{R}$ be a function of two variables, as defined above. The *partial derivative* of f, with respect to the first variable, is obtained by treating the second variable as a constant and taking the derivative with respect to the first variable. For example, suppose

$$f(x, y) = x^2y + 7xy + y^2.$$

If we give y a fixed value c, we obtain a function of x only:

$$x^2c + 7xc + c^2,$$

and the derivative of this with respect to x is $2xc + 7c$. We replaced y by c only to emphasise that it is kept fixed, so reverting to the name y, we get $2xy + 7y$.

The partial derivative of f with respect to the first variable is denoted by

$$\frac{\partial f}{\partial x} \quad \text{or} \quad f_1(x, y).$$

It is important to remember that it is a function of x and y, although the ∂-notation makes it difficult to display the values. For the particular function defined in the previous paragraph, we obtained

$$\frac{\partial f}{\partial x} = f_1(x, y) = 2xy + 7y.$$

We can define the partial derivative of f with respect to the second variable in a similar way. In our example, we get

$$\frac{\partial f}{\partial y} = f_2(x, y) = x^2 + 7x + 2y.$$

Since f_1 and f_2 are themselves functions of (x, y), we can define their partial derivatives with respect to x and y. For example, we may form $(f_1)_1$, the partial derivative of f_1 with respect to x. This is usually denoted

$$\frac{\partial^2 f}{\partial x^2} \quad \text{or} \quad f_{11}(x, y).$$

(In the first of these notations, the 2s occurring as superscripts signify that f has been differentiated twice.) In a similar way, the second partial derivative $(f_2)_2$ may be formed. This is usually denoted

$$\frac{\partial^2 f}{\partial y^2} \quad \text{or} \quad f_{22}(x, y).$$

Now, in general, the derivative f_1 will be a function of both x and y, and so we may form its partial derivative, $(f_1)_2$, with respect to y. The notations used for this second partial derivative are f_{12} and

$$\frac{\partial^2 f}{\partial y \partial x}.$$

Analogously, the derivative of f_2 with respect to x can be calculated and we denote it by f_{21} or

$$\frac{\partial^2 f}{\partial x \partial y}.$$

There are therefore four *second* partial derivatives of f.

For the function $f(x, y) = x^2y + 7xy + y^2$ considered above we get

$$f_{11}(x,y) = 2y, \qquad f_{12}(x,y) = 2x + 7,$$
$$f_{21}(x,y) = 2x + 7, \qquad f_{22}(x,y) = 2.$$

You will notice that the *mixed* second partial derivatives f_{12} and f_{21} turn out to be equal in this case. Fortunately, this will always happen provided f is a 'well-behaved' function, such as those which are presumed to occur in economic models.

Example A function often used in economics is the *Cobb–Douglas* production function

$$q(k,l) = Ak^{\alpha}l^{\beta}.$$

Here A, α and β are constants which can be chosen to reflect the characteristics of a particular situation. For example, we might consider a specific firm with production function $6k^{1/4}l^{3/4}$, which is the case $A = 6$, $\alpha = 1/4$, $\beta = 3/4$.

For a given production function, the partial derivative $q_1 = \partial q/\partial k$ measures the rate of change of q with respect to k when l is kept fixed. Using the same reasoning as for an ordinary derivative (Section 6.3), we can interpret this as a 'marginal' function. In this case it is approximately equal to the change in production when one additional unit of capital is available, and so it is known as the *marginal product of capital*. Similarly, $q_2 = \partial q/\partial l$ is the *marginal product of labour*. For the general Cobb–Douglas production function we have

$$q_1(k,l) = A\alpha k^{\alpha-1}l^{\beta} \quad \text{and} \quad q_2(k,l) = A\beta k^{\alpha}l^{\beta-1}.$$

We can also work out the second partial derivatives:

$$q_{11}(k,l) = A\alpha(\alpha-1)k^{\alpha-2}l^{\beta}, \qquad q_{12}(k,l) = A\alpha\beta k^{\alpha-1}l^{\beta-1},$$
$$q_{21}(k,l) = A\alpha\beta k^{\alpha-1}l^{\beta-1}, \qquad q_{22}(k,l) = A\beta(\beta-1)k^{\alpha}l^{\beta-2}.$$

Note the equality of the mixed derivatives. ■

11.3 The Chain Rule

For the sake of motivation, we continue to look at the production function. Suppose that both k and l change over a period of time in some known way, so that we have formulae for $k(t)$ and $l(t)$, where t is a parameter measuring time. For example, we might have

$$k(t) = 4 + 0.1t \quad \text{and} \quad l(t) = 9 - 0.05t,$$

which means that k increases linearly while l decreases linearly, as functions of time. If we know the production function q in terms of k and l, then we can also work out the level of production in terms of t, and so we can see how it will be affected by the changes in k and l with time. For example, if q is the Cobb–Douglas function $q(k,l) = kl$, and k and l change as above, we get the formula

$$q(k,l) = kl = (4 + 0.1t)(9 - 0.05t) = 36 + 0.7t - 0.005t^2$$

for the output in terms of t.

More generally, suppose we are given a function f of two variables (x, y), both of which are themselves functions of t. As illustrated in Figure 11.2, we can think of this situation as defining a *composite function* $F(t) = f(x(t), y(t))$.

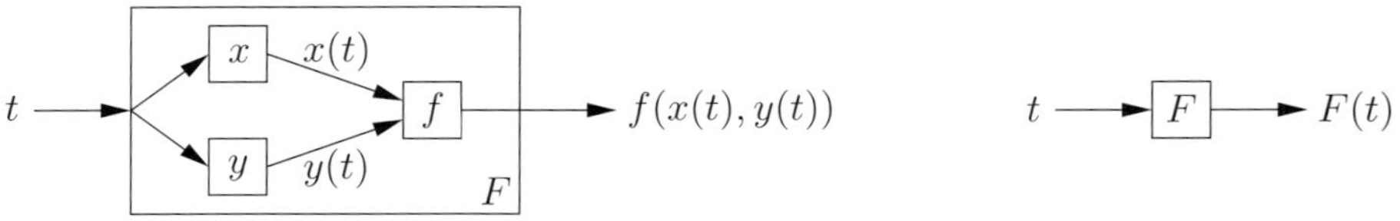

Figure 11.2 The composite function F.

In the case of a single variable we have a rule, the 'function of a function' rule, which enables us to work out the derivative of a composite function. There is a similar rule here, known as the *chain rule*:

$$\frac{dF}{dt} = \frac{\partial f}{\partial x}\frac{dx}{dt} + \frac{\partial f}{\partial y}\frac{dy}{dt}.$$

Example Suppose $f(x, y) = xy$, $x(t) = 3 + 4t$ and $y(t) = 6 - 2t$. Then, using the chain rule:

$$\frac{dF}{dt} = y \times 4 + x \times (-2) = (6 - 2t)4 + (3 + 4t)(-2) = 18 - 16t.$$

In this case we can check the result explicitly, because we can work out a formula for $F(t)$:

$$F(t) = x(t)y(t) = (3 + 4t)(6 - 2t) = 18 + 18t - 8t^2.$$

Differentiating this in the usual way we get $dF/dt = 18 - 16t$, as before. ■

We can justify the chain rule by recalling the approximation formula, which tells us that when t changes by a small amount Δt, $F(t) = f(x(t), y(t))$ changes by an amount $\Delta F \simeq F'(t)\Delta t$. Here there are two reasons why changing the value of t affects the value of F: changing t changes both $x = x(t)$ and $y = y(t)$, and these changes in turn affect F.

Consider first the effect on x. The change in x resulting from a small change in t is approximated by $\Delta x \simeq x'(t)\Delta t$. Similarly, the change in $f(x, y)$ resulting from a change Δx in x (keeping y constant) is approximately $(\partial f/\partial x)\Delta x$. Thus the change in F is

$$\Delta_x F \simeq \frac{\partial f}{\partial x}\Delta x \simeq \frac{\partial F}{\partial x}\frac{dx}{dt}\Delta t.$$

A similar argument shows that the change in F which results from a change in t and its effect on y may be approximated by

$$\Delta_y F \simeq \frac{\partial f}{\partial y}\frac{dy}{dt}\Delta t.$$

The total change in F is therefore the sum of these two quantities:

$$\Delta F = \Delta_x F + \Delta_y F \simeq \frac{\partial f}{\partial x}\frac{dx}{dt}\Delta t + \frac{\partial f}{\partial y}\frac{dy}{dt}\Delta t,$$

and we have

$$\frac{dF}{dt} \simeq \frac{\Delta F}{\Delta t} \simeq \frac{\partial f}{\partial x}\frac{dx}{dt} + \frac{\partial f}{\partial y}\frac{dy}{dt}.$$

There are generalisations of the chain rule, covering the case where f is a function of several variables, each of which is itself a function of several other variables. The simplest case will suffice to explain how it goes. Suppose f is a function of x and y, which are both functions of u and v. We have a function F of u and v defined by $F(u,v) = f(x(u,v), y(u,v))$, and the partial derivatives of F are given by

$$\frac{\partial F}{\partial u} = \frac{\partial f}{\partial x}\frac{\partial x}{\partial u} + \frac{\partial f}{\partial y}\frac{\partial y}{\partial u} \quad \text{and} \quad \frac{\partial F}{\partial v} = \frac{\partial f}{\partial x}\frac{\partial x}{\partial v} + \frac{\partial f}{\partial y}\frac{\partial y}{\partial v}.$$

Example Suppose that a firm produces two goods, and that its revenue R is a function $R(q_1, q_2)$ of the quantities q_1 and q_2 produced. (*Caution:* the subscripts here refer to the two goods, *not* the partial derivatives as in Section 11.2.) We may suppose that q_1 and q_2 are themselves functions of capital k and labour l. The chain rule formulae for $\partial R/\partial k$ and $\partial R/\partial l$ are

$$\frac{\partial R}{\partial k} = \frac{\partial R}{\partial q_1}\frac{\partial q_1}{\partial k} + \frac{\partial R}{\partial q_2}\frac{\partial q_2}{\partial k},$$

$$\frac{\partial R}{\partial l} = \frac{\partial R}{\partial q_1}\frac{\partial q_1}{\partial l} + \frac{\partial R}{\partial q_2}\frac{\partial q_2}{\partial l}.$$

Suppose, for the sake of example, that

$$R(q_1, q_2) = q_1^2 q_2, \quad q_1 = 5k + l \quad \text{and} \quad q_2 = 6k + 2l.$$

Then the partial derivatives of R are

$$\frac{\partial R}{\partial q_1} = 2q_1 q_2 \quad \text{and} \quad \frac{\partial R}{\partial q_2} = q_1^2,$$

whereas the partial derivatives of q_1 and q_2 are

$$\frac{\partial q_1}{\partial k} = 5, \quad \frac{\partial q_1}{\partial l} = 1, \quad \frac{\partial q_2}{\partial k} = 6 \quad \text{and} \quad \frac{\partial q_2}{\partial l} = 2.$$

Thus the partial derivative of R with respect to k is

$$\begin{aligned}\frac{\partial R}{\partial k} &= 5(2q_1 q_2) + 6(q_1^2) \\ &= 10q_1 q_2 + 6q_1^2 = 10(5k + l)(6k + 2l) + 6(5k + l)^2 \\ &= 10(30k^2 + 16kl + 2l^2) + 6(25k^2 + 10kl + l^2) \\ &= 450k^2 + 220kl + 26l^2.\end{aligned}$$

Calculating the other partial derivative in the same way:

$$\frac{\partial R}{\partial l} = 2q_1 q_2 + 2(q_1^2) = 110k^2 + 52kl + 6l^2.$$

■

Of course, in this example there is another way of obtaining these results, by first working out the explicit formula for R in terms of k and l:

$$\begin{aligned} R = q_1^2 q_2 &= (5k + l)^2(6k + 2l) = (25k^2 + 10kl + l^2)(6k + 2l) \\ &= 150k^3 + 110k^2 l + 26kl^2 + 2l^3. \end{aligned}$$

From this we can work out the partial derivatives directly.

The reader might think that the direct approach is easier, and wonder why we bother with the chain rule. There are two reasons. First, for some problems it is easier: in Example 11.4 below, to find $F'(2)$ by first working out the formula for F in terms of t is a more lengthy process than the method given. Second, the chain rule has important theoretical consequences. In the next chapter we shall use it to develop an important technique known as 'implicit differentiation'.

Worked Examples

Example 11.1 Find the partial derivatives and second partial derivatives of the function

$$f(x, y) = 3x^2 + 4xy + y^2.$$

Solution

The partial derivatives of f are

$$\frac{\partial f}{\partial x} = f_1 = 6x + 4y \quad \text{and} \quad \frac{\partial f}{\partial y} = f_2 = 4x + 2y.$$

Then the second partial derivatives are

$$\begin{aligned} \frac{\partial^2 f}{\partial x^2} &= f_{11} = 6, & \frac{\partial^2 f}{\partial x \partial y} &= f_{12} = 4, \\ \frac{\partial^2 f}{\partial y \partial x} &= f_{21} = 4, & \frac{\partial^2 f}{\partial y^2} &= f_{22} = 2. \end{aligned}$$

Note the equality of the mixed derivatives.

Example 11.2 Find the partial derivatives of the function $g(x, y) = y^{\sqrt{x}}$.

Solution

We recall that the way to deal with variable powers is to use the rule $a^x = \exp(x \ln a)$. Thus here we have

$$g(x, y) = \exp(\sqrt{x} \ln y).$$

Taking y as fixed, g is the composite of the function 'exp' and the function $A\sqrt{x}$, where $A = \ln y$ is constant. Using the ordinary function of a function rule we get

$$\frac{\partial g}{\partial x} = \exp(\sqrt{x} \ln y) \frac{\partial}{\partial x} (\sqrt{x} \ln y) = y^{\sqrt{x}} \left(\frac{1}{2} x^{-1/2} \ln y \right) = \frac{\ln y}{2\sqrt{x}} y^{\sqrt{x}}.$$

It is far easier to determine $\partial g / \partial y$. Taking x as fixed, $\sqrt{x}$ is a constant and the derivative (with respect to y) of $y^{\sqrt{x}}$ is $\sqrt{x} y^{\sqrt{x}-1}$. Alternatively, writing $y^{\sqrt{x}} = \exp(\sqrt{x} \ln y)$ and differentiating with respect to y:

$$\frac{\partial g}{\partial y} = \exp(\sqrt{x} \ln y) \frac{\partial}{\partial y} (\sqrt{x} \ln y) = y^{\sqrt{x}} \left(\frac{\sqrt{x}}{y} \right) = \frac{\sqrt{x} y^{\sqrt{x}}}{y} = \sqrt{x} y^{\sqrt{x}-1}.$$

Example 11.3 Suppose that a firm has production function $q(k, l) = Ak^{\alpha} l^{1-\alpha}$ where $A > 0$ and $0 < \alpha < 1$. Show that the marginal product of labour $\partial q / \partial l$ is positive, and that it is a decreasing function of l when k is fixed.

Solution

The marginal product of labour is

$$\frac{\partial q}{\partial l} = A(1 - \alpha) k^{\alpha} l^{-\alpha} = A(1 - \alpha) \left(\frac{k}{l} \right)^{\alpha}.$$

This is clearly positive, since we are given that $A > 0$ and $1 - \alpha > 0$. Furthermore, as l increases then, for fixed k, k/l decreases. Since $\alpha > 0$, it follows that the marginal product of labour decreases with l.

There is another way of verifying that this is so. The rate of change of the marginal product of labour with respect to l is its derivative, in other words, the *second* derivative $\partial^2 q / \partial l^2$. By the usual rules we get

$$\frac{\partial^2 q}{\partial l^2} = \frac{\partial}{\partial l} \left(A(1 - \alpha) k^{\alpha} l^{-\alpha} \right) = A(1 - \alpha)(-\alpha) k^{\alpha} l^{-\alpha-1} = -A\alpha(1 - \alpha) k^{\alpha} l^{-\alpha-1}.$$

Because $A > 0$, $\alpha > 0$ and $1 - \alpha > 0$, this is negative, from which it follows that the marginal product of labour is a decreasing function of l.

In economic language, we have shown that a firm with production function $Ak^{\alpha} l^{1-\alpha}$ has 'diminishing marginal product of labour'. Roughly speaking, this means that, as the workforce increases in size, production increases also, but at a slower rate.

Example 11.4 Suppose that $f(x, y) = x^2 y + y^2$. Let $x(t) = 3t^2 + 3$ and $y(t) = t^3 - 7$ and let $F(t) = f(x(t), y(t))$. Use the chain rule to find $F'(2)$.

Solution

The chain rule tells us that

$$F'(t) = \frac{\partial f}{\partial x} x'(t) + \frac{\partial f}{\partial y} y'(t).$$

Now, $f_1(x,y) = 2xy, f_2(x,y) = x^2 + 2y$ and $x'(t) = 6t$, $y'(t) = 3t^2$. Therefore

$$F'(t) = (2xy)(6t) + (x^2 + 2y)(3t^2).$$

If our aim was to find a general formula for $F'(t)$ in terms of t, we would now substitute into this the explicit expressions for x and y as functions of t (as in the first example in Section 11.3). However, to calculate the value of $F'(2)$, it is not necessary to do this. When $t = 2$, the corresponding values of x and y are $x = 15$ and $y = 1$, and hence $F'(2) = 2(15)(1)(12) + (15^2 + 2)(12) = 3084$.

Main Topics

- functions of more than one variable (particularly of two variables)
- partial derivatives and second partial derivatives
- the chain rule

Key Terms, Notations and Formulae

- function of n variables, $f(x_1, x_2, \dots, x_n), f : \mathbb{R}^n \to \mathbb{R}$
- capital, k; labour, l; production function, $q(k,l)$
- partial derivatives, $f_1 = \dfrac{\partial f}{\partial x}, f_2 = \dfrac{\partial f}{\partial y}$
- second partial derivatives, $f_{11} = \dfrac{\partial^2 f}{\partial x^2}$, $f_{22} = \dfrac{\partial^2 f}{\partial y^2}$, $f_{12} = \dfrac{\partial^2 f}{\partial y \partial x}$, $f_{21} = \dfrac{\partial^2 f}{\partial x \partial y}$
- $f_{12} = f_{21}$ for well-behaved f
- Cobb–Douglas production function, $q(k,l) = Ak^{\alpha} l^{\beta}$
- chain rule, $F(t) = f(x(t), y(t))$, $F'(t) = \dfrac{\partial f}{\partial x}\dfrac{dx}{dt} + \dfrac{\partial f}{\partial y}\dfrac{dy}{dt}$

Exercises

11.1 Suppose that

$$f(x,y) = x^3 y - \frac{x}{y}.$$

Find the partial derivatives of f.

11.2 The function f is given by $f(x,y) = xy \ln\left(\dfrac{x}{y}\right)$ for $x, y > 0$.

Find the partial derivatives $\dfrac{\partial f}{\partial x}$ and $\dfrac{\partial f}{\partial y}$.

Verify that $x\dfrac{\partial f}{\partial x} + y\dfrac{\partial f}{\partial y} = 2f(x,y)$.

11.3 Suppose that $f(x, y) = x^2y$ and that $x(t) = 2 + 3t$ and $y(t) = t^2 + 1$. If $F(t) = f(x(t), y(t))$, find the derivative dF/dt by using the chain rule.

Problems

11.1 Find the first and second partial derivatives of the functions

$$f(x, y) = x^2y + xy^3, \quad g(x, y) = 5x^{2/3}y^{1/4} \quad \text{and} \quad h(x, y) = x^2(x^2 + y^3)^{2/3}.$$

11.2 If $f(x, y) = x^{y^2}$, find $\frac{\partial f}{\partial x}$ and $\frac{\partial f}{\partial y}$.

[*Hint:* to help calculate $\frac{\partial f}{\partial y}$, note that $x^{y^2} = \exp\left(y^2 \ln x\right)$.]

11.3 Suppose that $f(x, y) = x^2y$. Let $x(t) = 2 - t$ and $y(t) = 3t + 7$ and let $F(t) = f(x(t), y(t))$. Use the chain rule to find an expression for $F'(t)$.

11.4 Suppose that a firm's capital and labour vary in time according to

$$k(t) = 2t^2 \quad \text{and} \quad l(t) = 2t + 5.$$

If the firm has production function $q(k, l) = kl^2$, determine the rate at which production changes with time.

11.5 Suppose that $f(x, y) = x^2y$ and that x and y are defined in terms of u and v by $x(u, v) = u^2 + v^2$ and $y(u, v) = u^3 - v^3$. Let $F(u, v) = f(x(u, v), y(u, v))$. Find the partial derivatives $\partial F/\partial u$ and $\partial F/\partial v$ using the chain rule.

11.6 Suppose that $f(x, y) = x^{1/2}y^{3/4}$ and that x and y are defined in terms of u and v by $x(u, v) = u^2 + v^2$ and $y(u, v) = uv$. Let $F(u, v) = f(x(u, v), y(u, v))$. Find the partial derivatives $\partial F/\partial u$ and $\partial F/\partial v$ using the chain rule.

12 Applications of Partial Derivatives

12.1 Functions Defined Implicitly

If we are given a function $g : \mathbb{R}^2 \to \mathbb{R}$ the equation $g(x, y) = 0$ can, in some cases, be solved to give 'y as a function of x'. For example, if $g(x, y)$ is $x^2 - 4y$ then the equation is

$$x^2 - 4y = 0, \quad \text{which gives} \quad y = \frac{x^2}{4}.$$

In general, we say that an equation $g(x, y) = 0$ defines y *implicitly* as a function of x if there is a function $y(x)$ which satisfies the equation, for a range of values of x; that is

$$g(x, y(x)) = 0 \quad \text{for all } x \in S,$$

where S is some appropriate set of values. This simply means that the 'solution' $y(x)$ satisfies the equation $g(x, y) = 0$. For example, when $g(x, y)$ is $x^2 - 4y$ the solution $y(x) = x^2/4$ satisfies

$$g(x, y(x)) = x^2 - 4y(x) = x^2 - 4\left(\frac{x^2}{4}\right) = 0 \quad \text{for all } x.$$

Often an equation $g(x, y) = 0$ defines y implicitly as a function of x even when it is difficult or impossible to solve the equation and find a formula for $y(x)$. The same observation holds for an equation $g(x, y) = c$, where c is any constant value.

Example Consider the equation

$$x^2 + 3xy + 2y^2 = 48.$$

Here it is possible to 'solve' for y in terms of x by writing the equation as a quadratic in y:

$$2y^2 + (3x)y + (x^2 - 48) = 0.$$

The usual formula (Section 2.4) gives two solutions:

$$y(x) = \frac{1}{4}\left(-3x + \sqrt{x^2 + 384}\right)$$

and

$$y(x) = \frac{1}{4}\left(-3x - \sqrt{x^2 + 384}\right).$$

These two functions are defined implicitly by the equation. The set of points (x, y) satisfying the equation $x^2 + 3xy + 2y^2 = 48$ is illustrated in Figure 12.1.

We see that this set has two distinct parts to it. These are sometimes called *branches*, and they correspond to the two functions described above. The upper branch is the graph of the function

$$y(x) = \frac{1}{4}\left(-3x + \sqrt{x^2 + 384}\right)$$

and the lower branch is the graph of the function

$$y(x) = \frac{1}{4}\left(-3x - \sqrt{x^2 + 384}\right).$$

■

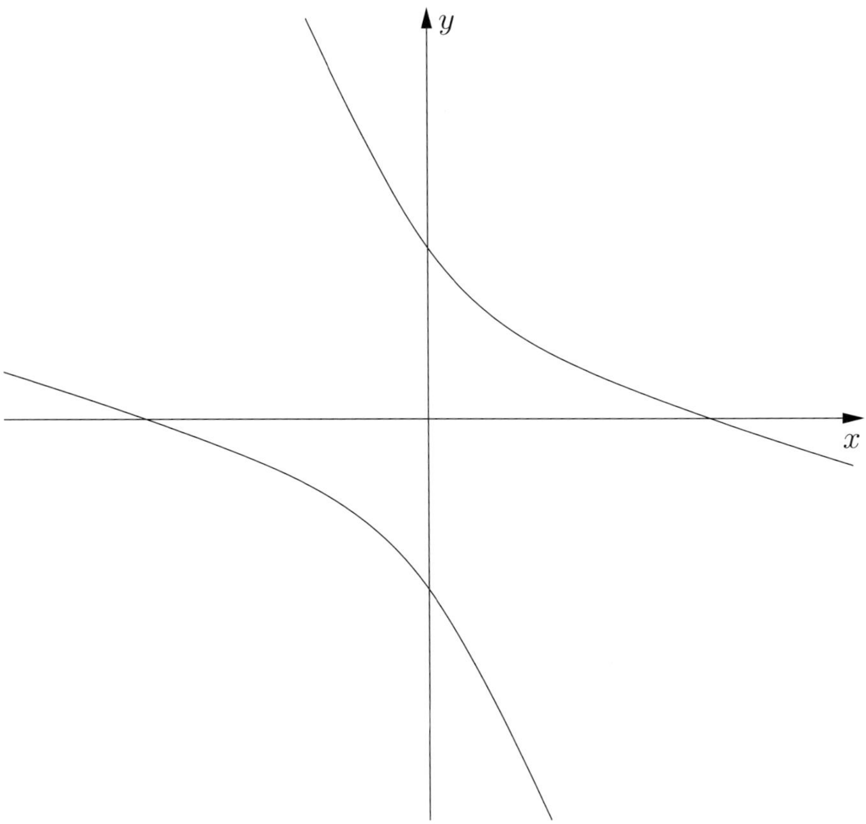

Figure 12.1 The set of points (x, y) satisfying $x^2 + 3xy + 2y^2 = 48$.

12.2 The Derivative of an Implicit Function

If $g(x,y) = c$ defines a function $y(x)$ implicitly, how can we work out the derivative dy/dx? Of course, if we can solve the equation explicitly then there is no problem (in theory). But as we have noted, that may not be possible, and even when it is the resulting formula may be complicated. In this section we shall explain how dy/dx can be found simply in terms of the partial derivatives of g, by the formula

$$\frac{dy}{dx} = -\frac{\partial g/\partial x}{\partial g/\partial y}.$$

The trick is to think of the equation $g(x, y(x)) = c$, which defines $y(x)$, in the following way. The given g is a function of two variables; call them X and Y for the moment. If we make both X and Y functions of x, specifically by putting $X = x$ and $Y = y(x)$, then the composite function $G(x) = g(X,Y) = g(x, y(x))$, a function of x, is *constant*. Now we can write down dG/dx by using the chain rule (Section 11.3):

$$\frac{dG}{dx} = \frac{\partial g}{\partial X}\frac{dX}{dx} + \frac{\partial g}{\partial Y}\frac{dY}{dx}.$$

In our case $X = x$ and $Y = y(x)$, so $dX/dx = 1$ and $dY/dx = dy/dx$. Also G is constant as a function of x, so $dG/dx = 0$. Thus the chain rule equation becomes

$$0 = \frac{\partial g}{\partial x}1 + \frac{\partial g}{\partial y}\frac{dy}{dx}.$$

Rearranging, we get the required formula for dy/dx:

$$\frac{dy}{dx} = -\frac{\partial g/\partial x}{\partial g/\partial y} \quad \text{or} \quad \frac{dy}{dx} = -\frac{g_1(x,y)}{g_2(x,y)}.$$

Example In the example in Section 12.1 we considered the equation $g(x,y) = 48$, where $g(x,y) = x^2 + 3xy + 2y^2$. Here we have

$$\frac{dy}{dx} = -\frac{g_1(x,y)}{g_2(x,y)} = -\frac{2x+3y}{3x+4y}.$$

Suppose we wish to find the rate of change of $y(x)$ when $x = 4$. First we must remember that there are *two* functions $y(x)$ in this case, corresponding to the $\pm$ sign in the formula

$$y(x) = \frac{1}{4}\left(-3x \pm \sqrt{x^2 + 384}\right).$$

If we choose the + sign, then we have $y = 2$ when $x = 4$, and in graphical terms we are looking at the branch whose graph contains the point $(4,2)$ (Figure 12.2); this is the upper branch of Figure 12.1. The slope of that branch at $x = 4$ is thus $-(2x + 3y)/(3x + 4y) = -14/20 = -7/10$. ■

Example Consider the equation

$$g(x,y) = \exp(xy) + 2(x + y) = 5.$$

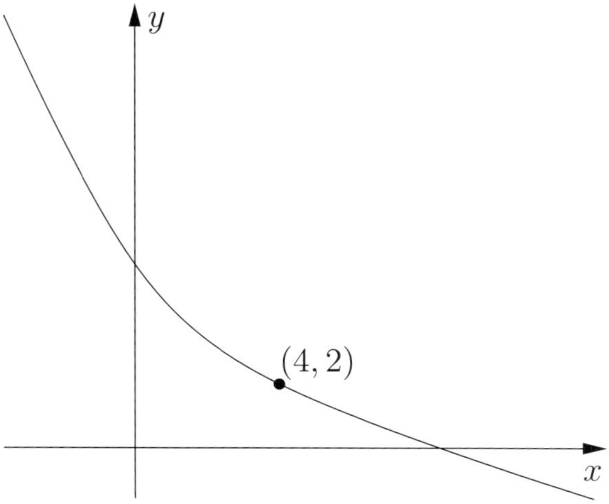

Figure 12.2 Graph of the upper branch determined by $x^2 + 3xy + 2y^2 = 48$.

Here we have

$$\frac{dy}{dx} = -\frac{g_1(x,y)}{g_2(x,y)} = -\frac{y\exp(xy) + 2}{x\exp(xy) + 2}.$$

For example, at the point $x = 0$, where $y = 2$, the derivative of $y(x)$ is $-(4/2) = -2$. ■

Finally, a word of caution. It is very easy to forget the minus sign in the formula for dy/dx. The correct formula is

$$\frac{dy}{dx} = -\frac{\partial g/\partial x}{\partial g/\partial y},$$

even though the minus sign may 'look wrong'. If we (rashly) cancel the ∂gs on the right-hand side, we obtain $-\partial y/\partial x$, whereas we might expect $\partial y/\partial x$. Of course the fact is that we cannot treat symbols such as ∂g, ∂x, ∂y, dx, dy as if they were numbers to which the laws of algebra apply.

12.3 Contours and Isoquants

It is often helpful to use geometrical ideas when we have to deal with functions of several variables. One of the most useful devices is derived from the way in which hills and dales are represented on a map, by means of contour lines. If $h(x, y)$ represents the height above sea level (in metres) at the point (x, y), then the 100 m contour consists of the points for which $h(x, y) = 100$.

Following this idea, for any function f of two variables we define a *contour* to be a set of the form

$$\{(x, y) \mid f(x, y) = c\}$$

for some constant c. In Figure 12.3 we sketch some typical contours for the functions defined by the formulae $x + y$, $x^2 + y^2$ and xy. For reasons of space, only the parts of the contours which lie in $\mathbb{R}^2_+$ are shown.

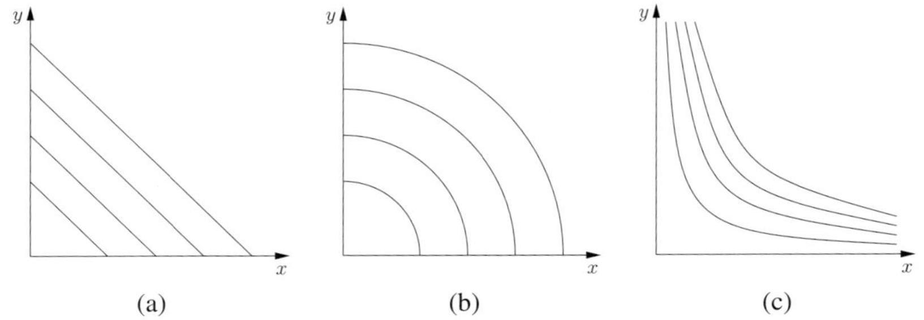

Figure 12.3 Some contours for $x + y$, $x^2 + y^2$ and xy.

Suppose we want to find the slope of the contour $f(x, y) = c$ at the point (a, b). This is a problem which can be solved by the method described in the previous section, because the equation $f(x, y) = c$ implicitly defines y as a function of x, and the graph of this function is just the contour. As we found, the slope is just

$$\frac{dy}{dx} = -\frac{f_1}{f_2},$$

where the partial derivatives are evaluated at the point (a, b).

Example Figure 12.1 is a sketch of what we now call the contour for the value $c = 48$ of the function $x^2 + 3xy + 2y^2$. Using the formula $dy/dx = -(2x + 3y)/(3x + 4y)$ we have already found that the slope of this contour at the point $(4, 2)$ is $-7/10$.

Suppose we wish to consider the contour of the same function which passes through the point $(2, 3)$. Then the formula immediately gives the slope as
$-(4 + 9)/(6 + 12) = -13/18$. ■

When the function under consideration is a production function $q(k, l)$, the contours are called *isoquants* ('iso-quant' meaning 'same quantity'). Thus the points (k, l) lying on a particular isoquant correspond to a set of inputs of capital and labour, all of which result in a given fixed output.

Example The slope of an isoquant of the function $q(k, l) = 6k^{1/2}l^{1/4}$ is given by

$$\frac{dl}{dk} = -\frac{q_1}{q_2} = -\frac{6\,(1/2)k^{-1/2}l^{1/4}}{6\,(1/4)k^{1/2}l^{-3/4}} = -\frac{2l}{k}.$$

For instance, the slope of the isoquant passing through the point $(5, 7)$ is $-14/5$ at that point. ■

12.4 Scale Effects and Homogeneous Functions

Suppose we have a function of two variables, such as $f(x, y) = 3x^2y + 7xy^2$, and we multiply the 'inputs' x and y by a constant c. In this case we get the 'output'

$$f(cx, cy) = 3(cx)^2(cy) + 7(cx)(cy)^2 = c^3(3x^2y + 7xy^2) = c^3f(x, y).$$

Thus, for this particular f, multiplying the inputs by c results in the 'output' being multiplied by c^3. In general, if a function h is such that

$$h(cx, cy) = c^D h(x, y),$$

then we say that h is *homogeneous of degree D*. The number D is called the *degree of homogeneity* of h. The function f given by the formula above is homogeneous of degree 3. *Note that many (indeed most) functions are not homogeneous.*

The notion of a homogeneous function is related to the idea of 'returns to scale' in economics. In the case of a production function, we say that there are *constant returns to scale* if a proportional increase in k and l results in the same proportional increase in $q(k, l)$; that is if

$$q(ck, cl) = cq(k, l).$$

This means, for example, that doubling both capital and labour doubles the production. Clearly, this is the same as saying that q is homogeneous of degree 1. If q is homogeneous of degree $D > 1$ then the proportional increase in $q(k, l)$ will be larger than that in k and l, and we say that there are *increasing returns to scale*. On the other hand, if $D < 1$ we say that there are *decreasing returns to scale*.

Example For the Cobb–Douglas production function $q(k, l) = Ak^\alpha l^\beta$, we have

$$q(ck, cl) = A(ck)^\alpha(cl)^\beta = c^{\alpha+\beta}(Ak^\alpha l^\beta) = c^{\alpha+\beta}q(k, l).$$

This means that the Cobb–Douglas function is homogeneous of degree $\alpha + \beta$; in particular, if $\alpha + \beta = 1$ then there are constant returns to scale. ■

There is a useful result about homogeneous functions known as *Euler's Theorem* (named after the Swiss mathematician Léonard Euler (1707–1783), whose name is pronounced 'Oiler'). For any homogeneous function h of degree D, it asserts that

$$x\frac{\partial h}{\partial x} + y\frac{\partial h}{\partial y} = Dh.$$

Example As above, let f be defined by $f(x, y) = 3x^2y + 7xy^2$, which we have noted is homogeneous of degree 3. Then

$$\frac{\partial f}{\partial x} = 6xy + 7y^2 \quad \text{and} \quad \frac{\partial f}{\partial y} = 3x^2 + 14xy,$$

so that

$$\begin{aligned} x\frac{\partial f}{\partial x} + y\frac{\partial f}{\partial y} &= x(6xy + 7y^2) + y(3x^2 + 14xy) = 9x^2y + 21xy^2 \\ &= 3(3x^2y + 7xy^2) \\ &= 3f(x, y), \end{aligned}$$

as predicted by Euler's Theorem. ■

We can derive an important economic insight from Euler's Theorem. Suppose that a production function $q(k, l)$ has constant returns to scale; that is $q(k, l)$ is homogeneous of degree 1. In that case the theorem says that

$$k\frac{\partial q}{\partial k} + l\frac{\partial q}{\partial l} = q.$$

Recall that $\partial q/\partial k$ is the marginal product of capital and $\partial q/\partial l$ is the marginal product of labour. Now it is a reasonable assumption that each factor of production, capital and labour, should be rewarded at a level equal to its marginal product. This is because, if one extra worker gives an increase in output of, say, 500 tins of catfood per week, then a reasonable weekly wage for that worker is (the money equivalent of) 500 tins of catfood. Under this assumption, the total reward to labour should be $l(\partial q/\partial l)$ measured in units of the product, such as tins of catfood. Similarly, $k(\partial q/\partial k)$ should be the total reward to capital. Euler's Theorem tells us that, in the case of constant returns to scale, these rewards add up exactly to the amount produced, which is as it should be.

Worked Examples

Example 12.1 Without solving the equation, show that $2x^2 + 5xy + y^2 = 19$ defines an implicit function $y(x)$ for which $y(2) = 1$, and find dy/dx when $x = 2$. Express the answer in geometrical terms.

Solution

Putting $x = 2$ and $y = 1$ we see that the equation is satisfied, since $2(2^2) + 5(2)(1) + 1^2 = 19$. Using the formula given in Section 12.2, we have

$$\frac{dy}{dx} = -\frac{4x + 5y}{5x + 2y} = -\frac{13}{12},$$

when $(x, y) = (2, 1)$. In geometrical terms, this means that the slope of the contour $2x^2 + 5xy + y^2 = c$ which passes through the point $(2, 1)$ is $-13/12$ at that point. (Of course, we know that the contour is the one for which $c = 19$.)

Example 12.2 The positive quantity y is defined implicitly as a function of x by the equation

$$x^4y^3 + 4x^2y^2 - 2x^5y = 3.$$

Find dy/dx when $x = 1$.

Solution

By the rule for implicit differentiation:

$$\frac{dy}{dx} = -\frac{g_1(x,y)}{g_2(x,y)} = -\frac{4x^3y^3 + 8xy^2 - 10x^4y}{3x^4y^2 + 8x^2y - 2x^5}.$$

We have to find the value (or values) of y when $x = 1$, noting that we are given that y must be positive. Putting $x = 1$ in the equation, we get $y^3 + 4y^2 - 2y - 3 = 0$. Now

$$y^3 + 4y^2 - 2y - 3 = (y - 1)(y^2 + 5y + 3),$$

and $y^2 + 5y + 3$ is never zero when $y > 0$. Thus we must have $y = 1$. Hence the required value of dy/dx is obtained from the general expression by substituting $x = 1$ and $y = 1$, which gives $dy/dx = -2/9$.

Example 12.3 Write down explicitly the Cobb–Douglas production function $Ak^\alpha l^\beta$ when

$$\text{(i) } A = 1, \alpha = 1/2, \beta = 1/2; \quad \text{(ii) } A = 1, \alpha = 1, \beta = -1.$$

In both cases sketch some typical isoquants in $\mathbb{R}^2_+$, and find the gradient of the isoquant at the point $(1, 1)$.

Solution

The functions are

$$\text{(i)} \quad u(k,l) = \sqrt{kl} \quad \text{and} \quad \text{(ii)} \quad v(k,l) = k/l.$$

A typical isoquant for u therefore has equation $\sqrt{kl} = c$, that is $kl = c^2$, or equivalently $l = c^2/k$. This represents a *hyperbola*, a curve of the form shown in Figure 12.4a, if $c \neq 0$. (If $c = 0$, the isoquant has equation $kl = 0$ and is therefore the k-axis together with the l-axis.)

A typical isoquant for v has equation $k/l = c$. This is just the straight line $k - cl = 0$, and as c varies we get the family of straight lines sketched in Figure 12.4b.

Recall that the gradient of an isoquant $f(k,l) = c$ at the point (k_0, l_0) is $-f_1/f_2$ evaluated at that point.

For u, we have $u(k,l) = \sqrt{kl}$, so the partial derivatives are

$$u_1 = \frac{1}{2}k^{-1/2}l^{1/2}, \quad u_2 = \frac{1}{2}k^{1/2}l^{-1/2},$$

and

$$\frac{dl}{dk} = -\frac{\frac{1}{2}k^{-1/2}l^{1/2}}{\frac{1}{2}k^{1/2}l^{-1/2}} = -\frac{l}{k}.$$

The gradient at $(1, 1)$ is therefore $-1/1 = -1$.

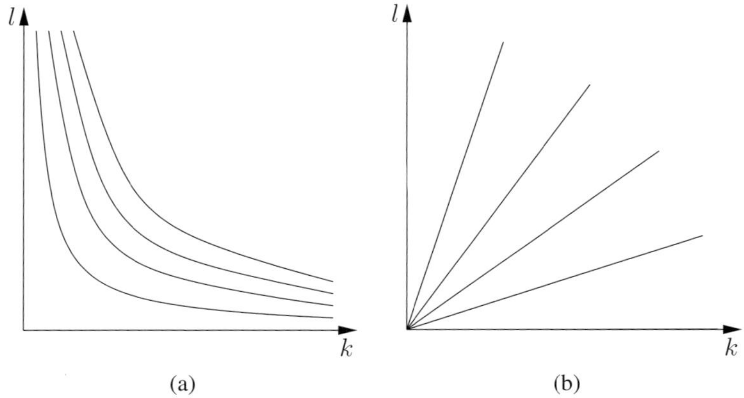

Figure 12.4 Typical isoquants for $\sqrt{kl}$ and k/l.

Similarly, the derivative dl/dk on an isoquant $v(k,l) = c$ is

$$\frac{dl}{dk} = -\frac{v_1}{v_2} = -\frac{1/l}{-k/l^2} = \frac{l}{k},$$

which is 1 at $(1,1)$.

Example 12.4 The notation $\min(x,y)$ stands for 'the smaller of the values of x and y'. Sketch typical contours for the function $m(x,y) = \min(x,y)$ in $\mathbb{R}^2_+$. Comment on the problem of finding the gradient at $(1,1)$ of the contour which passes through that point.

Solution

A contour for m has equation $\min(x,y) = c$. Now if $x < y$, $\min(x,y) = x$. This means that in the region where $x < y$ (the part of $\mathbb{R}^2_+$ which lies above the line $x = y$), a contour is just a line $x = c$. Similarly, if $x > y$ (the region below the line $x = y$), a contour is a line $y = c$. The contours therefore have the form shown in Figure 12.5.

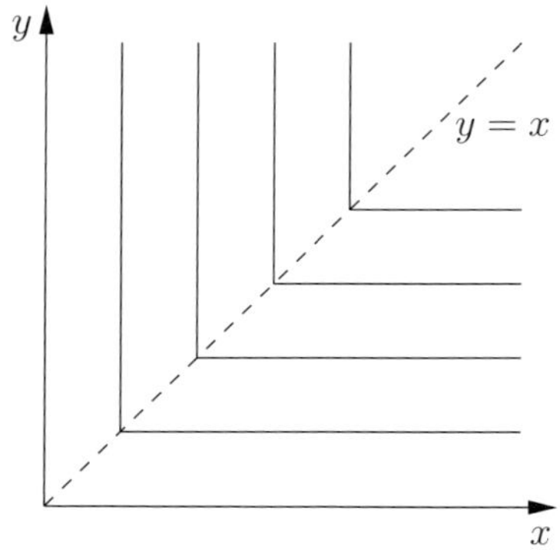

Figure 12.5 The contours $\min(x,y) = c$.

It is clear that the gradient of the contour $\min(x, y) = 1$ cannot be defined at $(1, 1)$. The contour is not smooth at that point and instead changes direction abruptly.

Example 12.5 Which of the following functions are homogeneous? For those that are homogeneous, what is the degree of homogeneity?

$$f(x, y) = \sqrt{x^2 + xy}, \quad g(x, y) = \sqrt{x + y^2},$$

$$h(x, y) = x(x^2 + xy)^{1/5} + y^{3/5}x^{4/5}.$$

Solution

To determine whether a function is homogeneous, we use the definition. That is, we calculate $f(cx, cy)$ and check whether it is of the form $c^D f(x, y)$. For the first function we have

$$f(cx, cy) = \sqrt{(cx)^2 + (cx)(cy)} = \sqrt{c^2(x^2 + xy)} = c\sqrt{x^2 + xy} = cf(x, y).$$

Thus f is homogeneous of degree 1.

Turning to g, we have

$$g(cx, cy) = \sqrt{cx + (cy)^2}.$$

Suppose that for some D this is equal to $c^D g(x, y)$. Then we should have

$$\sqrt{cx + (cy)^2} = c^D\sqrt{x + y^2},$$

for all (x, y). Putting $x = 0$ this implies that $\sqrt{(c^2y^2)} = c^D\sqrt{y^2}$, so $D = 1$. On the other hand, putting $y = 0$ we must have $\sqrt{cx} = c^D\sqrt{x}$, so $D = 1/2$. This contradiction shows that g is not homogeneous.

Finally, for h we have

$$\begin{aligned} h(cx, cy) &= cx((cx)^2 + (cx)(cy))^{1/5} + (cy)^{3/5}(cx)^{4/5} \\ &= c^{7/5}\left(x(x^2 + xy)^{1/5} + y^{3/5}x^{4/5}\right) \\ &= c^{7/5}h(x, y). \end{aligned}$$

So h is homogeneous of degree $7/5$.

Example 12.6 Show that the function $f(x, y) = (x^2 + y^2)^{3/2}x^{1/2}y^{1/2}$ is homogeneous of degree 4, and verify that

$$x\frac{\partial f}{\partial x} + y\frac{\partial f}{\partial y} = 4f.$$

Solution

To show that the function is homogeneous of degree 4, we observe that

$$\begin{aligned} f(cx, cy) &= ((cx)^2 + (cy)^2)^{3/2}(cx)^{1/2}(cy)^{1/2} \\ &= (c^2)^{3/2}(x^2 + y^2))^{3/2}c^{1/2}x^{1/2}c^{1/2}y^{1/2} \\ &= c^4 f(x, y). \end{aligned}$$

Now

$$\frac{\partial f}{\partial x} = \frac{3}{2}2x(x^2 + y^2)^{1/2}(x^{1/2}y^{1/2}) + (x^2 + y^2)^{3/2}\left(\frac{1}{2}x^{-1/2}y^{1/2}\right),$$

$$\frac{\partial f}{\partial y} = \frac{3}{2}2y(x^2 + y^2)^{1/2}(x^{1/2}y^{1/2}) + (x^2 + y^2)^{3/2}\left(\frac{1}{2}x^{1/2}y^{-1/2}\right).$$

It follows that

$$\begin{aligned} x\frac{\partial f}{\partial x} + y\frac{\partial f}{\partial y} &= (3x^2 + 3y^2)(x^2 + y^2)^{1/2}(x^{1/2}y^{1/2}) + (x^{1/2}y^{1/2})(x^2 + y^2)^{3/2} \\ &= 3(x^2 + y^2)^{3/2}(x^{1/2}y^{1/2}) + (x^2 + y^2)^{3/2}(x^{1/2}y^{1/2}) \\ &= 4(x^2 + y^2)^{3/2}(x^{1/2}y^{1/2}) = 4f(x, y), \end{aligned}$$

as predicted by Euler's Theorem.

Main Topics

- implicitly defined functions
- finding the derivative of a function defined implicitly
- contours and isoquants and their slopes
- homogeneous functions and Euler's Theorem

Key Terms, Notations and Formulae

- if $g(x, y) = c$ defines y implicitly, then $\dfrac{dy}{dx} = -\dfrac{\partial g/\partial x}{\partial g/\partial y} = -\dfrac{g_1}{g_2}$
- contour of $f(x, y)$: $\{(x, y) \mid f(x, y) = c\}$
- isoquant: contour of production function
- $h(x, y)$ homogeneous of degree D if $h(cx, cy) = c^D h(x, y)$
- constant, increasing and decreasing returns to scale
- Euler's Theorem: if $h(x, y)$ homogeneous of degree D, then $x\dfrac{\partial h}{\partial x} + y\dfrac{\partial h}{\partial y} = Dh$

Exercises

12.1 Suppose the quantity y is defined as a function of x through the equation

$$x^2y^3 - 6x^3y^2 + 2xy = 1.$$

Find $\dfrac{dy}{dx}$ when $x = 1/2$.

12.2 Show that the function $f(x,y) = (x^4+y^4)^{1/2}(x^2y+x^3)$ is homogeneous of degree 5. Verify that

$$x\frac{\partial f}{\partial x} + y\frac{\partial f}{\partial y} = 5f.$$

12.3 Suppose that the function

$$f(x,y) = xy^a \sin\left(\frac{x}{y}\right) + \frac{x^b}{y}$$

is homogeneous, where a and b are numbers. Determine a formula for b in terms of a. Also find the degree of homogeneity of the function in terms of a and verify that the function satisfies Euler's Theorem.

Problems

12.1 Write down all the first-order and second-order partial derivatives for the Cobb–Douglas function

$$f(x,y) = Ax^{\alpha}y^{\beta}.$$

Explain what is meant by saying that the function is homogeneous of degree $\alpha + \beta$ and comment on the relationship between this property and the notion of 'returns to scale'. For the particular Cobb–Douglas function with $A = 1$, $\alpha = 2$, $\beta = 1$, sketch some typical contours $f(x,y) = c$ with $c > 0$. Find the gradient at the point $(3,2)$ of the contour which passes through that point.

12.2 Let $g(x,y) = x^2y^{3/2}$ for $x, y > 0$. Sketch some typical contours $g(x,y) = c$ with $c > 0$. Calculate the first-order partial derivatives of g. Find the slope at the point $(1,2)$ of the contour which passes through that point.

12.3 Which of the following functions are homogeneous? For those that are, what are their degrees of homogeneity?

$$f(x,y) = \sqrt{xy} + \frac{x^2}{y}, \quad g(x,y) = \left(x^2 + \frac{2y^3}{x}\right)^{1/3} + xy^{1/3},$$

$$h(x,y) = (y^2 + 2xy)^{1/9} + x^{1/18}y^{3/18}.$$

12.4 Show that the function

$$f(x,y) = 2y^3x + 5y^4 - (y^{3/4} - 2x^{3/4})^4x$$

is homogeneous of degree 4. Verify that

$$x\frac{\partial f}{\partial x} + y\frac{\partial f}{\partial y} = 4f.$$

12.5 Suppose that production q, capital k and labour l satisfy $g(q,k,l) = 0$. In other words, the production function $q(k,l)$ is defined implicitly, and it satisfies

$g(q(k,l),k,l) = 0$ identically. How do you think we might calculate the partial derivatives $\partial q/\partial k$ and $\partial q/\partial l$ in a manner similar to that developed in this chapter for functions g of only two variables?

Illustrate your method by working out the partial derivatives when q is defined by the equation $q^3k^2 + l^3 + qkl = 0$.

13 Optimisation in Two Variables

13.1 Profit Maximisation Again

Problems in which we need to optimise a function of several variables occur frequently and naturally in economics. A simple example occurs when we consider the profit maximisation problem for a firm which makes two goods.

Example Let us suppose that a firm known as the All Purpose Outfit (APO) makes two goods, Brand X and Brand Y. It will be convenient to let x and y denote the quantities of X and Y produced, respectively. Our first aim is to discover how revenue and profit depend on x and y.

Suppose that the selling price of each unit of X is fixed at $p^X = 4$ and that the price of each unit of Y is fixed at $p^Y = 1$. Then the revenue obtained when APO produces x units of X and y units of Y is clearly

$$R(x, y) = 4x + y.$$

On the other hand, the production of x units of X and y units of Y will involve a cost $C(x, y)$. Suppose, for the sake of example, that this *joint cost function* for APO is

$$C(x, y) = 5 + x^2 - xy + y^2.$$

Then the profit function is

$$\begin{aligned} \Pi(x, y) &= R(x, y) - C(x, y) \\ &= 4x + y - (5 + x^2 - xy + y^2) \\ &= 4x + y - 5 - x^2 + xy - y^2. \end{aligned}$$

In order to determine the levels of production of each good that will maximise the profit, we shall therefore have to find the point (x, y) which gives the maximum value of the function $\Pi(x, y)$. ■

13.2 How Prices Are Related to Quantities

As in the case where there is only one good (Chapter 9), we have to decide what assumptions to make about the selling prices of X and Y. Generally, if p^X and p^Y

are the selling prices of one unit of X and one unit of Y, then the revenue obtained by producing amounts x and y is

$$R(x,y) = xp^{\mathrm{X}} + yp^{\mathrm{Y}},$$

simply because xp^{X} is the revenue from X and yp^{Y} is the revenue from Y. That much is plain. However, in order to make progress we must try to understand how the prices p^{X} and p^{Y} are affected by x and y.

In fact, we shall consider three ways in which the prices and quantities may be related.

- Case 1: p^{X} and p^{Y} are constants.

This is what happens when there is perfect competition, or equivalently when we have an 'efficient small firm', as discussed in Chapter 10. The simple example given in Section 13.1 belongs to Case 1.

- Case 2: p^{X} depends only on x and p^{Y} depends only on y.

This is the case when the firm has a monopoly in both X and Y, and there is no interaction between the markets for the two goods; then p^{X} and p^{Y} are the respective inverse demand functions.

Example Suppose that the firm APO in the example above is a monopoly, not an efficient small firm. Then, according to the discussion in Section 9.3, the prices paid for its products are given by the respective inverse demand functions. If, for example, the demand sets for X and Y are

$$D_{\mathrm{X}} = \{(x,p) \mid x + 2p = 10\} \quad \text{and} \quad D_{\mathrm{Y}} = \{(y,p) \mid y + p = 24\},$$

then the inverse demand functions for X and Y are given by

$$p^{\mathrm{X}}(x) = 5 - 0.5x \quad \text{and} \quad p^{\mathrm{Y}}(y) = 24 - y.$$

The profit function is

$$\begin{aligned}
\Pi(x,y) &= xp^{\mathrm{X}} + yp^{\mathrm{Y}} - C(x,y) \\
&= x(5 - 0.5x) + y(24 - y) - (5 + x^2 - xy + y^2) \\
&= 5x - 0.5x^2 + 24y - y^2 - 5 - x^2 + xy - y^2 \\
&= 5x + 24y - 1.5x^2 - 2y^2 + xy - 5.
\end{aligned}$$

■

Finally we have the most interesting case.

- Case 3: each of p^{X} and p^{Y} depends on *both* x and y.

Here we are assuming that the amount of X on the market affects the demand for Y, and vice versa. Equivalently, this is the case when the price of X affects the demand for Y, as well as the demand for X, and vice versa. For instance, a firm may make two different types of chocolate bar and consumers may switch between them depending on their relative prices. In this case, the two chocolate bars are competing products, often referred to by economists as 'substitutes'. As another such example, consider the market for e-books and printed books. As the price of e-books falls, the demand for them increases and the demand for printed books decreases, because people generally buy only one version of the latest novel. Thus the price of e-books affects not only the demand for e-books, but also the demand for printed books. Alternatively, instead of purchase of one of the goods suppressing demand for the other, it could be the case that the opposite relationship holds; for example, X could be a necklace and Y a matching bracelet. As here, where purchase of either good stimulates demand for the other, the goods are known as 'complements'.

Example Suppose now that APO is the only firm producing X and Y and that the demand for X is given by

$$x = 2 - 2p^{\mathrm{X}} + p^{\mathrm{Y}},$$

and the demand for Y is given by

$$y = 13 + p^{\mathrm{X}} - 2p^{\mathrm{Y}}.$$

Note that the general form of these equations tells us how the various quantities are linked. If the price of X is fixed and the price of Y is increased, then the demand for X rises and the demand for Y falls. This is the behaviour one might expect if X and Y are two different types of chocolate bar.

We may rearrange the equations to find expressions for p^{X} and p^{Y}. Multiplying the first equation by 2 and adding it to the second, we obtain

$$2x + y = 2(2 - 2p^{\mathrm{X}} + p^{\mathrm{Y}}) + 13 + p^{\mathrm{X}} - 2p^{\mathrm{Y}} = 17 - 4p^{\mathrm{X}} + p^{\mathrm{X}} = 17 - 3p^{\mathrm{X}},$$

from which we get p^{X} as a function of x and y:

$$p^{\mathrm{X}}(x, y) = (17 - 2x - y)/3.$$

Using this expression for p^{X}, together with the first equation, we can obtain a similar expression for p^{Y}:

$$p^{\mathrm{Y}} = x - 2 + 2p^{\mathrm{X}} = x - 2 + \frac{2}{3}(17 - 2x - y) = \frac{1}{3}(28 - x - 2y).$$

The profit function in this case is

$$\begin{aligned}
\Pi(x, y) &= xp^{\mathrm{X}} + yp^{\mathrm{Y}} - C(x, y) \\
&= \frac{x}{3}(17 - 2x - y) + \frac{y}{3}(28 - x - 2y) - (5 + x^2 - xy + y^2) \\
&= -5 + \frac{17}{3}x + \frac{28}{3}y - \frac{5}{3}x^2 - \frac{5}{3}y^2 + \frac{1}{3}xy.
\end{aligned}$$

■

Of course, all three cases discussed above are covered by the blanket assumption that p^{X} and p^{Y} are functions of x and y, provided we allow the functions to be constant with respect to one or both variables. Making the reasonable assumption that the joint cost function is also a function $C(x, y)$ of two variables, we have a general expression for the profit from making and selling x units of X and y units of Y:

$$\Pi(x, y) = R(x, y) - C(x, y) = xp^{\mathrm{X}}(x, y) + yp^{\mathrm{Y}}(x, y) - C(x, y).$$

In the rest of this chapter we shall explain how to find the maximum value of such a function.

13.3 Critical Points

When we say that the point (a, b) is a *local maximum* of a function f we mean that

$$f(x, y) \leq f(a, b)$$

for all points (x, y) in the neighbourhood of (a, b). An equivalent statement is that, at the point (a, b), small changes h and k in the variables always result in a negative or zero change in f. In our usual notation for small changes this condition is

$$\Delta f = f(a + h, b + k) - f(a, b) \leq 0.$$

Let B denote the function of one variable defined by $B(x) = f(x, b)$. In other words, we fix $y = b$ and let x vary. Then we have

$$B(x) = f(x, b) \leq f(a, b) = B(a),$$

for all x in the neighbourhood of a. This simply means that $B(a)$ is a maximum value of B and, by our rules for functions of one variable, we must have $B'(a) = 0$. But $B(x) = f(x, b)$, so the derivative of B is just the *partial derivative* of f with respect to the first variable. That is, $B'(a) = f_1(a, b) = 0$.

Similarly, let A denote the function of one variable defined by $A(y) = f(a, y)$. Here we have $A'(y) = f_2(a, y)$, and the condition for a maximum at (a, b) is that $A'(b) = f_2(a, b) = 0$.

It follows that the following two conditions hold at a point (a, b) where f is a maximum:

$$\frac{\partial f}{\partial x} = 0 \quad \text{and} \quad \frac{\partial f}{\partial y} = 0.$$

We call these the *first-order conditions*, and say that a point (x, y) is a *critical point* if the first-order conditions hold at the point. In other words, a critical point is one where both partial derivatives are zero.

We have shown that a local maximum is a critical point. If we define a *local minimum* to be a point (a, b) for which Δf is positive for any small change in the variables, then it is clear that a local minimum is also a critical point. So we have established that the first step in the search for maxima is to find the critical points, but we must remember that *not every critical point is a maximum*. Indeed, the situation for functions of two variables is rather more complicated than for functions of one variable, as we shall see in the next section.

Example In order to find the critical points of the function

$$g(x,y) = x^4 + 2x^2y + 2y^2 + y$$

we work out the partial derivatives of g:

$$\frac{\partial g}{\partial x} = 4x^3 + 4xy \quad \text{and} \quad \frac{\partial g}{\partial y} = 2x^2 + 4y + 1.$$

The critical points are given by the first-order conditions

$$4x^3 + 4xy = 0 \quad \text{and} \quad 2x^2 + 4y + 1 = 0.$$

The first of these equations says that $x(x^2 + y) = 0$, and so either (i) $x = 0$ or (ii) $y = -x^2$.

(i) When $x = 0$, the second equation is $4y + 1 = 0$, giving $y = -1/4$. Thus one critical point is $(0, -1/4)$. (ii) When $y = -x^2$, the second equation becomes $2x^2 - 4x^2 + 1 = 0$ and so x is either $1/\sqrt{2}$ or $-1/\sqrt{2}$. The corresponding values of y are determined by $y = -x^2$, and are both $-1/2$. Thus, there are three critical points: $(0, -1/4)$, $(1/\sqrt{2}, -1/2)$ and $(-1/\sqrt{2}, -1/2)$. ■

13.4 Maxima, Minima and Saddle Points

We begin by introducing a useful way of visualising the behaviour of functions of two variables. It is analogous to the use of graph sketching in the case of functions of one variable.

Given a function f of two variables, we think of the points (x, y) as lying in a horizontal plane. For each point (x, y) we represent the value $f(x, y)$ by the point at 'height' $f(x, y)$ lying directly above (x, y). All such points form a surface, which we regard as the 'graph' of f. See Figure 13.1.

At a critical point (a, b), both partial derivatives of f are zero. Geometrically, this means that the surface is horizontal at (a, b).

We can be more specific. If the surface has a 'peak' at the value $f(a, b)$ (Figure 13.2a), this means that the point (a, b) is a maximum. As we have already observed, a maximum of f is characterised by the fact that the change Δf is negative for any small changes in the variables. Similarly, if the surface has a 'pit' (Figure 13.2b), the critical point is a minimum, and Δf is positive.

It is most important to note that a critical point of a function of two variables may be neither a local maximum nor a local minimum. For example, Figure 13.3 depicts the surface representing $f(x, y) = x^2 - y^2$ in the neighbourhood of the critical point $(0, 0)$. We can see that, for some points (h, k) in the neighbourhood of $(0, 0)$, $f(h, k)$ is greater than $f(0, 0)$, while for other such points it is less. Equivalently, Δf takes both positive and negative values in the neighbourhood of $(0, 0)$. Such points are known as *saddle points*.

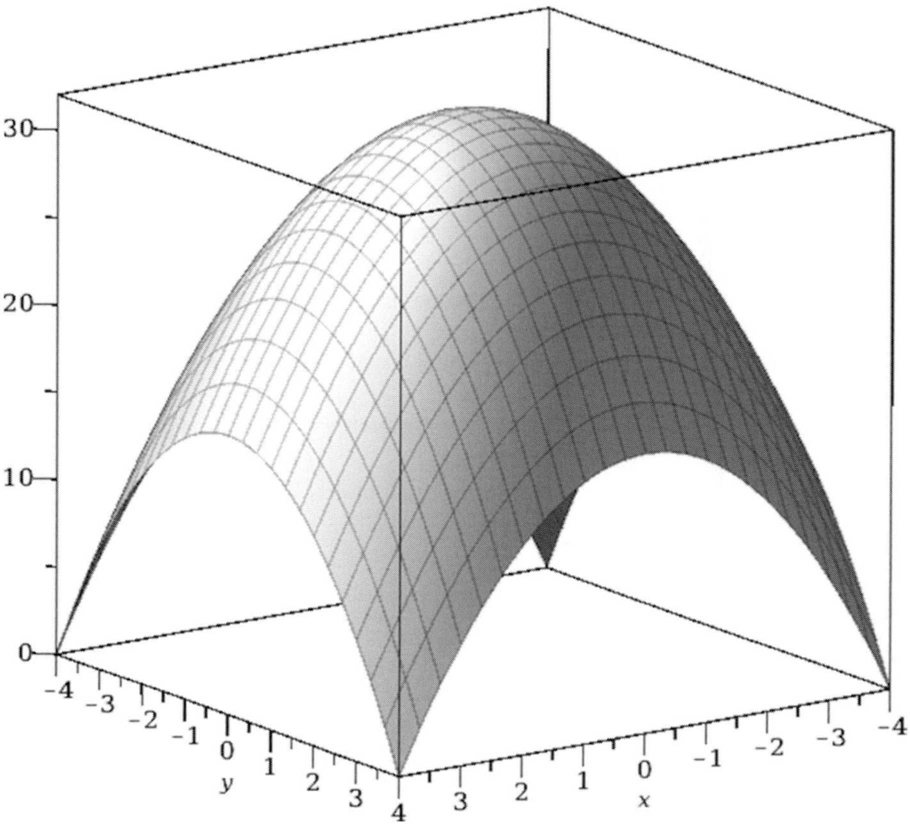

Figure 13.1 A part of the surface representing $f(x, y)$.

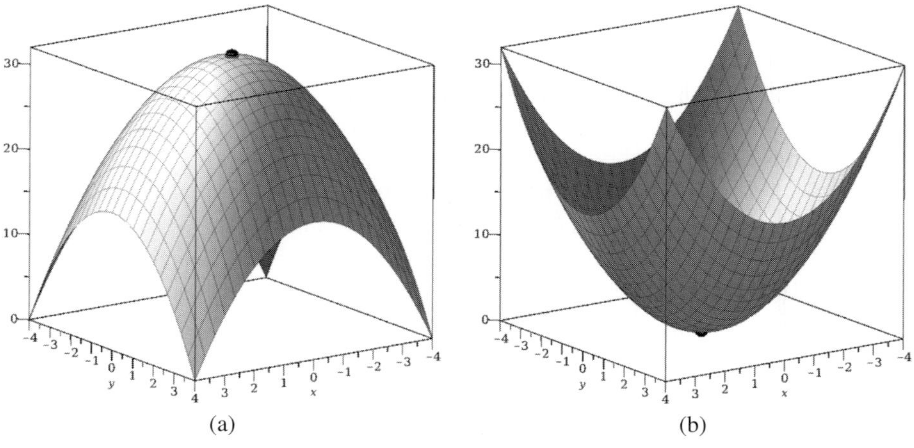

Figure 13.2 A local maximum and a local minimum.

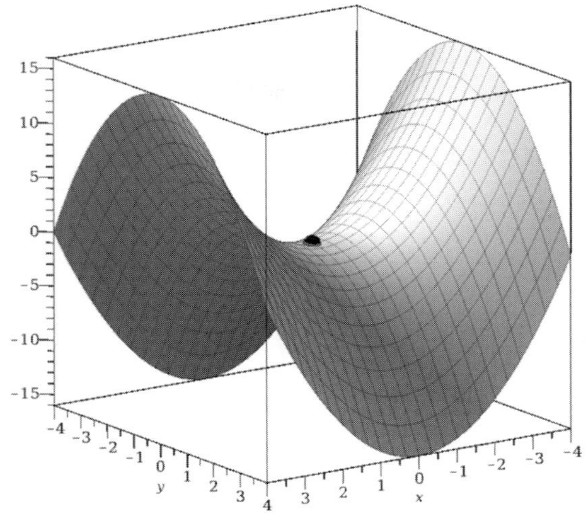

Figure 13.3 A saddle point.

13.5 Classification of Critical Points – Introduction

Our task is to *classify* critical points; that is to determine whether a critical point is a maximum, a minimum, a saddle point or possibly some other kind of point. For functions of one variable, we showed in Chapter 8 that the sign of the *second* derivative is a useful guide. For functions of two variables, there are similar conditions involving the second partial derivatives.

We begin by looking at a function of the form

$$Q(x, y) = Ax^2 + Bxy + Cy^2 + Lx + My + N.$$

One immediate justification for this is that all the profit functions obtained in Sections 13.1 and 13.2 are of this form. But we shall explain in due course how the problem of classifying critical points for a function of two variables can be reduced to this case, provided certain simple conditions are satisfied.

First we need to work out the critical points of Q. They are given by the first-order conditions

$$Q_1(x, y) = 2Ax + By + L = 0 \quad \text{and} \quad Q_2(x, y) = Bx + 2Cy + M = 0.$$

These two simultaneous equations in x and y can be solved by elementary algebra. It is helpful to use the abbreviation D for $4AC - B^2$; then the solution is

$$x_0 = \frac{BM - 2CL}{D}, \quad y_0 = \frac{BL - 2AM}{D}.$$

We conclude that Q has just one critical point (x_0, y_0), provided of course that $D \neq 0$. It remains to classify that point.

We recall that the nature of the critical point is determined by the behaviour of ΔQ in its neighbourhood. Given the values of x_0 and y_0 determined above, ΔQ can be worked out by elementary algebra. Omitting the details, the result is reassuringly simple:

$$\begin{aligned}\Delta Q &= Q(x_0 + h, y_0 + k) - Q(x_0, y_0)\\ &= Ah^2 + Bhk + Ck^2.\end{aligned}$$

Now, assuming that $A \neq 0$, we use the standard technique of completing the square (see Section 2.4), which gives

$$\begin{aligned}\Delta Q &= A\left(h + \frac{B}{2A}k\right)^2 + \left(C - \frac{B^2}{4A}\right)k^2\\ &= A\left(h + \frac{B}{2A}k\right)^2 + \left(\frac{D}{4A}\right)k^2.\end{aligned}$$

Putting $X = h + (B/2A)k$ and $Y = k/2$, we have

$$\Delta Q = AX^2 + \left(\frac{D}{A}\right)Y^2.$$

The reason for putting ΔQ into this form is that X^2 and Y^2 are never negative. So the behaviour of ΔQ is determined by the signs of the coefficients A and D/A, or what comes to the same thing, the signs of A and D.

Suppose first that D is positive and A is negative, so that D/A is negative. This means that ΔQ is negative and the critical point is a maximum.

Next, if D and A are both positive, then D/A is positive. This means that ΔQ is positive and the critical point is a minimum.

Finally, we ask what happens when D is negative. In this case the coefficients A and D/A have opposite signs, whatever the sign of A. So ΔQ takes one sign when $X = 0$ and the other when $Y = 0$. In other words, we can find arbitrarily small values of h and k for which ΔQ takes either sign. Thus the critical point is a saddle point.

Example In the example discussed in Section 13.1 we found the profit function $\Pi(x, y)$ to be $4x + y - 5 - x^2 + xy - y^2$. Rearranging the terms we see that this is a function of the same general form as Q. The critical point is given by

$$\frac{\partial \Pi}{\partial x} = 4 - 2x + y = 0 \quad \text{and} \quad \frac{\partial \Pi}{\partial y} = 1 + x - 2y = 0.$$

That is

$$2x - y = 4 \quad \text{and} \quad -x + 2y = 1.$$

Multiplying the first equation by 2 and adding this to the second equation, we have

$$2(2x - y) + (-x + 2y) = 2(4) + 1, \quad \text{or} \quad 3x = 9.$$

So $x = 3$ and, since $y = 2x - 4$, we have $y = 2 \times 3 - 4 = 2$. So the profit function Π has a critical point at $(3, 2)$. (Of course, we could also have used the general formula for (x_0, y_0) given above to find the critical point.)

We need to be sure that this gives a maximum of Π, rather than any other kind of critical point. For this, we note that here the coefficient of x^2 is $A = -1$, and similarly $B = 1$ and $C = -1$, so

$$A < 0 \quad \text{and} \quad D = 4AC - B^2 = 3 > 0.$$

As shown above, this means that the critical point is indeed a maximum and the maximum profit is $\Pi(3, 2) = 2$. ■

In this example we assumed that the profit function achieves its largest value (or its *global maximum*) at the maximum point $(3, 2)$. Although it is true in this example that the function has its largest value at the maximum point, it is possible for a function to take arbitrarily large values, in which case it has no largest value at all. When this happens, a maximum point is merely a *local* maximum. (This is analogous to the discussion in Chapter 8, for functions of one variable.) Having planted these seeds of caution, we shall blithely assume from now on, for the sake of simplicity, that if there is only one local maximum, then the function has a global maximum there. (This will hold in all our examples.)

The two cases of the profit function discussed in Section 13.2 are also functions of the form Q, and their critical points could be classified using the method in the example above. However, we shall now extend the method, and in practice it is preferable to work with the method in its general form.

13.6 The Classification of Critical Points in General

In the case of functions of the form Q discussed in the previous section, the nature of the critical point is determined by the signs of A and $D = 4AC - B^2$. In that case the second partial derivatives of Q are constants: $Q_{11} = 2A$, $Q_{12} = Q_{21} = B$, $Q_{22} = 2C$. So A and D are simply related to the second partial derivatives:

$$A = Q_{11}/2 \qquad \text{and} \quad D = Q_{11}Q_{22} - Q_{12}^2.$$

This means that the conditions for a maximum, minimum and saddle point of Q can be expressed in terms of the signs of Q_{11} and $D = Q_{11}Q_{22} - Q_{12}^2$.

It turns out that precisely the same results hold for *any* function f of two variables: the critical points of f can be classified using f_{11} and $D = f_{11}f_{22} - f_{12}^2$, according to the following *second-order conditions*.

> Suppose that (a, b) is a critical point of f.
>
> - If $f_{11}f_{22} - f_{12}^2 > 0$ and $f_{11} < 0$, it is a maximum.
> - If $f_{11}f_{22} - f_{12}^2 > 0$ and $f_{11} > 0$, it is a minimum.
> - If $f_{11}f_{22} - f_{12}^2 < 0$, it is a saddle point.

Essentially, the same method works because Δf can be approximated in the neighbourhood of a critical point by an expression involving only terms in h^2, k^2 and kh, and the coefficients of these terms are just the second partial derivatives, as was the case for ΔQ. We have to take this result for granted here, but we hope that we have done enough to justify the use of the second-order conditions to classify the critical points of functions of two variables in general. The reader is advised to commit these conditions to memory.

It is important to realise that in some cases the second-order conditions do not tell us what kind of critical point we have; for instance, if the quantity $D = f_{11}f_{22} - f_{12}^2$ turns out to be zero at the critical point, then the second-order conditions do not enable us to classify the nature of the point.

Example In Section 13.3 we showed that the critical points of the function $g(x, y) = x^4 + 2x^2y + 2y^2 + y$ are

$$(0, -1/4), \quad (1/\sqrt{2}, -1/2), \quad (-1/\sqrt{2}, -1/2).$$

To classify them we need the second partial derivatives of g:

$$g_{11}(x, y) = 12x^2 + 4y, \quad g_{12}(x, y) = 4x, \quad g_{22}(x, y) = 4.$$

Note that, as is generally the case, these are not constant, so we have to look at each critical point individually.
At $(0, -1/4)$: we have $g_{11} = -1$, $g_{12} = -1$, $g_{22} = 4$, so $g_{11} < 0$ and $D = (-1)(4) - 1 < 0$. Therefore $(0, -1/4)$ is a saddle point.

At $(1/\sqrt{2}, -1/2)$: we have $g_{11} = 4$, $g_{12} = 2\sqrt{2}$, $g_{22} = 4$, so $g_{11} > 0$ and $D = (4)(4) - 8 > 0$. Therefore $(1/\sqrt{2}, -1/2)$ is a minimum.

At $(-1/\sqrt{2}, -1/2)$: we have $g_{11} = 4$, $g_{12} = -2\sqrt{2}$, $g_{22} = 4$, so $g_{11} > 0$ and $D = (4)(4) - 8 > 0$. Therefore $(-1/\sqrt{2}, -1/2)$ is also a minimum. ■

Worked Examples

Example 13.1 Show that the profit function

$$\Pi(x, y) = 5x + 24y - 1.5x^2 - 2y^2 + xy - 5$$

obtained in Section 13.2 for the firm APO – when it is assumed to have a monopoly on X and Y but the markets do not interact – has a critical point at $(4, 7)$, and verify that it is a maximum.

Solution

The first-order conditions are

$$\Pi_1(x, y) = 5 - 3x + y = 0 \quad \text{and} \quad \Pi_2(x, y) = 24 - 4y + x = 0.$$

That is $3x - y = 5$ and $-x + 4y = 24$. Multiplying the first equation by 4 and adding this to the second, we obtain $11x = 44$, so $x = 4$. Since $y = 3x - 5$, the value of y is $3 \times 4 - 5 = 7$. Therefore the profit function has a critical point at $(4, 7)$.

To classify the critical point, we use the second-order conditions. The second partial derivatives are

$$\Pi_{11} = -3, \quad \Pi_{12} = 1, \quad \Pi_{22} = -4.$$

(Note that they are constant here.) We see that Π_{11} is negative and $D = \Pi_{11}\Pi_{22} - \Pi_{12}^2 = 12 - 1 = 11$ is positive. It follows that the critical point is a maximum, and the maximum profit is $\Pi(4, 7) = 89$.

Example 13.2 If $f(x, y) = x^3 - y^3 - 2xy + 1$, find and classify the critical points of f.

Solution

The first-order conditions are

$$f_1(x, y) = 3x^2 - 2y = 0 \quad \text{and} \quad f_2(x, y) = -3y^2 - 2x = 0.$$

From the first of these, $y = 3x^2/2$. Substituting for y in the second, we get

$$3\left(\frac{3}{2}x^2\right)^2 + 2x = 0, \quad \text{that is} \quad \frac{27}{4}x^4 + 2x = 0.$$

On multiplying by 4 and factorising, this becomes $x(27x^3 + 8) = 0$, which has solutions $x = 0$ and $x = -2/3$. The corresponding values of y, given by $y = 3x^2/2$, are $y = 0$ and $y = 2/3$. Therefore there are two critical points, $(0, 0)$ and $(-2/3, 2/3)$.

To classify them, we apply the second-order conditions. We have

$$f_{11}(x, y) = 6x, \quad f_{12}(x, y) = -2, \quad f_{22}(x, y) = -6y.$$

At $(0, 0)$, $D = -(-2)^2 < 0$, so $(0, 0)$ is a saddle point.

At $(-2/3, 2/3)$, $f_{11} = -4 < 0$ and $D = (-4)(-4) - (-2)^2 > 0$, so the point $(-2/3, 2/3)$ is a maximum.

Example 13.3 A firm has a monopoly for the manufacture of two goods, X and Y, for which the inverse demand functions are

$$p^{\mathrm{X}} = 6 - x, \quad p^{\mathrm{Y}} = 16 - 2y,$$

where x and y are the quantities of X and Y, and p^{X} and p^{Y} are the respective prices. The firm's cost function is $C(x, y) = \frac{1}{2}x^2 + \frac{1}{2}y^2 + xy$. Determine the output quantities which will maximise the firm's profit, and calculate the maximum profit.

Solution

The profit function is

$$\begin{aligned}\Pi(x,y) &= xp^{X} + yp^{Y} - C(x,y)\\ &= x(6-x) + y(16-2y) - \left(\frac{x^2}{2} + \frac{y^2}{2} + xy\right)\\ &= 6x + 16y - \frac{3}{2}x^2 - xy - \frac{5}{2}y^2.\end{aligned}$$

The first-order conditions for a critical point are

$$\Pi_1(x,y) = 6 - 3x - y = 0 \quad \text{and} \quad \Pi_2(x,y) = 16 - x - 5y = 0.$$

Multiplying the first equation by 5 and subtracting the second, we have

$$30 - 15x - 5y - 16 + x + 5y = 0, \quad \text{that is} \quad 14 - 14x = 0.$$

Thus, $x = 1$ and there is only one critical point, $(1,3)$.

To verify that $(1,3)$ is a maximum, we need the second partial derivatives:

$$\Pi_{11}(x,y) = -3, \quad \Pi_{12}(x,y) = -1, \quad \Pi_{22}(x,y) = -5.$$

Thus $\Pi_{11} < 0$ and $D = \Pi_{11}\Pi_{22} - \Pi_{12}^2 = 15 - 1 > 0$, from which it follows that the critical point is a maximum. The maximum profit is $\Pi(1,3) = 27$.

Example 13.4 A monopoly manufactures two commodities, X and Y, the markets for which interact. The demand functions are given by

$$x = 800(p^{Y} - p^{X}), \quad y = 400(9 + 2p^{X} - 4p^{Y}).$$

How much of each commodity should be manufactured to maximise the profit, given that it costs \$1 to produce one unit of X and \$1.50 to produce one unit of Y?

Solution

We can eliminate p^{X} by adding the demand functions, which gives $x + y = 3600 - 800p^{Y}$. Thus

$$p^{Y} = \frac{3600 - x - y}{800} \quad \text{and} \quad p^{X} = \frac{800p^{Y} - x}{800} = \frac{3600 - 2x - y}{800}.$$

The cost function is $x + \frac{3}{2}y$ and so the profit function is

$$\begin{aligned}\Pi(x,y) &= xp^{X} + yp^{Y} - C(x,y)\\ &= \frac{1}{800}\left(3600x - 2x^2 - xy + 3600y - xy - y^2\right) - x - \frac{3}{2}y\\ &= \frac{1}{800}\left(2800x - 2x^2 - 2xy + 2400y - y^2\right).\end{aligned}$$

The first-order conditions for a critical point are

$$\Pi_1(x,y) = \frac{2800 - 4x - 2y}{800} = 0 \quad \text{and} \quad \Pi_2(x,y) = \frac{2400 - 2y - 2x}{800} = 0,$$

so that $2x + y = 1400$ and $x + y = 1200$. Solving these yields $x = 200$ and $y = 1000$.

We must check that the critical point (200, 1000) is a maximum. The second partial derivatives are

$$\Pi_{11}(x,y) = -1/200, \quad \Pi_{12}(x,y) = -1/400, \quad \Pi_{22}(x,y) = -1/400.$$

So $\Pi_{11} < 0$ and $D = \Pi_{11}\Pi_{22} - \Pi_{12}^2 > 0$, which implies that the critical point is a maximum and that the optimal levels of production are $x = 200$ and $y = 1000$.

Main Topics

- finding critical points of two-variable functions
- classifying critical points as maxima, minima or saddle points
- profit maximisation for a firm producing two goods

Key Terms, Notations and Formulae

- joint cost function, $C(x,y)$
- revenue, $R(x,y) = xp^{\mathrm{X}} + yp^{\mathrm{Y}}$
- maximum, minimum, saddle point
- critical point
- first-order conditions: $\dfrac{\partial f}{\partial x} = 0$ and $\dfrac{\partial f}{\partial y} = 0$
- second-order conditions: at a critical point
  - if $f_{11}f_{22} - f_{12}^2 > 0$ and $f_{11} < 0$, we have a maximum
  - if $f_{11}f_{22} - f_{12}^2 > 0$ and $f_{11} > 0$, we have a minimum
  - if $f_{11}f_{22} - f_{12}^2 < 0$, we have a saddle point

Exercises

13.1 Find and classify (i.e., determine the nature of) the critical points of the function

$$f(x,y) = x^2 - 2xy + 2y^2 - 2y + 2.$$

13.2 The function f is given, for $x, y \neq 0$, by

$$f(x,y) = \frac{4}{x} - \frac{2x}{y} + y.$$

Show that f has one critical point and determine what type of critical point this is.

13.3 The function f is given by

$$f(x,y) = 10 - 2x^2 + y^2 - y^4 + 2xy^2.$$

Find the critical points of f and determine the nature of each.

13.4 A firm is the only producer of two goods, X and Y. If the firm produces quantities x and y of X and Y, respectively, then the corresponding prices p^X and p^Y are related by the equations

$$p^Y - p^X + 13 - 6x = 0 \quad \text{and} \quad p^X - 4p^Y + 26 - 6y = 0.$$

The firm's total cost for producing x of X and y of Y is $2x + y$. Find an expression in terms of x and y for the firm's profit function. Determine the quantities x and y that maximise the firm's profit.

Problems

13.1 Find the maximum value of the function

$$f(x, y) = 6 + 4x - 3x^2 + 4y + 2xy - 3y^2.$$

13.2 Show that the function

$$g(x, y) = 3x^2 + 2xy + 2y^2 - 160x - 120y + 18$$

has only one critical point, and classify it.

13.3 For the firm APO, regarded as a monopoly producing two goods whose markets are interrelated, we obtained in Section 13.2 the profit function

$$\Pi(x, y) = -5 + \frac{17}{3}x + \frac{28}{3}y - \frac{5}{3}x^2 - \frac{5}{3}y^2 + \frac{1}{3}xy.$$

Show that this function has a maximum at $(2, 3)$.

13.4 Find the critical points of the functions

$$u(x, y) = y^3 + 3xy - x^3 \quad \text{and} \quad v(x, y) = x^3 - 3xy^2 + y^4,$$

and classify them.

13.5 A small firm manufactures two goods, X and Y, and the market price of these goods is unaffected by the level of the firm's production. If the firm's cost function is $C(x, y) = 2x^2 + xy + 2y^2$ and the market price of X is \$12 per unit and the market price of Y is \$18 per unit, determine the number of units of each that the firm should produce to maximise its profit.

13.6 A monopoly manufactures two goods, X and Y, with demand functions

$$x = 12 - p^X \quad \text{and} \quad y = 18 - p^Y.$$

The firm's cost function is $C(x, y) = x^2 + y^2 + 2xy$. Find the maximum profit achievable, and the quantities produced of each of X and Y in order to achieve this.

13.7 A firm manufactures two products, X and Y, and sells these in related markets. Suppose that the firm is the only producer of X and Y and that the inverse demand functions for X and Y are

$$p^X = 13 - 2x - y \quad \text{and} \quad p^Y = 13 - x - 2y.$$

Determine the production levels that maximise profit, given that the cost function is $C(x, y) = x + y$.

14 Vectors, Preferences and Convexity

14.1 Vectors and Bundles

An *n-vector* $\mathbf{v}$ is a list of n numbers, written either as a *row vector*

$$(v_1, v_2, \ldots, v_n)$$

or a *column vector*

$$\begin{pmatrix} v_1 \\ v_2 \\ \vdots \\ v_n \end{pmatrix}.$$

The numbers v_1, v_2, and so on are known as the *components*, *entries* or *coordinates* of $\mathbf{v}$. In economics, a vector often stands for a *bundle* of commodities, so that the ith component v_i represents a quantity of commodity i. For example, a bundle of five apples and seven bananas can be denoted by the row-vector $(5, 7)$. Clearly, vectors with $n = 2$ components correspond to points in $\mathbb{R}^2$, and so we can use plane diagrams to represent them in the usual way. Similarly, a vector with $n > 2$ components can be thought of as a point in n-dimensional space.

It is most important to understand that only certain kinds of algebraic operations can be performed with vectors. Specifically, we can define *addition* of two n-vectors by the rule

$$(w_1, w_2, \ldots, w_n) + (v_1, v_2, \ldots, v_n) \;=\; (w_1 + v_1, w_2 + v_2, \ldots, w_n + v_n).$$

Also, we can multiply a vector by any single number α (usually called a *scalar* in this context) using the following rule:

$$\alpha(v_1, v_2, \ldots, v_n) = (\alpha v_1, \alpha v_2, \ldots, \alpha v_n).$$

For example

$$(4, \,-1, 5) + (3, 2, 1) = (7, 1, 6) \quad \text{and} \quad 3(4, \,-1, 5) = (12, \,-3, 15).$$

The operations of addition and multiplication by a scalar may be combined. For example

$$2(1, 2, 3) + 3(1, 2, 1) = (2, 4, 6) + (3, 6, 3) = (5, 10, 9).$$

These operations have obvious interpretations when we think of vectors as bundles of goods.

Example Consider a family of seven, two parents and five children. Suppose that each of the five children has six pairs of dirty jeans and one dirty shirt, while the parents each have two pairs of dirty jeans and four dirty shirts. Then the family's total bundle of dirty jeans and dirty shirts is

$$5(6, 1) + 2(2, 4) = (30, 5) + (4, 8) = (34, 13).$$

■

The definitions of vector addition and multiplication by a scalar lead to some simple rules for calculating with vectors, such as

$$(\mathbf{w} + \mathbf{v}) + \mathbf{x} = \mathbf{w} + (\mathbf{v} + \mathbf{x}) \quad \text{and} \quad (\alpha + \beta)\mathbf{v} = \alpha\mathbf{v} + \beta\mathbf{v}.$$

However, it must be stressed that there is no useful and natural way of 'multiplying' two vectors to obtain another vector. But there is a very useful way of combining two vectors to obtain a *number*. The *dot product* of two vectors is defined by

$$\mathbf{v}.\mathbf{w} = v_1 w_1 + v_2 w_2 + \cdots + v_n w_n.$$

For example, if $\mathbf{v} = (4, \ -1, 5)$ and $\mathbf{w} = (3, 2, 1)$ then

$$\mathbf{v}.\mathbf{w} = (4 \times 3) + ((-1) \times 2) + (5 \times 1) = 15.$$

Note that the dot product is not analogous to the ordinary multiplication of real numbers. In fact it is a special case of 'matrix multiplication', which we shall discuss in the next chapter. (Sometimes the dot product is known as the 'inner product' or 'scalar product'.)

14.2 Prices and Budgets

In this section we shall use vector notation to write down the total cost of a bundle of different items, and we shall look at the relation between the cost and the amount of money which can be spent.

Example Suppose I purchase x_1 apples, x_2 bananas and x_3 oranges, for which the prices per item are 15 cents, 10 cents and 25 cents, respectively. Then the cost of the apples, in cents, is $15x_1$, the cost of the bananas is $10x_2$ and the cost of the oranges is $25x_3$. So the total cost is $15x_1 + 10x_2 + 25x_3$ cents.

In terms of the *price vector* $\mathbf{p} = (15, 10, 25)$, the cost of the bundle $\mathbf{x} = (x_1, x_2, x_3)$ is the dot product $\mathbf{p}.\mathbf{x} = 15x_1 + 10x_2 + 25x_3$. ■

In general, suppose there are n goods and the price per unit of the ith good is p_i, so that the price vector is $\mathbf{p} = (p_1, p_2, \ldots, p_n)$. Then the cost of a bundle $\mathbf{x} = (x_1, x_2, \ldots, x_n)$ is

$$\mathbf{p}.\mathbf{x} = p_1x_1 + p_2x_2 + \cdots + p_nx_n.$$

So if I have a given amount M to spend, I can purchase any bundle $\mathbf{x}$ for which the cost $\mathbf{p}.\mathbf{x}$ does not exceed M.

The condition $\mathbf{p}.\mathbf{x} \leq M$ is known as the *budget constraint*, and the set of bundles $\mathbf{x}$ satisfying the budget constraint is the *budget set*. When we are talking about bundles of goods, we implicitly assume that no component of $\mathbf{x}$ is negative, and so strictly speaking the budget set consists of all those $\mathbf{x}$ which satisfy $\mathbf{p}.\mathbf{x} \leq M$ and which have no negative components. In the case of two goods, the budget set may be illustrated graphically, as in the following example.

Example Suppose, as before, that the price of an apple is 15 cents and the price of a banana is 10 cents. If I have \$1.50 to spend on apples and bananas, then the budget constraint is the condition that total expenditure must not exceed \$1.50. Since the cost of a bundle (x_1, x_2) consisting of x_1 apples and x_2 bananas is $15x_1 + 10x_2$, the budget constraint is

$$15x_1 + 10x_2 \leq 150 \quad \text{or} \quad 3x_1 + 2x_2 \leq 30.$$

The budget set, the set of all (x_1, x_2) satisfying this inequality, consists of points on and below the line with equation $3x_1 + 2x_2 = 30$. As we do not consider negative quantities in this context, the budget set is the shaded region in Figure 14.1. ■

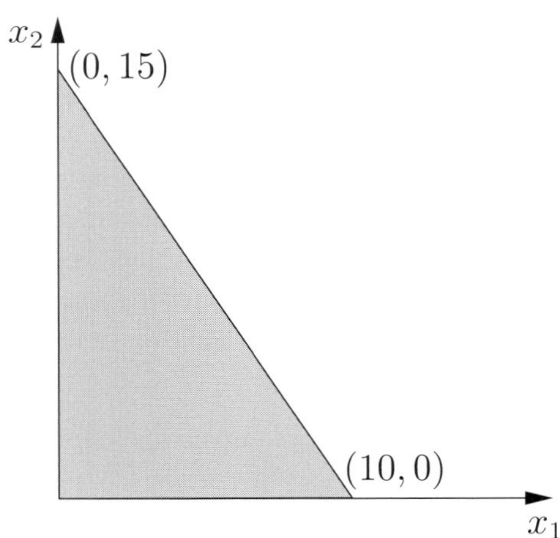

Figure 14.1 The budget set $3x_1 + 2x_2 \leq 30$.

14.3 Preferences, Utility and Indifference Curves

It is a basic assumption of economics that consumers 'prefer more to less'; in other words, six apples are always preferred to five apples. This can be taken as defining what we mean when we say that apples are a *good*. In the case of one good considered in isolation, a 'bundle' is just a single number and any reasonable consumer will rank two bundles in order of preference simply by comparing the corresponding numbers and choosing the larger. However, when we have two goods, such as apples and bananas, there is no unique way of ranking two bundles, such as $(4,6)$ and $(6,3)$. The ranking will depend upon the individual preferences of the consumer.

Example Suppose Ann regards apples and bananas as equally desirable; in other words, she is always prepared to exchange one apple for one banana. Then if she has to choose between $(4,6)$ and $(6,3)$ she will choose the first bundle, because it contains a total of 10 equally desirable items, whereas the second bundle contains only 9.

On the other hand, suppose Bill considers one apple to be equivalent to three bananas. Then for him the first bundle provides $3 \times 4 + 6 = 18$ 'banana-units', while the second provides $3 \times 6 + 3 = 21$. So he will prefer the second bundle. ■

In order to discuss individual preferences more carefully, we introduce the idea of a *utility function*. For a given consumer and a given bundle (x_1,x_2), we denote by $u(x_1,x_2)$ the 'utility' which the consumer derives from the bundle (x_1,x_2). It is convenient to use vector notation, so we often write $\mathbf{x}$ for (x_1,x_2) and $u(\mathbf{x})$ for $u(x_1,x_2)$.

Loosely speaking, $u(x_1,x_2)$ is a measure of how valuable or desirable the consumer considers the bundle (x_1,x_2) to be, or how much enjoyment the bundle provides, but we have to be careful with this interpretation. The utility function enables us to rank bundles in order of desirability to the consumer. To do this, we compare two bundles $\mathbf{x}$ and $\mathbf{y}$ by computing $u(\mathbf{x})$ and $u(\mathbf{y})$. If we find that $u(\mathbf{x})$ is greater than $u(\mathbf{y})$, then we may deduce that the consumer prefers $\mathbf{x}$ to $\mathbf{y}$. However, we should *not* attach any meaning to *how much larger* $u(\mathbf{x})$ is than $u(\mathbf{y})$. For example, if $u(\mathbf{x})$ is twice $u(\mathbf{y})$, we cannot deduce that '$\mathbf{x}$ is twice as good as $\mathbf{y}$'.

It is very helpful to have a diagrammatic method of representing u, and we can do this by considering the contours of u. You will recall that a contour consists of the set of points (x_1,x_2) for which $u(x_1,x_2)$ takes a given constant value. Thus two points on the same contour represent bundles which have the same utility, and the consumer is indifferent between them. For this reason, a contour of a utility function is known as an *indifference curve*.

Example (continued) We consider the indifference curves for four consumers: Ann and Bill (whom we have already met) and two new characters, Chas and Dave.

Suppose we are given a bundle $\mathbf{x} = (x_1,x_2)$ of apples and bananas. Ann considers apples and bananas to be of equal worth, so her utility function u_A is just the total number of items: $u_A(x_1,x_2) = x_1 + x_2$. On the other hand, Bill considers one apple to be equivalent to three bananas, so the utility which he derives from x_1 apples is $3x_1$ 'banana-units' and his utility function is $u_B(x_1,x_2) = 3x_1 + x_2$.

Ann's indifference curves are given by $x_1 + x_2 = c$, where c is a constant. These are straight lines with gradient -1 (Figure 14.2). The indifference curves for Bill are given by $3x_1 + x_2 = c$; these are straight lines with gradient -3.

Suppose Chas's preferences are described by the utility function $x_1^2+2x_2^2$ and Dave's preferences are described by the utility function $x_1^2x_2$. The equation $x_1^2 + 2x_2^2 = c$

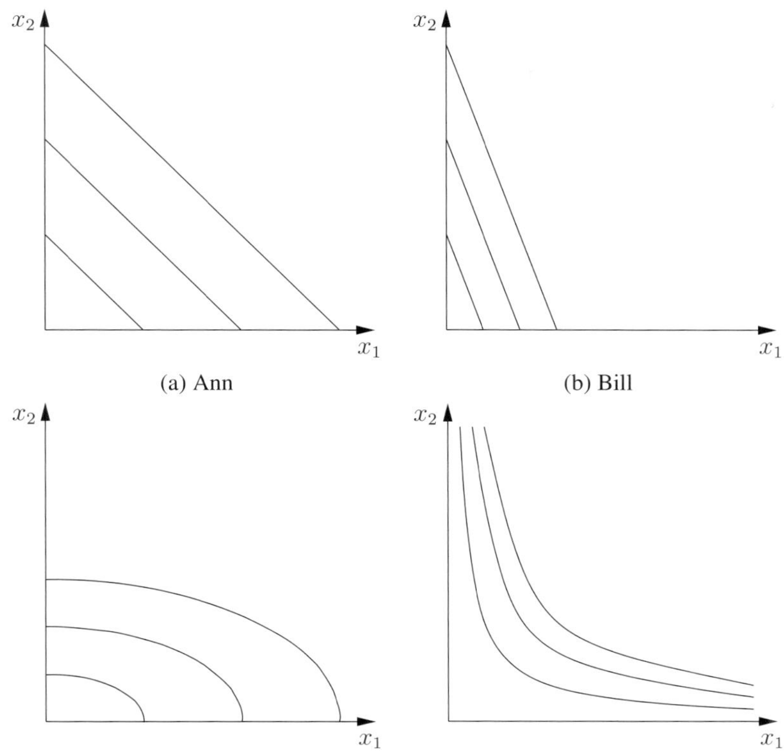

Figure 14.2 Indifference curves for Ann, Bill, Chas and Dave.

describing an indifference curve for Chas is the equation of an ellipse or (if you prefer) a squashed circle. An indifference curve for Dave has equation $x_1^2x_2 = c$. Typical examples are shown in Figure 14.2. ∎

One useful remark illustrated by Figure 14.2 is that the indifference curves $u(\mathbf{x}) = c$ move in a 'north-easterly' direction away from the origin as the constant c increases. In other words, given two bundles, a consumer will prefer the one which is on the indifference curve further from the origin in this sense. This is the fact which corresponds, in this context, to the economic assumption that consumers prefer more to less.

Example (continued) Suppose that the three bundles

$$\mathbf{r} = (10, 0), \quad \mathbf{s} = (8, 3), \quad \mathbf{t} = (6, 6)$$

are available. The utility of these bundles for our four consumers is as follows:

$$\begin{array}{llll} \text{Ann:} & u_A(\mathbf{r}) = 10 & u_A(\mathbf{s}) = 11 & u_A(\mathbf{t}) = 12; \\ \text{Bill:} & u_B(\mathbf{r}) = 30 & u_B(\mathbf{s}) = 27 & u_B(\mathbf{t}) = 24; \end{array}$$

$$\text{Chas:} \quad u_C(\mathbf{r}) = 100 \quad u_C(\mathbf{s}) = 82 \quad u_C(\mathbf{t}) = 108;$$

$$\text{Dave:} \quad u_D(\mathbf{r}) = 0 \quad u_D(\mathbf{s}) = 192 \quad u_D(\mathbf{t}) = 216.$$

Thus, Ann prefers **t** to **s** which, in turn, she prefers to **r**. We may say that she ranks the bundles in the order $\mathbf{t}, \mathbf{s}, \mathbf{r}$. Similarly, Bill ranks the bundles $\mathbf{r}, \mathbf{s}, \mathbf{t}$; Chas ranks them $\mathbf{t}, \mathbf{r}, \mathbf{s}$; and Dave ranks them $\mathbf{t}, \mathbf{s}, \mathbf{r}$.
Note that the same bundles are ranked differently by the consumers. This is because preferences are determined by the individual's utility functions. ■

Note that, as mentioned earlier, we should attach no economic meaning to the numerical values of the utility functions other than the preference ordering they imply. Thus, with the information calculated in the above example, we may say that Ann prefers **t** to **s** and **s** to **r**, but we may *not* say that she likes **t** 12/11 times as much as she likes **s**. Along the same lines, it is not possible to make 'inter-personal comparisons'. For example, although $u_A(\mathbf{r}) = 10$ and $u_B(\mathbf{r}) = 30$, we cannot deduce that Bill derives more enjoyment from bundle **r** than Ann does.

14.4 Linear and Convex Combinations

Later in this book we shall meet an important part of microeconomic theory which depends upon a very specific assumption about the shape of indifference curves. We shall introduce this topic here, working with two-vectors, or bundles of two commodities, considered as points in $\mathbb{R}^2$.

Given two vectors **v** and **w**, we say that any vector of the form $\lambda\mathbf{v} + \mu\mathbf{w}$, for some numbers λ and μ, is a *linear combination* of **v** and **w**. Here we are particularly interested in the linear combinations for which $\lambda \geq 0$, $\mu \geq 0$ and $\lambda + \mu = 1$; any point of this form is said to be a *convex combination* of **v** and **w**. The reason why the convex combinations are interesting is because they are precisely the points lying between **v** and **w** on the straight-line segment joining them. For example, the midpoint of the segment is the point $\frac{1}{2}\mathbf{v} + \frac{1}{2}\mathbf{w}$, and the point one-third of the way along from **v** towards **w** is $\frac{2}{3}\mathbf{v} + \frac{1}{3}\mathbf{w}$.

We say that a set X of points in $\mathbb{R}^2$ is *convex* if, whenever two points are in X, so is every convex combination of them. So in a convex set X, the straight-line segment joining any two points in X lies wholly in X. In Figure 14.3 a convex set and a non-convex set are illustrated; the second set is not convex because points such as **z** on the line segment joining **x** to **y** are not in the set.

Given a utility function u and a positive constant c, we define the set

$$U_c = \{\mathbf{x} \mid u(\mathbf{x}) \geq c\}.$$

This is the set of bundles having utility at least c. Geometrically, U_c consists of all the points on the indifference curve $u(\mathbf{x}) = c$, together with those points **x** which are on indifference curves 'further from the origin', as indicated in Figure 14.4.

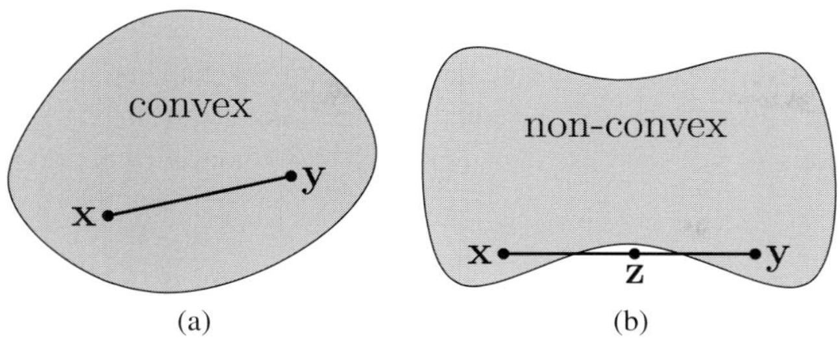

Figure 14.3 A convex set and a non-convex set.

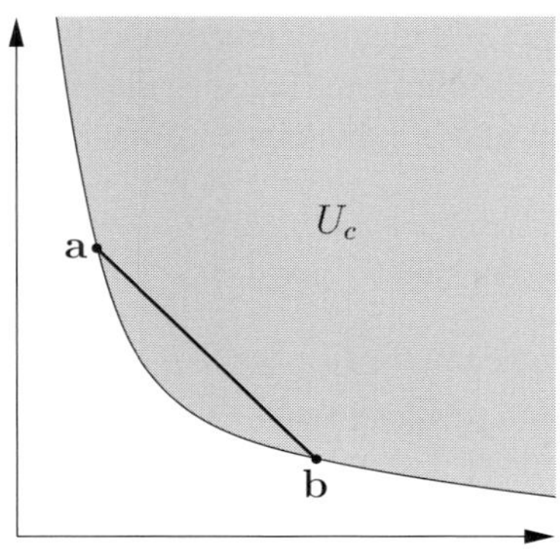

Figure 14.4 Convex preferences.

A fundamental assumption about consumers' preferences frequently invoked in microeconomic theory is the following:

- *the sets U_c are convex.*

In order to see what this means, consider Figure 14.4. Here **a** and **b** are bundles on the same indifference curve $u(\mathbf{x}) = c$, and the assumption requires that every point on the straight-line segment joining them is in U_c. This means that such points have utility which is at least equal to c. In other words, the indifference curves must bulge towards the origin in the manner shown. Roughly speaking, the assumption is equivalent to the statement that consumers prefer a convex combination (or 'mixture') of **a** and **b** to either of the extreme bundles **a** and **b**.

This assumption usually applies when we are considering consumers whose consumption is 'large'. For example, a manufacturer or retailer will normally prefer to have a mixture of two items in stock, rather than just one or the other.

Of course, it is possible to think of situations where the assumption is false. For instance, let us suppose that an office worker likes both coffee and tea, so that he considers them both to be 'goods'. If his utility function is $u(x_1, x_2) = x_1^2 + x_2^2$, then a typical indifference curve $x_1^2 + x_2^2 = c$ is part of a circle. The fact that the sets U_c are not convex in this case indicates that the worker does *not* prefer mixtures of the two different types of drink. For example, the utility of the bundle $(1, 1)$ is 2, whereas the utility of each of the bundles $(2,0)$ and $(0,2)$ is 4. In plain language, the office worker would rather have two cups of coffee or two cups of tea than one of each.

Despite such exceptions, the preference for mixtures seems to be a valid assumption in many situations. We shall use it, as appropriate, without further comment.

14.5 Choosing Optimal Bundles

We now turn to the problem of finding the best bundle, subject to a given budget constraint. Specifically, we suppose that the price vector $\mathbf{p} = (p_1, p_2)$ for two goods is given, and we know the consumer's budget M and utility function u. The problem is to determine the bundle $\mathbf{x}$ which maximises the utility and satisfies the budget constraint. This is the 'best buy' from the viewpoint of the individual consumer. In symbols, we require to solve the problem

$$\text{maximise} \quad u(\mathbf{x}) \quad \text{such that} \quad \mathbf{p}.\mathbf{x} \leq M.$$

We begin by thinking about this problem geometrically. Since p_1, p_2 and M are given, the budget set is a fixed triangular region, as illustrated in Figure 14.5. We shall be particularly interested in the *budget line* $p_1x_1 + p_2x_2 = M$, which we shall denote by L. The budget line represents the bundles which cost exactly M, the budget limit, and we would expect the best bundle to be one of them.

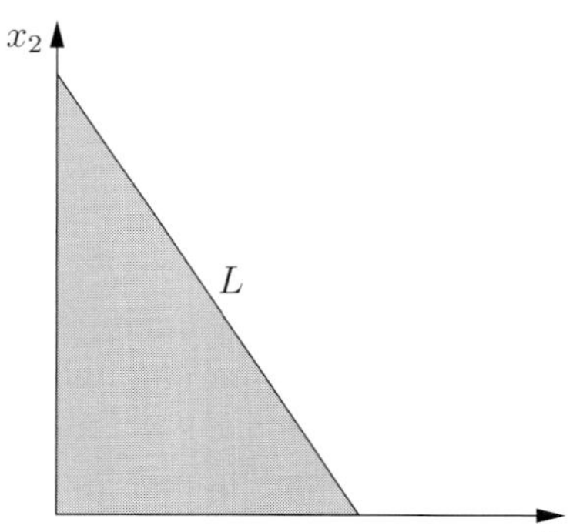

Figure 14.5 The budget set and line.

How do the indifference curves $u(\mathbf{x}) = c$ behave with respect to L? First, recall that the indifference curves move away from the origin as the value of c increases. For small values of c, the curves intersect L in two points **a** and **b** (Figure 14.6a). By the convexity assumption, points on the segment of L joining **a** and **b** lie in U_c. Such points have greater utility than **a** and **b**, and so the corresponding value of c is not optimal.

On the other hand, for large values of c, the indifference curves $u(\mathbf{x}) = c$ do not intersect L (Figure 14.6b). So these values of c are not feasible: there is no bundle in the budget set which provides that level of utility.

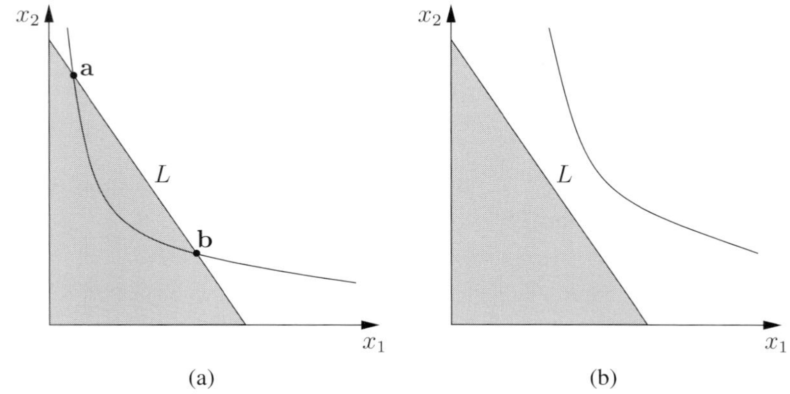

Figure 14.6 The budget set and indifference curves in two cases.

Now we can see how to characterise the optimal level of utility. It occurs when the indifference curve $u(\mathbf{x}) = c$ intersects L in just one point, in other words *when the budget line is a tangent to an indifference curve*. This is illustrated in Figure 14.7, where the point $\mathbf{x}^*$ represents the optimal bundle.

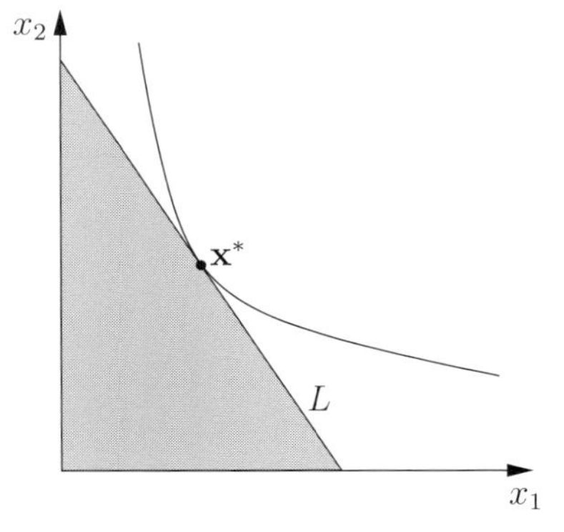

Figure 14.7 The budget set and the optimal indifference curve.

Not only does the geometrical method provide a useful insight, it also tells us how to work out the optimal bundle $\mathbf{x}^*$ in any particular case.

Example Suppose that Dave has \$1.50 to spend on apples and bananas, which cost $15c$ and $10c$, respectively, so that his budget constraint is

$$3x_1 + 2x_2 \leq 30.$$

We already know that his utility function is $u_D(x_1, x_2) = x_1^2 x_2$.

His optimal bundle $\mathbf{x}^* = (x_1^*, x_2^*)$ of apples and bananas is determined by the condition that the indifference curve is tangent to the budget line at $\mathbf{x}^*$. This means that the gradient, or slope, of the indifference curve at $\mathbf{x}^*$ is the same as that of the budget line $3x_1 + 2x_2 = 30$, which is $-3/2$.

Now Dave's indifference curves are $x_1^2 x_2 = c$, and we can calculate the slope of such a curve at the point (x_1, x_2) by the formula obtained in Section 12.3:

$$\frac{dx_2}{dx_1} = -\frac{\partial u_D/\partial x_1}{\partial u_D/\partial x_2} = -\frac{2x_1x_2}{x_1^2} = -\frac{2x_2}{x_1}.$$

At the optimal point (x_1^*, x_2^*) this must be equal to $-3/2$, the slope of the budget line. So x_1^* and x_2^* must satisfy

$$-\frac{2x_2^*}{x_1^*} = -\frac{3}{2} \quad \text{or} \quad x_2^* = \frac{3}{4}x_1^*.$$

Since $\mathbf{x}^*$ lies on the budget line, it also satisfies the equation $3x_1^* + 2x_2^* = 30$. So we have two equations for x_1^* and x_2^*, which give

$$3x_1^* + 2\left(\frac{3}{4}x_1^*\right) = 30, \quad \text{that is} \quad x_1^* = 20/3.$$

It follows that $x_2^* = 5$, and the optimal bundle for Dave is $(20/3, 5)$. ■

We shall return to this topic, in a slightly more general setting, in Chapter 22.

Worked Examples

Example 14.1 Denote by **a**, **b**, **c** the column vectors

$$\mathbf{a} = \begin{pmatrix} 1 \\ 2 \\ 3 \end{pmatrix}, \quad \mathbf{b} = \begin{pmatrix} -2 \\ 1 \\ -3 \end{pmatrix}, \quad \mathbf{c} = \begin{pmatrix} -2 \\ -1 \\ 1 \end{pmatrix}.$$

Calculate

$$2\mathbf{a} - 5\mathbf{b}, \quad 2\mathbf{a} - 5\mathbf{b} + \mathbf{c}, \quad \mathbf{a}.\mathbf{b}, \quad \mathbf{a}.\mathbf{c}.$$

Solution

We have

$$2\mathbf{a} - 5\mathbf{b} = 2\begin{pmatrix} 1 \\ 2 \\ 3 \end{pmatrix} - 5\begin{pmatrix} -2 \\ 1 \\ -3 \end{pmatrix} = \begin{pmatrix} 2 \\ 4 \\ 6 \end{pmatrix} + \begin{pmatrix} 10 \\ -5 \\ 15 \end{pmatrix} = \begin{pmatrix} 12 \\ -1 \\ 21 \end{pmatrix}.$$

It follows that

$$2\mathbf{a} - 5\mathbf{b} + \mathbf{c} = \begin{pmatrix} 12 \\ -1 \\ 21 \end{pmatrix} + \begin{pmatrix} -2 \\ -1 \\ 1 \end{pmatrix} = \begin{pmatrix} 10 \\ -2 \\ 22 \end{pmatrix}.$$

The dot product **a**.**b** is

$$\mathbf{a}.\mathbf{b} = 1(-2) + 2(1) + 3(-3) = -2 + 2 - 9 = -9$$

and $\mathbf{a}.\mathbf{c}$ is

$$\mathbf{a}.\mathbf{c} = 1(-2) + 2(-1) + 3(1) = -2 - 2 + 3 = -1.$$

Example 14.2 The preferences of John, Paul, George and Ringo for bundles $\mathbf{x} = (x_1, x_2)$ of apples and bananas can be described as follows:

- John is always willing to exchange apples for bananas on the basis of two apples for one banana;
- Paul's utility function is $7x_1 + 3x_2$;
- George has utility function $2x_1^2 + x_2^2$;
- Ringo has utility function $x_1 x_2^2$.

Sketch some typical indifference curves for each consumer and in each case say whether or not the sets $U_c = \{\mathbf{x} \mid u(\mathbf{x}) \geq c\}$ are convex.

Solution

Since John considers two apples to be worth one banana, or one apple to be worth half a banana, the utility of the bundle (x_1, x_2) of x_1 apples and x_2 bananas is $x_1/2 + x_2$, when measured in 'banana-units'. Equally well we could measure it in 'apple-units', in which case the formula would be $x_1 + 2x_2$. This second formula is simpler, so we shall take John's utility function to be $u_J(x_1, x_2) = x_1 + 2x_2$. A typical indifference curve for John has equation $x_1 + 2x_2 = c$, which is a straight line with gradient $-1/2$.

Paul's indifference curves $u_P(x_1, x_2) = 7x_1 + 3x_2 = c$ are straight lines with gradient $-7/3$ (See Figure 14.8.).

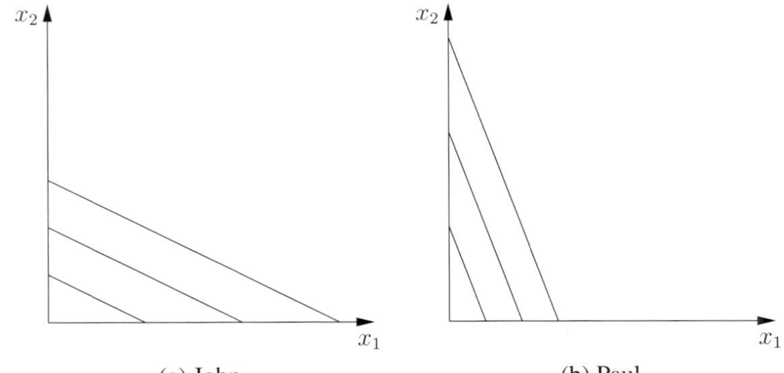

Figure 14.8 Indifference curves for John and Paul.

A typical indifference curve for George has equation $2x_1^2 + x_2^2 = c$, which describes an ellipse, longer in the x_2-direction than in the x_1-direction. Ringo's indifference curves are $x_1 x_2^2 = c$; that is the typical 'Cobb–Douglas' form. (See Figure 14.9.)

For John, Paul and Ringo, the sets $U_c = \{\mathbf{x} \mid u(\mathbf{x}) \geq c\}$ are convex. However, this is clearly not so for George.

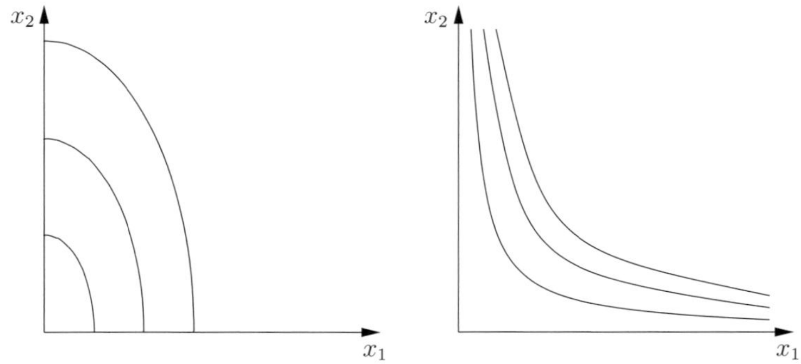

(a) George (b) Ringo

Figure 14.9 Indifference curves for George and Ringo.

Example 14.3 Rank the following bundles in order of preference for John, Paul, George and Ringo, with the utility functions given in the previous question:

$$\mathbf{r} = (8, 0), \quad \mathbf{s} = (6, 3), \quad \mathbf{t} = (2, 7).$$

Solution

For each consumer, we calculate the utility of each of the three bundles on offer.

John:	$u_J(\mathbf{r}) = 8$	$u_J(\mathbf{s}) = 12$	$u_J(\mathbf{t}) = 16;$
Paul:	$u_P(\mathbf{r}) = 56$	$u_P(\mathbf{s}) = 51$	$u_P(\mathbf{t}) = 35;$
George:	$u_G(\mathbf{r}) = 128$	$u_G(\mathbf{s}) = 81$	$u_G(\mathbf{t}) = 57;$
Ringo:	$u_R(\mathbf{r}) = 0$	$u_R(\mathbf{s}) = 54$	$u_R(\mathbf{t}) = 98.$

Therefore, John ranks them in the order **t**, **s**, **r**; Paul ranks them **r**, **s**, **t**; George ranks them **r**, **s**, **t**; and Ringo ranks them **t**, **s**, **r**.

Example 14.4 Suppose the price of an apple is \$0.30 and the price of a banana is \$0.10. If John, Paul, George and Ringo each have \$2.10 to spend, show on a diagram the budget set, the set of all possible bundles of apples and bananas which they can purchase. If each of them chooses the bundle in the budget set which maximises his utility (as given in Example 14.2), determine how many apples and bananas each buys. (Assume that fractions of apples and bananas may be purchased.)

Solution

The budget constraint is $0.3x_1 + 0.1x_2 \leq 2.10$, which may be written as

$$3x_1 + x_2 \leq 21.$$

So the budget set consists of points on and below the budget line $3x_1 + x_2 = 21$.

For each consumer, the optimal bundle is the one in the budget set with the highest utility for that consumer. As usual, we start with the graphical approach. Figure 14.10 illustrates, for each of the four consumers, the budget set, together with typical

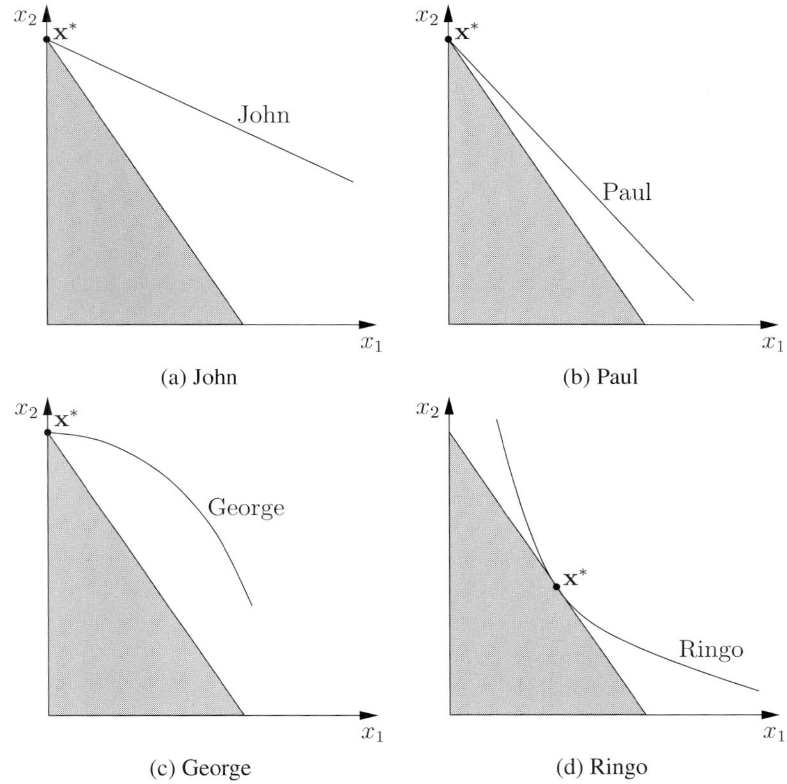

Figure 14.10 The optimal bundles for John, Paul, George and Ringo.

indifference curves and the optimal bundle $\mathbf{x}^*$. From the diagrams, the optimal bundle for each of John, Paul and George is the bundle $(0, 21)$ at the extreme top-left of the budget set. These three will therefore buy no apples and 21 bananas each.

To determine Ringo's optimal bundle $\mathbf{x}^*$, we use the fact that the budget line is tangent to the indifference curve through $\mathbf{x}^*$. At that point, the indifference curve therefore has the same gradient, -3, as the budget line. Now the gradient of the indifference curve $u_R(x_1, x_2) = x_1 x_2^2 = c$ is

$$\frac{dx_2}{dx_1} = -\frac{\partial u_R/\partial x_1}{\partial u_R/\partial x_2} = -\frac{x_2^2}{2x_1x_2} = -\frac{x_2}{2x_1}.$$

Since this must equal -3 at (x_1^*, x_2^*), we have $x_2^*/(2x_1^*) = 3$, or $x_2^* = 6x_1^*$. Since $\mathbf{x}^*$ lies on the budget line, it satisfies the equation $3x_1^* + x_2^* = 21$. Given that $x_2^* = 6x_1^*$, this yields

$$3x_1^* + 6x_1^* = 21; \quad \text{that is} \quad x_1^* = 21/9 = 7/3,$$

from which it follows that $x_2^* = 14$. The optimal bundle for Ringo is therefore $(7/3, 14)$: he should buy $7/3$ apples and 14 bananas.

Main Topics

- vectors and their properties
- the budget constraint
- utility functions and indifference curves
- linear and convex combinations, and convex sets
- convexity and utility functions
- ranking bundles in order of preference
- finding the optimal bundle subject to a given budget constraint

Key Terms, Notations and Formulae

- vector, row vector, column vector
- components, coordinates
- commodity bundle
- dot product, $\mathbf{v}.\mathbf{w} = v_1w_1 + v_2w_2 + \cdots + v_nw_n$
- budget constraint, $p_1x_1 + p_2x_2 + \cdots + p_nx_n \leq M$, or $\mathbf{p}.\mathbf{x} \leq M$
- utility function, $u(\mathbf{x}) = u(x_1, x_2)$; $u(\mathbf{x}) > u(\mathbf{y})$ means $\mathbf{x}$ preferable to $\mathbf{y}$
- indifference curve
- linear combination, $\lambda\mathbf{v} + \mu\mathbf{w}$; convex combination if $\lambda, \mu \geq 0$, $\lambda + \mu = 1$
- convex set
- convexity assumption: for all c, $U_c = \{\mathbf{x} \mid u(\mathbf{x}) \geq c\}$ convex
- budget line
- at optimal bundle, budget line is tangent to indifference curve

Exercises

14.1 Let $\mathbf{x} = (1,2,3)$ and $\mathbf{y} = (3,2,1)$. Find $\mathbf{x}+\mathbf{y}$ and $\mathbf{x}.\mathbf{y}$.

14.2 Use the geometrical approach to determine the optimal bundles for the consumers Ann, Bill and Chas, subject to the budget constraint $3x_1 + 2x_2 \leq 30$. (Use the utility functions given in Section 14.3.)

14.3 Susan has utility function $x_1^2x_2^2$ where the bundle (x_1, x_2) comprises x_1 kilograms of flour and x_2 kilograms of sugar. If one kilogram of flour costs \$2 and one kilogram of sugar costs \$1, determine her optimal bundle if she has \$4 to spend.

Problems

14.1 The row vectors $\mathbf{x}, \mathbf{y}, \mathbf{z}$ are

$$\mathbf{x} = (1,\ -1, 2)\,,\ \mathbf{y} = (3,\ -3,\ -3)\,,\ \mathbf{z} = (2, 7, 6)\,.$$

Calculate (i) $3\mathbf{x} + 5\mathbf{y}$, (ii) $2\mathbf{x} - \mathbf{y} + 3\mathbf{z}$, (iii) $\mathbf{x}.\mathbf{y}$ and $\mathbf{x}.\mathbf{z}$.

Show by direct calculation that $\mathbf{x}.(\mathbf{y}+\mathbf{z}) = \mathbf{x}.\mathbf{y} + \mathbf{x}.\mathbf{z}$.

14.2 Mick, Keith, Bill and Charlie, four health-conscious consumers, like both freshly squeezed orange juice and sunflower seeds. Their preferences for bundles $\mathbf{x} = (x_1, x_2)$ comprising x_1 glasses of orange juice and x_2 grams of sunflower seeds are described as follows:

- Mick considers a glass of orange juice to be as desirable as 10 grams of sunflower seeds.
- Keith is always willing to exchange two glasses of orange juice for 5 grams of sunflower seeds.
- Bill has utility function $x_1^2 + x_2^2$.
- Charlie has utility function $x_1 x_2$.

Sketch some typical indifference curves for each consumer and in each case say whether or not the set $\{\mathbf{x} \mid u(\mathbf{x}) \geq c\}$ is convex.

14.3 Rank the following bundles in order of preference for each of the consumers Mick, Keith, Bill and Charlie of the previous question:

$$\mathbf{r} = (4, 0), \quad \mathbf{s} = (2, 15), \quad \mathbf{t} = (1, 20).$$

14.4 Suppose that a glass of freshly squeezed orange juice costs \$0.50 and that sunflower seeds cost \$0.10 per gram. If Mick, Keith, Bill and Charlie each have \$26 to spend, show on a diagram the budget set, the set of all possible bundles which they can purchase. Assuming that each consumer chooses the bundle in the budget set which maximises their utility, determine what each purchases.

15 Matrix Algebra

15.1 What is a Matrix?

Many readers of this book will already have some elementary knowledge of matrices. The purpose of this chapter is to summarise the important facts in the generality required for more advanced work.

A *matrix* is an array of numbers

$$\begin{pmatrix} a_{11} & a_{12} & \dots & a_{1n} \\ a_{21} & a_{22} & \dots & a_{2n} \\ \vdots & \vdots & \ddots & \vdots \\ a_{m1} & a_{m2} & \dots & a_{mn} \end{pmatrix}.$$

We denote this array by the single letter A, or by (a_{ij}), and we say that A has m *rows* and n *columns*, or that it is an $m \times n$ matrix. We also say that A is a matrix of *size* $m \times n$. The number a_{ij} is known as the (i,j)th entry of A. The row vector $\begin{pmatrix} a_{i1} & a_{i2} & \cdots & a_{in} \end{pmatrix}$ is *row i* of A, or the *ith row* of A, and the column vector

$$\begin{pmatrix} a_{1j} \\ a_{2j} \\ \vdots \\ a_{mj} \end{pmatrix}$$

is *column j* of A, or the *jth column* of A.

Matrices are useful because they provide a compact notation, and because we can 'do algebra' with them. If A and B are two matrices of the same size, then we define $A+B$ to be the matrix whose elements are the sums of the corresponding elements in A and B. Formally, the (i,j)th entry of the matrix $A+B$ is $a_{ij}+b_{ij}$, where a_{ij} and b_{ij} are the (i,j)th entries of A and B, respectively. For example

$$\begin{pmatrix} 3 & 8 & 1 & 3 \\ 4 & 1 & 3 & 5 \\ 2 & 5 & 2 & 3 \end{pmatrix} + \begin{pmatrix} 2 & 1 & 7 & 4 \\ 4 & 3 & 2 & 1 \\ 2 & 1 & -3 & 6 \end{pmatrix} = \begin{pmatrix} 5 & 9 & 8 & 7 \\ 8 & 4 & 5 & 6 \\ 4 & 6 & -1 & 9 \end{pmatrix}.$$

Also, if c is a number, we define cA to be the matrix whose elements are c times those of A; that is cA has (i,j)th entry ca_{ij}. For example

$$3\begin{pmatrix} 3 & 8 & 1 & 3 \\ 4 & 1 & 3 & 5 \\ 2 & 5 & 2 & 3 \end{pmatrix} = \begin{pmatrix} 9 & 24 & 3 & 9 \\ 12 & 3 & 9 & 15 \\ 6 & 15 & 6 & 9 \end{pmatrix}.$$

We note that there are obvious algebraic consequences of these definitions, such as the familiar rules for dealing with brackets: $A + (B + C) = (A + B) + C$ and $c(A + B) = cA + cB$.

15.2 Matrix Multiplication

Suppose A and B are matrices such that the number (say n) of columns of A is equal to the number of rows of B. Then each row of A is a vector with n components, as is each column of B, and we can form the dot product of two such vectors. This observation allows us to define the product $C = AB$ to be the matrix whose elements are

$$c_{ij} = a_{i1}b_{1j} + a_{i2}b_{2j} + \cdots + a_{in}b_{nj}.$$

Although this formula looks bad in raw algebraic terms, it is quite easy to use in practice. What it says is that the element in row i and column j of the product is obtained by taking the dot product of row i of A and column j of B.

Example In the following product the element in row 2 and column 3 is found by taking the dot product of the row and column printed in bold type:

$$\begin{pmatrix} 3 & 8 & 1 \\ \mathbf{4} & \mathbf{1} & \mathbf{3} \end{pmatrix} \begin{pmatrix} 1 & 3 & \mathbf{2} & 4 \\ 2 & 2 & \mathbf{1} & 5 \\ 2 & 5 & \mathbf{2} & 3 \end{pmatrix} = \begin{pmatrix} 21 & 30 & 16 & 55 \\ 12 & 29 & \mathbf{15} & 30 \end{pmatrix}.$$

The answer is 15 because

$$15 = 4 \times 2 + 1 \times 1 + 3 \times 2.$$

The other elements of the product can be worked out in the same way. ■

It must be stressed that when A has n columns then B must have n rows if AB is to be defined. In any other case, the product is not defined. Given that A is an $m \times n$ matrix and B is an $n \times p$ matrix, it follows that AB is an $m \times p$ matrix.

The definition of matrix multiplication allows us to use some more familiar algebraic rules, but care is needed. Among the rules which we can use are

$$A(BC) = (AB)C \quad \text{and} \quad A(B + C) = AB + AC.$$

On the other hand, it is most important to remember that AB and BA are not usually equal. Indeed it is quite possible that one of the products is defined but the other is not. Even if both are defined, they are generally not equal. For example, if

$$A = \begin{pmatrix} 5 & 1 \\ 2 & 3 \end{pmatrix} \quad \text{and} \quad B = \begin{pmatrix} 1 & 2 \\ 3 & 4 \end{pmatrix},$$

then

$$AB = \begin{pmatrix} 8 & 14 \\ 11 & 16 \end{pmatrix} \quad \text{and} \quad BA = \begin{pmatrix} 9 & 7 \\ 23 & 15 \end{pmatrix},$$

which are not equal.

A particularly useful matrix is the *identity matrix*

$$I = \begin{pmatrix} 1 & 0 & \dots & 0 \\ 0 & 1 & \dots & 0 \\ \vdots & \vdots & \ddots & \vdots \\ 0 & 0 & \dots & 1 \end{pmatrix},$$

which has the number 1 in each of the positions on the 'main diagonal' and 0 elsewhere. Note that I is a *square* matrix; the number of rows is equal to the number of columns. Note also that there is an identity matrix of any size $n \times n$.

The identity matrix has the property that, whenever the products AI and IA can be defined, we have

$$IA = AI = A.$$

It is easy to check this result using the explicit definition of matrix multiplication.

15.3 How to Make Money with Matrices

Many practical situations can be expressed simply in matrix form, and we shall give a number of examples in the next few chapters. In particular, financial economists nowadays employ a form of language which is easy to express in matrix terms, and which we shall introduce here.

The basic idea is that an investor can choose between a number of different *assets*; these may be thought of as shares in companies, holdings of foreign exchange, old-master paintings, or whatever. After a fixed time period, say 1 year, the assets will have a new value, and this will depend on what has happened in the meantime. We model this uncertainty by assuming that there are a number of *states* which may occur.

Example The citizens of the island of Apathia can invest in land, bonds (which yield a fixed return) and stocks (which yield an uncertain return). A general election is due, after which one of the two political parties, the Liberal-Conservatives or the Conservative-Liberals, will be in power. There are three assets: Asset 1 = land, Asset 2 = bonds, Asset 3 = stocks, and there are two states: State 1 = LibCons in power, State 2 = ConLibs in power.

Clearly, the value of an investment will depend upon which of the states occurs. For example, government by LibCons may result in an increase in the value of land and a decrease in the value of stocks, while government by ConLibs may have the opposite effect. We can describe this by writing down the value in a year's time of the amount of each asset which costs one dollar now, under the assumption that a given state occurs. In this example, we could get a table like the following:

	LibCon	ConLib
Land	1.25	0.95
Bonds	1.05	1.05
Stocks	0.90	1.15.

In other words, we have a 3×2 matrix $R = (r_{ij})$ which determines the *return* of each asset in the various states. In general, if there are m assets and n states, then the returns matrix R will be an $m \times n$ matrix.

Suppose an Apathian investor decides, before the election, to invest \$5000 in land, \$1000 in bonds and \$4000 in stocks. Then the row vector

$$\begin{pmatrix} 5000 & 1000 & 4000 \end{pmatrix}$$

is her *portfolio*. Its value in 1 year's time depends upon which party wins the election. If the LibCons win, the value will be

$$5000 \times 1.25 + 1000 \times 1.05 + 4000 \times 0.90 = 10\,900.$$

If the ConLibs win, the value will be

$$5000 \times 0.95 + 1000 \times 1.05 + 4000 \times 1.15 = 10\,400.$$

This is just an example of matrix multiplication. The portfolio can be considered as a 1×3 matrix Y, the table of returns R is a 3×2 matrix and their product

$$YR = \begin{pmatrix} 5000 & 1000 & 4000 \end{pmatrix} \begin{pmatrix} 1.25 & 0.95 \\ 1.05 & 1.05 \\ 0.90 & 1.15 \end{pmatrix} = \begin{pmatrix} 10\,900 & 10\,400 \end{pmatrix}$$

is a 1×2 matrix. When R is given, YR tells us the possible values of the portfolio Y in the various states.

The terminology introduced above is very useful when we try to model investment decisions. For example, a cautious investor might ask if there is a *riskless* portfolio; that is one which has the same value whichever party wins the election.

On the other hand, an investor might look for a portfolio which costs nothing. In the Apathian context, this could be a portfolio such as

$$\begin{pmatrix} 5000 & -10\,000 & 5000 \end{pmatrix},$$

which means that the investor *borrows* \$10 000 from the bank and uses it to buy land and stocks. At the end of the year she owes the bank more, $10\,000 \times 1.05$ dollars, but the changes in value of land and stocks may offset this. In fact, we calculate that

$$YR = \begin{pmatrix} 5000 & -10\,000 & 5000 \end{pmatrix} \begin{pmatrix} 1.25 & 0.95 \\ 1.05 & 1.05 \\ 0.90 & 1.15 \end{pmatrix} = \begin{pmatrix} 250 & 0 \end{pmatrix}.$$

Thus if the LibCons win she has gained \$250 without having to 'invest' anything, in real terms, while if the ConLibs win she is no worse off.

A portfolio like the one just described, which costs nothing, cannot lose and in at least one state yields a profit, is called an *arbitrage* portfolio. The question of the existence of riskless and arbitrage portfolios leads us inevitably to the study of linear equations, which we begin in the next chapter. ■

Worked Examples

Example 15.1 The matrices A, B, C are defined as follows:

$$A = \begin{pmatrix} 2 & 5 \\ 0 & 1 \end{pmatrix}, \quad B = \begin{pmatrix} 1 & 2 & 1 \\ 2 & 0 & 0 \end{pmatrix}, \quad C = \begin{pmatrix} 2 & 2 \\ 0 & 1 \\ 5 & 0 \end{pmatrix}.$$

(i) Why is AB defined? Why is BA not defined? Calculate AB.
(ii) Why are both BC and CB defined? Is it true that $BC = CB$?
(iii) Calculate $A(BC)$ and $(AB)C$. In general, should the answers be equal?

Solution

(i) A is a 2×2 matrix and B is a 2×3 matrix. The number of columns of A is the same as the number of rows of B, so the matrix product AB can be defined. In fact

$$AB = \begin{pmatrix} 2 & 5 \\ 0 & 1 \end{pmatrix} \begin{pmatrix} 1 & 2 & 1 \\ 2 & 0 & 0 \end{pmatrix} = \begin{pmatrix} 12 & 4 & 2 \\ 2 & 0 & 0 \end{pmatrix}.$$

Since B has 3 columns, whereas A has only 2 rows, BA is not defined.

(ii) B is a 2×3 matrix and C is a 3×2 matrix. The number of columns of B equals the number of rows of C and so BC is defined. The number of columns of C equals the number of rows of B and so CB is also defined. The product BC is a 2×2 matrix and the product CB is a 3×3 matrix. These cannot be equal because they have different sizes. (There is no need to calculate BC and CB to show that they are different.)

(iii) To calculate $A(BC)$, we first calculate BC. We have

$$BC = \begin{pmatrix} 1 & 2 & 1 \\ 2 & 0 & 0 \end{pmatrix} \begin{pmatrix} 2 & 2 \\ 0 & 1 \\ 5 & 0 \end{pmatrix} = \begin{pmatrix} 7 & 4 \\ 4 & 4 \end{pmatrix}.$$

Then

$$A(BC) = \begin{pmatrix} 2 & 5 \\ 0 & 1 \end{pmatrix} \begin{pmatrix} 7 & 4 \\ 4 & 4 \end{pmatrix} = \begin{pmatrix} 34 & 28 \\ 4 & 4 \end{pmatrix}.$$

From part (i)

$$AB = \begin{pmatrix} 12 & 4 & 2 \\ 2 & 0 & 0 \end{pmatrix}.$$

Therefore

$$(AB)C = \begin{pmatrix} 12 & 4 & 2 \\ 2 & 0 & 0 \end{pmatrix} \begin{pmatrix} 2 & 2 \\ 0 & 1 \\ 5 & 0 \end{pmatrix} = \begin{pmatrix} 34 & 28 \\ 4 & 4 \end{pmatrix} = A(BC).$$

As we noted in Section 15.2, it is true in general that $A(BC) = (AB)C$ whenever the products can be defined.

Example 15.2 $A = \begin{pmatrix} 2 & 1 \\ 0 & 1 \end{pmatrix}$ and let A^n be the nth power of A. For example

$$A^2 = AA = \begin{pmatrix} 2 & 1 \\ 0 & 1 \end{pmatrix} \begin{pmatrix} 2 & 1 \\ 0 & 1 \end{pmatrix} = \begin{pmatrix} 4 & 3 \\ 0 & 1 \end{pmatrix}$$

and

$$A^3 = A(AA) = AA^2 = \begin{pmatrix} 2 & 1 \\ 0 & 1 \end{pmatrix} \begin{pmatrix} 4 & 3 \\ 0 & 1 \end{pmatrix} = \begin{pmatrix} 8 & 7 \\ 0 & 1 \end{pmatrix}.$$

Suppose that

$$A^n = \begin{pmatrix} a_n & b_n \\ c_n & d_n \end{pmatrix}.$$

By using the fact that, for $n \geq 2$, $A^n = AA^{n-1}$, find recurrence equations for a_n, b_n, c_n, d_n. Solve these to determine an explicit formula for A^n.

Solution

We have

$$\begin{pmatrix} a_n & b_n \\ c_n & d_n \end{pmatrix} = A^n = A(A^{n-1}) = \begin{pmatrix} 2 & 1 \\ 0 & 1 \end{pmatrix} \begin{pmatrix} a_{n-1} & b_{n-1} \\ c_{n-1} & d_{n-1} \end{pmatrix}$$

$$= \begin{pmatrix} 2a_{n-1} + c_{n-1} & 2b_{n-1} + d_{n-1} \\ c_{n-1} & d_{n-1} \end{pmatrix}.$$

So we have the following equations:

$$a_n = 2a_{n-1} + c_{n-1};$$
$$b_n = 2b_{n-1} + d_n;$$
$$c_n = c_{n-1};$$
$$d_n = d_{n-1}.$$

Since $c_1 = 0$ and $d_1 = 1$, it follows that, for all n, $c_n = 0$ and $d_n = 1$. The equations for a_n and b_n then reduce to

$$a_n = 2a_{n-1} \quad \text{and} \quad b_n = 2b_{n-1} + 1.$$

Clearly, given that $a_1 = 2$, the solution for a_n is $a_n = 2^n$. If we let $y_n = b_{n+1}$ for $n \geq 0$, then $y_0 = b_1 = 1$ and the sequence $y_0, y_1, \ldots$ satisfies the recurrence

$$y_n = 2y_{n-1} + 1$$

for $n \geq 1$. Using the standard method from Chapter 3:

$$y^* = \frac{1}{1-2} = -1$$

and

$$y_n = y^* + (y_0 - y^*)2^n = -1 + 2(2^n) = 2^{n+1} - 1.$$

Therefore, $b_n = y_{n-1} = 2^n - 1$. It follows that

$$A^n = \begin{pmatrix} 2^n & 2^n - 1 \\ 0 & 1 \end{pmatrix}$$

for all $n \geq 1$.

Main Topics

- addition and scalar multiplication of matrices
- matrix multiplication
- modelling investment portfolios with matrices

Key Terms, Notations and Formulae

- $m \times n$ matrix
- (i,j)th entry or element a_{ij}
- rows and columns of matrix
- AB defined only if A is $m \times n$ and B $n \times p$; then $C = AB$ is $m \times p$ and

$$c_{ij} = a_{i1}b_{1j} + a_{i2}b_{2j} + \cdots + a_{in}b_{nj}$$

- AB need not equal BA
- square matrix
- identity matrix, I: $AI = IA = A$
- portfolio Y; states; returns matrix R
- riskless and arbitrage portfolios

Exercises

15.1 Let A and B be the matrices

$$A = \begin{pmatrix} 1 & 2 & 3 \\ 3 & 2 & 5 \end{pmatrix} \quad \text{and} \quad B = \begin{pmatrix} 2 & 1 \\ 0 & -1 \\ 1 & 2 \end{pmatrix}.$$

Show that

$$AB = \begin{pmatrix} 5 & 5 \\ 11 & 11 \end{pmatrix}.$$

15.2 Suppose that the matrix of returns for an investor is

$$R = \begin{pmatrix} 1.05 & 0.95 \\ 1.05 & 1.05 \\ 1.37 & 1.42 \end{pmatrix}.$$

Why might we say that the portfolio $Y = \begin{pmatrix} 500 & 10\,000 & 1000 \end{pmatrix}$ is 'riskless'? What return is the investor guaranteed?

Show that $Z = \begin{pmatrix} 1000 & -2000 & 1000 \end{pmatrix}$ is an arbitrage portfolio.

Problems

15.1 Let A and B be the matrices

$$A = \begin{pmatrix} 1 & 2 & -2 & -4 \\ -1 & 1 & 3 & -7 \end{pmatrix} \quad \text{and} \quad B = \begin{pmatrix} 1 \\ -2 \\ 3 \\ 8 \end{pmatrix}.$$

Calculate AB.

15.2 The matrices A and B are

$$A = \begin{pmatrix} 1 & 1 & 1 \\ 1 & 3 & 2 \\ 1 & 3 & 3 \end{pmatrix} \quad \text{and} \quad B = \begin{pmatrix} 1 & 0 & 2 \\ 2 & 3 & 5 \\ 3 & 1 & 5 \end{pmatrix}.$$

Why are both AB and BA defined? Are they equal?

15.3 The matrices A, B, C are as follows:

$$A = \begin{pmatrix} 1 & 0 & 2 \\ 2 & -1 & 1 \\ -3 & 0 & 2 \end{pmatrix}, \quad B = \begin{pmatrix} 1 & 0 \\ 1 & 1 \\ 0 & 2 \end{pmatrix}, \quad C = \begin{pmatrix} 0 & 1 \\ 2 & 1 \\ 3 & 4 \end{pmatrix}.$$

(i) Why is AB defined and BA not defined? Calculate AB.

(ii) Calculate AC.

(iii) Calculate $A(B + C)$ and verify that $A(B + C) = AB + AC$.

15.4 Suppose that A is the matrix

$$A = \begin{pmatrix} 3 & 1 \\ 0 & 1 \end{pmatrix}$$

and, for $n \geq 1$, suppose that

$$A^n = \begin{pmatrix} a_n & b_n \\ c_n & d_n \end{pmatrix}.$$

Find recurrence equations for a_n, b_n, c_n, d_n and solve them to determine an explicit formula for A^n.

15.5 Find a general formula for B^n when $B = \begin{pmatrix} 1 & -1 \\ 0 & 1 \end{pmatrix}$.

16 Linear Equations I

16.1 A Two-Industry 'Economy'

Example Suppose the 'economy' produces only two commodities, grommets and widgets. The grommet-making machines use widgets, and these wear out at the rate of s widgets per grommet produced, where s is a (hopefully) small positive number. Similarly, production of widgets consumes t grommets per widget. In addition, the needs of the population require a monthly net production of 5000 grommets and 2000 widgets. What should be the production schedule for the grommet and widget industries?

Suppose that the answer to this question is x_1 thousand grommets and x_2 thousand widgets per month. The total demand for grommets is 5000 for the population and tx_2 thousand for the widget industry, so we have the equation $x_1 = 5 + tx_2$. Similarly, working out the total demand for widgets we get $x_2 = 2 + sx_1$. Rearranging, we have the equations

$$\begin{aligned} x_1 - tx_2 &= 5, \\ -sx_1 + x_2 &= 2. \end{aligned}$$

Of course, the solution depends upon s and t. Using elementary algebra, we get

$$x_1 = \frac{5+2t}{1-st} \quad \text{and} \quad x_2 = \frac{2+5s}{1-st}.$$

So we have a *unique solution*, given by the above expressions, provided that $st \neq 1$. Given that s and t have the meanings assigned to them in this context, it is a reasonable assumption that st is a small positive number, so that $st < 1$.

But what if st does equal 1? In this case we can substitute $t = 1/s$ in the first equation, getting $x_1 - x_2/s = 5$. Dividing the second equation by s we get $x_1 - x_2/s = -2/s$. Since these two equations have the same left-hand sides, we see that when $5 \neq -2/s$ they are inconsistent and there is *no solution*.

But when $5 = -2/s$, that is, when $s = -2/5$, both equations reduce to the same equation $x_1 + 5x_2/2 = 5$, which has *infinitely many solutions*: for any α, the values $x_1 = \alpha$ and $x_2 = 2 - \frac{2}{5}\alpha$ give a solution. ■

Summarising, we see that a system of two linear equations in two unknowns may have

- a unique solution, or

- no solution, or
- infinitely many solutions.

If there are more than two equations and two unknowns the situation might conceivably become much more complicated, so we need to develop a method for dealing with such problems.

16.2 Linear Equations in Matrix Form

The problem discussed in the preceding section is an example of a system of linear equations. In its most general form, a *system of m linear equations in n unknowns* $x_1, x_2, \ldots, x_n$ is a set of m equations of the form

$$\begin{aligned}
a_{11}x_1 + a_{12}x_2 + \cdots + a_{1n}x_n &= b_1 \\
a_{21}x_1 + a_{22}x_2 + \cdots + a_{2n}x_n &= b_2 \\
\vdots \qquad\qquad & \quad \vdots \\
a_{m1}x_1 + a_{m2}x_2 + \cdots + a_{mn}x_n &= b_m.
\end{aligned}$$

The numbers a_{ij} are usually known as the *coefficients* of the system. We say that $(x_1^*, x_2^*, \ldots, x_n^*)$ is a *solution* of the system if *all* m equations hold true when $x_1 = x_1^*$, $x_2 = x_2^*$, and so on. Sometimes a system of linear equations is known as a set of *simultaneous* equations; such terminology emphasises that a solution is an assignment of values to each of the n unknowns such that *each and every* equation holds with this assignment.

In order to deal with large systems of linear equations, we usually write them in matrix form. First we observe that vectors are just special cases of matrices: a row vector or list of n numbers is simply a matrix of size $1 \times n$, and a column vector is a matrix of size $n \times 1$. The rule for multiplying matrices tells us how to calculate the product $A\mathbf{x}$ of an $m \times n$ matrix A and an $n \times 1$ column vector $\mathbf{x}$. According to the rule, $A\mathbf{x}$ is

$$\begin{pmatrix} a_{11} & a_{12} & \ldots & a_{1n} \\ a_{21} & a_{22} & \ldots & a_{2n} \\ \vdots & \vdots & \ddots & \vdots \\ a_{m1} & a_{m2} & \ldots & a_{mn} \end{pmatrix} \begin{pmatrix} x_1 \\ x_2 \\ \vdots \\ x_n \end{pmatrix} = \begin{pmatrix} a_{11}x_1 + a_{12}x_2 + \cdots + a_{1n}x_n \\ a_{21}x_1 + a_{22}x_2 + \cdots + a_{2n}x_n \\ \vdots \qquad\qquad \vdots \\ a_{m1}x_1 + a_{n2}x_2 + \cdots + a_{mn}x_n \end{pmatrix}.$$

Note that $A\mathbf{x}$ is a column vector with m rows, these being the left-hand sides of our system of linear equations. If we define another column vector $\mathbf{b}$, whose components are the right-hand sides b_i, the system is equivalent to the matrix equation

$$A\mathbf{x} = \mathbf{b}.$$

Obviously there are many advantages in this very compact way of writing the system. Indeed, the simplicity of the matrix form provides ample justification for the definition of matrix multiplication given in the previous chapter, which might have looked strange at first sight.

16.3 Solution of Linear Equations by Row Operations

It is a simple observation that the set of solutions of a system of linear equations is unaltered by the following three operations:

- Multiply both sides of an equation by a non-zero constant.
- Add a multiple of one equation to another.
- Interchange two equations.

We shall call these *elementary operations*. This observation is the foundation of an extremely useful technique for solving linear equations, to be described in general terms in the next chapter. The idea is that we use a sequence of elementary operations to reduce a system to a simpler one, whose solution is obvious. For instance, if we have the system

$$\begin{aligned} x_1 + 2x_2 &= 4, \\ 2x_1 - x_2 &= 5 \end{aligned}$$

of two equations in two unknowns, then we may 'eliminate' x_1 as follows. We multiply the first equation by 2 and then subtract the resulting equation from the second equation, giving

$$(2x_1 - x_2) - 2(x_1 + 2x_2) = 5 - 2 \times 4; \text{ that is } -5x_2 = -3.$$

Thus $x_2 = 3/5$. Then x_1 is found by substitution: $x_1 + 2x_2 = 4$, so $x_1 = 4 - 2(3/5) = 14/5$.

Consider now the following system of three equations in three unknowns:

$$\begin{aligned} x_1 + x_2 + x_3 &= 6, && (\bullet 1) \\ 2x_1 + 4x_2 + x_3 &= 5, && (\bullet 2) \\ 2x_1 + 3x_2 + x_3 &= 6. && (\bullet 3) \end{aligned}$$

Subtracting twice ($\bullet$1) from ($\bullet$2) in order to eliminate x_1 from ($\bullet$2), we obtain a new equation, which is

$$(2x_1 + 4x_2 + x_3) - 2(x_1 + x_2 + x_3) = 5 - 2 \times 6;$$

that is $2x_2 - x_3 = -7$, or $x_2 - (1/2)x_3 = -7/2$. We replace equation ($\bullet$2) by this equation, obtaining the system

$$\begin{aligned} x_1 + x_2 + x_3 &= 6, && (\bullet 1) \\ x_2 - \frac{1}{2}x_3 &= -\frac{7}{2}, && (\bullet 2') \\ 2x_1 + 3x_2 + x_3 &= 6. && (\bullet 3) \end{aligned}$$

This system has been obtained from the original by elementary operations, and therefore has the same solutions as the original. Now, in the same manner, to eliminate x_1 from equation (•3), we subtract twice (•1) from it. The resulting system is

$$\begin{aligned} x_1 + x_2 + x_3 &= 6, & (\bullet 1)\\ x_2 - \frac{1}{2}x_3 &= -\frac{7}{2}, & (\bullet 2')\\ x_2 - x_3 &= -6. & (\bullet 3') \end{aligned}$$

The last two equations of this system involve only the unknowns x_2 and x_3. These two equations, in isolation, form a system of two equations in two unknowns to which, in turn, we apply elementary operations.

Subtracting (•2′) from (•3′) gives

$$(x_2 - x_3) - \left(x_2 - \frac{1}{2}x_3\right) = -6 + \frac{7}{2} = -\frac{5}{2},$$

so that $-x_3/2 = -5/2$, or $x_3 = 5$. Now we have the system

$$\begin{aligned} x_1 + x_2 + x_3 &= 6,\\ x_2 - \frac{1}{2}x_3 &= -\frac{7}{2},\\ x_3 &= 5. \end{aligned}$$

This is a system which we can solve by working backwards. Since $x_2 - (1/2)x_3 = -7/2$, it follows that $x_2 = -7/2 + (1/2)(5) = -1$. Similarly, x_1 can now be determined. Since $x_1 + x_2 + x_3 = 6$ and $x_2 = -1, x_3 = 5$, we have $x_1 = 6 + 1 - 5 = 2$.

There is therefore precisely one solution to the original system of equations: $x_1 = 2$, $x_2 = -1, x_3 = 5$.

The foregoing procedure can be carried out more compactly by omitting the x_is. If we write the equations in matrix form $A\mathbf{x} = \mathbf{b}$, with

$$A = \begin{pmatrix} 1 & 1 & 1\\ 2 & 4 & 1\\ 2 & 3 & 1 \end{pmatrix} \quad \text{and} \quad \mathbf{b} = \begin{pmatrix} 6\\ 5\\ 6 \end{pmatrix},$$

then

$$(A \mid \mathbf{b}) = \left(\begin{array}{ccc|c} 1 & 1 & 1 & 6\\ 2 & 4 & 1 & 5\\ 2 & 3 & 1 & 6 \end{array}\right)$$

is called the *augmented matrix* of the system.

We use this form because of the important fact that elementary operations on the equations of the system correspond to the same operations on the rows of the augmented matrix. For that reason we shall now refer to them as *elementary row operations*. In the following sequence, the elementary row operations performed are just those corresponding to the operations on the system described above.

$$\left(\begin{array}{ccc|c}1&1&1&6\\2&4&1&5\\2&3&1&6\end{array}\right)\to\left(\begin{array}{ccc|c}1&1&1&6\\0&2&-1&-7\\2&3&1&6\end{array}\right)\to\left(\begin{array}{ccc|c}1&1&1&6\\0&1&-\frac{1}{2}&-\frac{7}{2}\\0&1&-1&-6\end{array}\right)$$

$$\to\left(\begin{array}{ccc|c}1&1&1&6\\0&1&-\frac{1}{2}&-\frac{7}{2}\\0&0&-\frac{1}{2}&-\frac{5}{2}\end{array}\right)\to\left(\begin{array}{ccc|c}1&1&1&6\\0&1&-\frac{1}{2}&-\frac{7}{2}\\0&0&1&5\end{array}\right).$$

The reader should check exactly which operations have been used at each step. For example, the transition from the first to the second matrix is effected by the elementary row operation which subtracts twice the first row from the second, giving the second row in the second matrix.

The final augmented matrix represents the system

$$\begin{aligned}x_1+x_2+x_3&=6,\\x_2-\frac{1}{2}x_3&=-\frac{7}{2},\\x_3&=5.\end{aligned}$$

As we have seen, the solution of this system can be found by 'working backwards', or by *back-substitution*, as it is officially known.

$$x_3=5,\quad x_2=\frac{1}{2}x_3-\frac{7}{2}=-1,\quad x_1=-x_2-x_3+6=2.$$

16.4 The Echelon Form in General

The example discussed in the previous section illustrates a quite general method used to solve systems of linear equations. This process is often known as *Gaussian elimination* (after the German mathematician Carl Friedrich Gauss (1777–1855)). We use elementary row operations to transform the augmented matrix of coefficients to what is known as the *echelon form*:

$$\left(\begin{array}{cccccc|c}1&*&*&*&\dots&*&*\\0&0&1&*&\dots&*&*\\0&0&0&1&\dots&*&*\\\vdots&\vdots&\vdots&\vdots&\ddots&\vdots&\vdots\\0&0&0&0&\dots&0&0\end{array}\right).$$

The characteristics of this form are that (a) the first non-zero entry in each row is 1 (we call this the *leading* 1); (b) the position of the leading 1 moves to the right as we go down the rows; (c) any rows which consist entirely of zeros are located at the bottom of the matrix.

In this chapter we concentrate on the special case where the echelon form is

$$\left(\begin{array}{cccccc|c} 1 & * & * & * & \dots & * & * \\ 0 & 1 & * & * & \dots & * & * \\ 0 & 0 & 1 & * & \dots & * & * \\ \vdots & \vdots & \vdots & \vdots & \ddots & \vdots & \vdots \\ 0 & 0 & 0 & 0 & \dots & 1 & * \end{array}\right).$$

Specifically, there are no rows consisting entirely of zeros, and the leading 1s move one step to the right as we go down the rows. It is clear that the solution of a system whose augmented matrix can be transformed to this form can be found by back-substitution.

Example Suppose that we have the following system of linear equations:

$$\begin{aligned} x_1 + 2x_2 + x_3 &= 1, \\ 2x_1 + 2x_2 &= 2, \\ 3x_1 + 5x_2 + 4x_3 &= 1. \end{aligned}$$

In matrix form, this is

$$\begin{pmatrix} 1 & 2 & 1 \\ 2 & 2 & 0 \\ 3 & 5 & 4 \end{pmatrix} \begin{pmatrix} x_1 \\ x_2 \\ x_3 \end{pmatrix} = \begin{pmatrix} 1 \\ 2 \\ 1 \end{pmatrix}.$$

We form the 3×4 augmented matrix

$$\left(\begin{array}{ccc|c} 1 & 2 & 1 & 1 \\ 2 & 2 & 0 & 2 \\ 3 & 5 & 4 & 1 \end{array}\right),$$

and use elementary row operations to reduce the augmented matrix to echelon form. We have

$$\begin{aligned} \left(\begin{array}{ccc|c} 1 & 2 & 1 & 1 \\ 2 & 2 & 0 & 2 \\ 3 & 5 & 4 & 1 \end{array}\right) &\to \left(\begin{array}{ccc|c} 1 & 2 & 1 & 1 \\ 0 & -2 & -2 & 0 \\ 0 & -1 & 1 & -2 \end{array}\right) \\ &\to \left(\begin{array}{ccc|c} 1 & 2 & 1 & 1 \\ 0 & 1 & 1 & 0 \\ 0 & -1 & 1 & -2 \end{array}\right) \\ &\to \left(\begin{array}{ccc|c} 1 & 2 & 1 & 1 \\ 0 & 1 & 1 & 0 \\ 0 & 0 & 2 & -2 \end{array}\right) \\ &\to \left(\begin{array}{ccc|c} 1 & 2 & 1 & 1 \\ 0 & 1 & 1 & 0 \\ 0 & 0 & 1 & -1 \end{array}\right). \end{aligned}$$

Therefore, the initial system has the same set of solutions as the system

$$\begin{aligned} x_1 + 2x_2 + x_3 &= 1, \\ x_2 + x_3 &= 0, \\ x_3 &= -1. \end{aligned}$$

But this is easy to solve by back-substitution. From the third equation, $x_3 = -1$. The second equation then gives $x_2 = 1$, and then the first gives $x_1 = 1 - 2x_2 - x_3 = 0$. ■

Worked Examples

Example 16.1 Let

$$C = \begin{pmatrix} 1 & 2 \\ 2 & 4 \end{pmatrix}, \quad \mathbf{x} = \begin{pmatrix} x_1 \\ x_2 \end{pmatrix}, \quad \mathbf{b} = \begin{pmatrix} 3 \\ 4 \end{pmatrix}.$$

Write out in full the systems of equations $C\mathbf{x} = \mathbf{b}$ and $C\mathbf{x} = \mathbf{0}$. Using only elementary algebra, explain why the first system has no solutions and the second system has infinitely many solutions.

Solution

The system $C\mathbf{x} = \mathbf{b}$ is

$$\begin{pmatrix} 1 & 2 \\ 2 & 4 \end{pmatrix} \begin{pmatrix} x_1 \\ x_2 \end{pmatrix} = \begin{pmatrix} 3 \\ 4 \end{pmatrix},$$

which is

$$\begin{pmatrix} x_1 + 2x_2 \\ 2x_1 + 4x_2 \end{pmatrix} = \begin{pmatrix} 3 \\ 4 \end{pmatrix}.$$

That is, $C\mathbf{x} = \mathbf{b}$ is the system of equations

$$\begin{aligned} x_1 + 2x_2 &= 3, \\ 2x_1 + 4x_2 &= 4. \end{aligned}$$

In a similar manner, $C\mathbf{x} = 0$ describes the system

$$\begin{aligned} x_1 + 2x_2 &= 0, \\ 2x_1 + 4x_2 &= 0. \end{aligned}$$

Consider the first of these two systems. The first equation is $x_1 + 2x_2 = 3$ and the second is easily seen to be equivalent, on division by 2, to $x_1 + 2x_2 = 2$. Since $2 \neq 3$, there are no solutions to this system.

The second system has two equations, but each equation conveys exactly the same information, namely that $x_1 + 2x_2 = 0$. (The second equation is just this multiplied by 2.) It follows that to obtain a solution we may let x_2 be any number at all and set

$x_1 = -2x_2$. Thus, for each real number r, $x_1 = -2r$ and $x_2 = r$ is a solution and, therefore, the system has infinitely many solutions.

Example 16.2 By performing elementary row operations on the augmented matrix, solve the following system of equations:

$$\begin{aligned} x_1 + 2x_2 + x_3 &= 1, \\ 2x_1 + 2x_2 + 5x_3 &= 2, \\ 3x_1 + 4x_2 + x_3 &= 3. \end{aligned}$$

Solution

The augmented matrix corresponding to the system is

$$\left(\begin{array}{ccc|c} 1 & 2 & 1 & 1 \\ 2 & 2 & 5 & 2 \\ 3 & 4 & 1 & 3 \end{array}\right).$$

We reduce this to echelon form, as follows:

$$\left(\begin{array}{ccc|c} 1 & 2 & 1 & 1 \\ 2 & 2 & 5 & 2 \\ 3 & 4 & 1 & 3 \end{array}\right) \to \left(\begin{array}{ccc|c} 1 & 2 & 1 & 1 \\ 0 & -2 & 3 & 0 \\ 0 & -2 & -2 & 0 \end{array}\right) \to \left(\begin{array}{ccc|c} 1 & 2 & 1 & 1 \\ 0 & -2 & -2 & 0 \\ 0 & -2 & 3 & 0 \end{array}\right)$$

$$\to \left(\begin{array}{ccc|c} 1 & 2 & 1 & 1 \\ 0 & 1 & 1 & 0 \\ 0 & -2 & 3 & 0 \end{array}\right) \to \left(\begin{array}{ccc|c} 1 & 2 & 1 & 1 \\ 0 & 1 & 1 & 0 \\ 0 & 0 & 5 & 0 \end{array}\right) \to \left(\begin{array}{ccc|c} 1 & 2 & 1 & 1 \\ 0 & 1 & 1 & 0 \\ 0 & 0 & 1 & 0 \end{array}\right).$$

This last, echelon, matrix represents the system

$$\begin{aligned} x_1 + 2x_2 + x_3 &= 1, \\ x_2 + x_3 &= 0, \\ x_3 &= 0 \end{aligned}$$

which, by virtue of the fact that elementary row operations do not change the solutions of a system of linear equations, must have the same solutions as the original system. But it is easy to determine the solution of this system by back-substitution. First, we have $x_3 = 0$, then $x_2 = -x_3 = 0$ and $x_1 = 1 - 2x_2 - x_3 = 1$. There is, therefore, one solution to the original system: $x_1 = 1, x_2 = 0, x_3 = 0$.

Example 16.3 Use the method of elementary row operations to solve the following system of equations:

$$\begin{aligned} 3x_1 - 3x_2 + 5x_3 &= 6, \\ x_1 + 7x_2 + 5x_3 &= 4, \\ 5x_1 + 10x_2 + 15x_3 &= 9. \end{aligned}$$

Solution

The augmented matrix corresponding to the system of equations is

$$\left(\begin{array}{ccc|c} 3 & -3 & 5 & 6 \\ 1 & 7 & 5 & 4 \\ 5 & 10 & 15 & 9 \end{array}\right),$$

which we reduce to echelon form using elementary row operations, as follows:

$$\left(\begin{array}{ccc|c} 3 & -3 & 5 & 6 \\ 1 & 7 & 5 & 4 \\ 5 & 10 & 15 & 9 \end{array}\right) \to \left(\begin{array}{ccc|c} 1 & 7 & 5 & 4 \\ 3 & -3 & 5 & 6 \\ 5 & 10 & 15 & 9 \end{array}\right) \to \left(\begin{array}{ccc|c} 1 & 7 & 5 & 4 \\ 0 & -24 & -10 & -6 \\ 0 & -25 & -10 & -11 \end{array}\right)$$

$$\to \left(\begin{array}{ccc|c} 1 & 7 & 5 & 4 \\ 0 & 1 & \frac{5}{12} & \frac{1}{4} \\ 0 & -25 & -10 & -11 \end{array}\right) \to \left(\begin{array}{ccc|c} 1 & 7 & 5 & 4 \\ 0 & 1 & \frac{5}{12} & \frac{1}{4} \\ 0 & 0 & \frac{5}{12} & -\frac{19}{4} \end{array}\right) \to \left(\begin{array}{ccc|c} 1 & 7 & 5 & 4 \\ 0 & 1 & \frac{5}{12} & \frac{1}{4} \\ 0 & 0 & 1 & -\frac{57}{5} \end{array}\right).$$

This last matrix, in echelon form, represents the system

$$\begin{aligned} x_1 + 7x_2 + 5x_3 &= 4, \\ x_2 + \frac{5}{12}x_3 &= \frac{1}{4}, \\ x_3 &= -\frac{57}{5}. \end{aligned}$$

Its solution (which is the same as that of the original system) may be determined by back-substitution:

$$x_3 = -\frac{57}{5}, \quad x_2 = \frac{1}{4} - \frac{5}{12}x_3 = 5, \quad x_1 = 4 - 7x_2 - 5x_3 = 26.$$

Example 16.4 Use the result of the previous example to express the vector $(6, 4, 9)$ as a linear combination of the vectors $(3, 1, 5)$, $(-3, 7, 10)$, $(5, 5, 15)$.

Solution

The system of equations in the previous example has solution $x_1 = 26, x_2 = 5, x_3 = -\frac{57}{5}$. It follows that

$$\begin{aligned} 3(26) - 3(5) + 5\left(-\frac{57}{5}\right) &= 6, \\ 1(26) + 7(5) + 5\left(-\frac{57}{5}\right) &= 4, \\ 5(26) + 10(5) + 15\left(-\frac{57}{5}\right) &= 9. \end{aligned}$$

This means that

$$26\begin{pmatrix} 3 \\ 1 \\ 5 \end{pmatrix} + 5\begin{pmatrix} -3 \\ 7 \\ 10 \end{pmatrix} - \frac{57}{5}\begin{pmatrix} 5 \\ 5 \\ 15 \end{pmatrix} = \begin{pmatrix} 6 \\ 4 \\ 9 \end{pmatrix}.$$

Thus, the row vector $(6, 4, 9)$ may be expressed as the linear combination

$$(6, 4, 9) = 26(3, 1, 5) + 5(-3, 7, 10) - \frac{57}{5}(5, 5, 15).$$

Example 16.5 Verify explicitly the identity $A(B\mathbf{y}) = (AB)\mathbf{y}$, where A and B are any 3×3 matrices, and $\mathbf{y}$ is a 3-vector.

Solution

Let $\mathbf{x} = B\mathbf{y}$ so that, for $j = 1, 2, 3$,

$$x_j = b_{j1}y_1 + b_{j2}y_2 + b_{j3}y_3.$$

Now $A\mathbf{x} = A(B\mathbf{y})$ and the ith component of $A\mathbf{x}$ is

$$a_{i1}x_1 + a_{i2}x_2 + a_{i3}x_3.$$

Substituting the expressions for the xs in terms of the ys, this is equal to

$$a_{i1}(b_{11}y_1 + b_{12}y_2 + b_{13}y_3) + a_{i2}(b_{21}y_1 + b_{22}y_2 + b_{23}y_3) + a_{i3}(b_{31}y_1 + b_{32}y_2 + b_{33}y_3).$$

Collecting together terms involving y_1, then terms involving y_2, and finally terms involving y_3, we obtain

$$(a_{i1}b_{11} + a_{i2}b_{21} + a_{i3}b_{31})y_1 + (a_{i1}b_{12} + a_{i2}b_{22} + a_{i3}b_{32})y_2 + (a_{i1}b_{13} + a_{i2}b_{23} + a_{i3}b_{33})y_3.$$

But, since

$$a_{i1}b_{1j} + a_{i2}b_{2j} + a_{i3}b_{3j}$$

is the (i,j)th entry of the matrix product AB, this is just the ith component of $(AB)\mathbf{y}$, from which the identity follows.

Example 16.6 The supply function for a commodity takes the form

$$q^S(p) = ap^2 + bp + c,$$

for some constants a, b, c. When $p = 1$, the quantity supplied is 5; when $p = 2$, the quantity supplied is 12; when $p = 3$, the quantity supplied is 23. Find the constants a, b, c.

Solution

The given information means that

$$q^S(1) = 5, \quad q^S(2) = 12, \quad q^S(3) = 23;$$

that is

$$a(1^2) + b(1) + c = 5,$$
$$a(2^2) + b(2) + c = 12,$$
$$a(3^2) + b(3) + c = 23.$$

So we have the following system of linear equations for a, b, c:

$$a + b + c = 5,$$
$$4a + 2b + c = 12,$$
$$9a + 3b + c = 23.$$

We solve this in the usual way, by reducing the augmented matrix to echelon form:

$$\left(\begin{array}{ccc|c} 1 & 1 & 1 & 5 \\ 4 & 2 & 1 & 12 \\ 9 & 3 & 1 & 23 \end{array}\right) \rightarrow \left(\begin{array}{ccc|c} 1 & 1 & 1 & 5 \\ 0 & -2 & -3 & -8 \\ 0 & -6 & -8 & -22 \end{array}\right) \rightarrow \left(\begin{array}{ccc|c} 1 & 1 & 1 & 5 \\ 0 & -2 & -3 & -8 \\ 0 & 0 & -1 & -2 \end{array}\right)$$

$$\rightarrow \left(\begin{array}{ccc|c} 1 & 1 & 1 & 5 \\ 0 & 1 & \frac{3}{2} & 4 \\ 0 & 0 & 1 & 2 \end{array}\right).$$

Therefore

$$\begin{aligned} a + b + c &= 5, \\ b + \frac{3}{2}c &= 4, \\ c &= 2, \end{aligned}$$

so that $c = 2, b = 1, a = 2$. The supply function is therefore given explicitly by $q^S(p) = 2p^2 + p + 2$.

Main Topics

- systems of linear equations
- linear equations in matrix form
- solving systems of linear equations using row operations

Key Terms, Notations and Formulae

- general system of m linear equations in n unknowns
- coefficients
- matrix form, $A\mathbf{x} = \mathbf{b}$
- augmented matrix $(A \mid \mathbf{b})$
- elementary row operations:
  - multiply a row by a non-zero constant
  - add a multiple of one row to another
  - interchange two rows
- back-substitution
- echelon form; leading 1s

Exercises

16.1 Solve the following system of equations:

$$\begin{aligned} x_1 + x_2 + x_3 &= 6, \\ 2x_1 + 4x_2 + x_3 &= 5, \\ 2x_1 + 3x_2 + x_3 &= 6. \end{aligned}$$

16.2 The function f is given, for $x > 0$, by

$$f(x) = \frac{a}{x} + bx + cx^2$$

for some constants a, b and c. The following facts about f and its derivative f' are known:

$$f(1) = 8,\ f(2) = 2,\ f'(1) = -9.$$

Show that a, b and c satisfy the following equations:

$$\begin{aligned} a + b + c &= 8, \\ a + 4b + 8c &= 4, \\ a - b - 2c &= 9. \end{aligned}$$

By solving this system of equations using elementary row operations, find the values of a, b and c.

16.3 For some numbers a, b and c, let f be the function of one variable given by

$$f(x) = ax^3 + \frac{b}{x} + c(1 + \ln x),$$

for $x > 0$. Given that

$$f(1) = 0,\ \ f'(1) = 20,\ \ f''(1) = 40,$$

find a system of linear equations for a, b and c. Find a, b and c by solving the system using elementary row operations.

16.4 Three goods are sold in the same market. If their prices are x_1, x_2, x_3, then the demand quantities y_1, y_2, y_3 and the supply quantities z_1, z_2, z_3 are given by the following equations:

$$\begin{aligned} y_1 &= 50 - x_1 + x_2 + x_3, \\ z_1 &= 2x_2 + x_3 - 20, \\ y_2 &= x_1 - x_2 - 2x_3 + 264, \\ z_2 &= 2x_1 + 4x_2 + 5x_3 - 30, \\ y_3 &= 2x_1 + 2x_2 - 2x_3 + 4, \\ z_3 &= 8x_1 + 2x_3 - 40. \end{aligned}$$

Non-negative numbers x_1^*, x_2^*, x_3^* are said to be equilibrium prices if, when the prices are $x_1 = x_1^*, x_2 = x_2^*$ and $x_3 = x_3^*$, then the supply and demand quantities for each good are equal; that is $y_1 = z_1$, $y_2 = z_2$ and $y_3 = z_3$. Using a matrix method, find the equilibrium prices.

Problems

16.1 Let

$$A = \begin{pmatrix} 2 & 4 & 4 \\ 4 & 8 & 8 \\ 1 & 2 & 2 \end{pmatrix}, \quad \mathbf{x} = \begin{pmatrix} x_1 \\ x_2 \\ x_3 \end{pmatrix}, \quad \mathbf{b} = \begin{pmatrix} 1 \\ 2 \\ 1 \end{pmatrix}.$$

Write out in full the systems of equations $A\mathbf{x} = \mathbf{b}$ and $A\mathbf{x} = \mathbf{0}$. Using only elementary algebra, explain why the first system has no solutions and the second system has infinitely many solutions.

16.2 By performing elementary row operations on the augmented matrix, solve the following system of equations:

$$\begin{aligned} 2x_1 + x_2 + x_3 &= 7, \\ x_1 + 2x_3 &= 7, \\ 2x_1 + 2x_2 - x_3 &= 4. \end{aligned}$$

16.3 By performing elementary row operations on the augmented matrix, solve the following system of equations:

$$\begin{aligned} 4x_1 + 2x_2 + 3x_3 &= 3, \\ 2x_1 + 3x_2 + 4x_3 &= 2, \\ 3x_1 + 4x_2 + 2x_3 &= 1. \end{aligned}$$

16.4 Use the method of elementary row operations to solve the following system of equations:

$$\begin{aligned} x_1 + 2x_2 + 2x_3 &= 7, \\ 4x_1 + 5x_2 + x_3 &= 11, \\ 7x_1 - 4x_2 - x_3 &= 1. \end{aligned}$$

16.5 Use the result of the previous example to express the vector $(7, 11, 1)$ as a linear combination of the vectors $(1, 4, 7)$, $(2, 5, -4)$, $(2, 1, -1)$.

16.6 The supply function for a good is

$$q^S(p) = ap^3 + bp^2 + c,$$

for some constants a, b, c. When $p = 1$, the quantity supplied is 1; when $p = 2$, the quantity supplied is 11; when $p = 3$, the quantity supplied is 35. Find the constants a, b, c.

16.7 The demand function for a commodity takes the form

$$q^D(p) = a + bp + \frac{c}{p},$$

for some constants a, b, c. When $p = 1$, the quantity demanded is 60; when $p = 2$, it is 40; and when $p = 4$, it is 15. Find the constants a, b, c.

16.8 In the 'grommets and widgets' problem considered in Section 16.1, discuss the economic significance (if any) of the cases where there is not a unique solution.

17 Linear Equations II

17.1 Consistent and Inconsistent Systems

In Chapter 16 we explained how to perform *elementary row operations* on the *augmented matrix* representing a linear system, and thereby reduce it to the standard *echelon form*.

All the systems considered in Chapter 16 had exactly one solution, and this was because the echelon form for those systems had a specific shape, as shown in Section 16.4. But we know that it is possible that a linear system has no solutions at all, or infinitely many. In this chapter we shall explain how these cases can also be determined by looking at the echelon form.

We begin with two examples. In the first one there are infinitely many solutions and in the second there are none.

Example Consider the system

$$\begin{aligned} x_1 + 2x_2 + x_3 &= 1, \\ 2x_1 + 2x_2 &= 2, \\ 3x_1 + 4x_2 + x_3 &= 3. \end{aligned}$$

As usual, we form the augmented matrix

$$\left(\begin{array}{ccc|c} 1 & 2 & 1 & 1 \\ 2 & 2 & 0 & 2 \\ 3 & 4 & 1 & 3 \end{array}\right)$$

and apply elementary row operations to reduce it to echelon form:

$$\left(\begin{array}{ccc|c} 1 & 2 & 1 & 1 \\ 2 & 2 & 0 & 2 \\ 3 & 4 & 1 & 3 \end{array}\right) \to \left(\begin{array}{ccc|c} 1 & 2 & 1 & 1 \\ 0 & 2 & 2 & 0 \\ 0 & 2 & 2 & 0 \end{array}\right) \to \left(\begin{array}{ccc|c} 1 & 2 & 1 & 1 \\ 0 & 1 & 1 & 0 \\ 0 & 2 & 2 & 0 \end{array}\right) \to \left(\begin{array}{ccc|c} 1 & 2 & 1 & 1 \\ 0 & 1 & 1 & 0 \\ 0 & 0 & 0 & 0 \end{array}\right).$$

This last matrix represents the system

$$\begin{aligned} x_1 + 2x_2 + x_3 &= 1, \\ x_2 + x_3 &= 0, \\ 0x_1 + 0x_2 + 0x_3 &= 0. \end{aligned}$$

The third of these equations conveys no information at all about x_1, x_2 and x_3, for it simply tells us that $0 = 0$. Consequently, the original system has the same solutions as the following system of two equations in three unknowns:

$$\begin{aligned} x_1 + 2x_2 + x_3 &= 1, \\ x_2 + x_3 &= 0. \end{aligned}$$

These equations tell us first that, given x_3, we have $x_2 + x_3 = 0$, so $x_2 = -x_3$. Then we have $x_1 = 1 - 2x_2 - x_3 = 1 + 2x_3 - x_3 = 1 + x_3$, so both x_1 and x_2 are determined in terms of x_3. Indeed, if we let x_3 be any real number s, then

$$x_1 = 1 + s, \quad x_2 = -s, \quad x_3 = s$$

is a solution to the system. So there are infinitely many solutions here. ■

Example Consider the system of equations

$$\begin{aligned} x_1 + 2x_2 + x_3 &= 1, \\ 2x_1 + 2x_2 &= 2, \\ 3x_1 + 4x_2 + x_3 &= 2. \end{aligned}$$

Using row operations to reduce the augmented matrix to echelon form, we obtain

$$\left(\begin{array}{ccc|c} 1 & 2 & 1 & 1 \\ 2 & 2 & 0 & 2 \\ 3 & 4 & 1 & 2 \end{array}\right) \to \left(\begin{array}{ccc|c} 1 & 2 & 1 & 1 \\ 0 & 2 & 2 & 0 \\ 0 & 2 & 2 & 1 \end{array}\right) \to \left(\begin{array}{ccc|c} 1 & 2 & 1 & 1 \\ 0 & 1 & 1 & 0 \\ 0 & 2 & 2 & 1 \end{array}\right) \to \left(\begin{array}{ccc|c} 1 & 2 & 1 & 1 \\ 0 & 1 & 1 & 0 \\ 0 & 0 & 0 & 1 \end{array}\right).$$

Thus the original system of equations is equivalent to the system

$$\begin{aligned} x_1 + 2x_2 + x_3 &= 1, \\ x_2 + x_3 &= 0, \\ 0x_1 + 0x_2 + 0x_3 &= 1. \end{aligned}$$

But this system has no solutions, since there are no values of x_1, x_2, x_3 which satisfy the last equation. It reduces to the false statement $0 = 1$, whatever values we give the unknowns. We deduce, therefore, that the original system has no solutions. ■

In the preceding example it turned out that the echelon form has a row of the kind $(0\ 0\ \ldots\ 0 \mid a)$, with $a \neq 0$. This means that the original system is equivalent to one in which there is an equation

$$0x_1 + 0x_2 + \cdots + 0x_n = a \qquad (a \neq 0).$$

Clearly this equation cannot be satisfied by any values of the x_is, and we say that such a system is *inconsistent*.

If there is no row of this kind, the system has at least one solution and we say that it is *consistent*.

17.2 The Rank of a Consistent System

In general, the rows of the echelon form of a consistent system are of two types. The first type, which we call a *zero* row, consists entirely of zeros: $(0\ 0\ \dots\ 0 \mid 0)$. The other type is a row in which at least one of the components, not the last one, is non-zero: $(*\ *\ \dots\ * \mid b)$, where at least one of the $*$s is not zero. We shall call this a *non-zero row*. The standard method of reduction to echelon form ensures that the zero rows (if any) come below the non-zero rows.

The number of non-zero rows in the echelon form is known as the *rank* of the system. As we shall see, the rank is of vital importance in determining the nature of the solution set.

Suppose first that the rank r is strictly less than n, the number of unknowns. Then the system in echelon form (and hence the original one) does not provide enough information to specify the values of $x_1, x_2, \dots, x_n$ uniquely.

Example Suppose we are given a system for which the augmented matrix reduces to the echelon form

$$\left(\begin{array}{cccccc|c} 1 & 3 & -2 & 0 & 2 & 0 & 0 \\ 0 & 0 & 1 & 2 & 0 & 3 & 1 \\ 0 & 0 & 0 & 0 & 0 & 1 & 5 \\ 0 & 0 & 0 & 0 & 0 & 0 & 0 \end{array}\right).$$

Here the rank (number of non-zero rows) is $r = 3$, which is strictly less than the number of unknowns, $n = 6$. The corresponding system is

$$\begin{aligned} x_1 + 3x_2 - 2x_3 + 2x_5 &= 0, \\ x_3 + 2x_4 + 3x_6 &= 1, \\ x_6 &= 5. \end{aligned}$$

These equations can be rearranged to give x_1, x_3 and x_6:

$$x_1 = -3x_2 + 2x_3 - 2x_5, \quad x_3 = 1 - 2x_4 - 3x_6, \quad x_6 = 5.$$

Using back-substitution to solve for x_1, x_3 and x_6 in terms of x_2, x_4 and x_5, we get

$$x_6 = 5, \quad x_3 = -14 - 2x_4, \quad x_1 = -28 - 3x_2 - 4x_4 - 2x_5.$$

The form of these equations tells us that we can assign any values to x_2, x_4 and x_5, and then the other variables will be determined. Explicitly, if we give x_2, x_4, x_5 the arbitrary values s, t, u, the solution is given by

$$x_1 = -28 - 3s - 4t - 2u,\ x_2 = s,\ x_3 = -14 - 2t,\ x_4 = t,\ x_5 = u,\ x_6 = 5.$$

Observe that there are *infinitely many solutions*, because the 'free unknowns' x_2, x_4, x_5 can take any values s, t, u. ■

Generally, we can describe what happens when the echelon form has $r < n$ non-zero rows $(0\ 0\ \dots\ 0\ 1\ *\ *\ \dots\ * \mid *)$. If the leading 1 is in the kth column, it is the coefficient of the unknown x_k. So if the rank is r and the leading 1s occur in columns

$c_1, c_2, \ldots, c_r$, then the general solution to the system can be expressed in a form where the unknowns $x_{c_1}, x_{c_2}, \ldots, x_{c_r}$ are given in terms of the other $n-r$ unknowns, and those $n-r$ unknowns are free to take any values. In the preceding example, we have $n = 6$ and $r = 3$, and the three unknowns x_1, x_3, x_6 can be expressed in terms of the $6-3 = 3$ free unknowns x_2, x_4, x_5.

The case $r = n$, where the number of non-zero rows r in the echelon form is equal to the number of unknowns n, was discussed at length in the previous chapter. The echelon form has no zero rows, and the leading 1s move one step to the right as we go down the rows. In this case there is a *unique solution* obtained by back-substitution from the echelon form. In fact, this can be thought of as a special case of the more general one discussed above: since $r = n$ there are $n - r = 0$ free unknowns and the solution is therefore unique.

We can now summarise our conclusions concerning a general linear system.

- If the echelon form has a row $(0\ 0\ \ldots\ 0 \mid a)$, with $a \neq 0$, the original system is inconsistent; it has *no solutions*.
- If the echelon form has no rows of the above type it is consistent, and the general solution involves $n - r$ free unknowns, where r is the rank. When $r < n$ there are *infinitely many solutions*, but when $r = n$ there are no free unknowns and so there is a *unique solution*.

17.3 The General Solution in Vector Notation

Consider again the example discussed in the previous section. We found that the general solution in terms of three free unknowns, or *parameters*, s, t, u is

$$x_1 = -28 - 3s - 4t - 2u,\ x_2 = s,\ x_3 = -14 - 2t,\ x_4 = t,\ x_5 = u,\ x_6 = 5.$$

It is instructive to write this in vector notation. If we write $\mathbf{x}$ as a column vector,

$$\mathbf{x} = \begin{pmatrix} x_1 \\ x_2 \\ x_3 \\ x_4 \\ x_5 \\ x_6 \end{pmatrix},$$

then

$$\mathbf{x} = \begin{pmatrix} -28 - 3s - 4t - 2u \\ s \\ -14 - 2t \\ t \\ u \\ 5 \end{pmatrix} = \begin{pmatrix} -28 \\ 0 \\ -14 \\ 0 \\ 0 \\ 5 \end{pmatrix} + \begin{pmatrix} -3s \\ s \\ 0 \\ 0 \\ 0 \\ 0 \end{pmatrix} + \begin{pmatrix} -4t \\ 0 \\ -2t \\ t \\ 0 \\ 0 \end{pmatrix} + \begin{pmatrix} -2u \\ 0 \\ 0 \\ 0 \\ u \\ 0 \end{pmatrix}.$$

That is, the general solution is

$$\mathbf{x} = \mathbf{v} + s\mathbf{u}_1 + t\mathbf{u}_2 + u\mathbf{u}_3,$$

where

$$\mathbf{v} = \begin{pmatrix} -28 \\ 0 \\ -14 \\ 0 \\ 0 \\ 5 \end{pmatrix}, \mathbf{u}_1 = \begin{pmatrix} -3 \\ 1 \\ 0 \\ 0 \\ 0 \\ 0 \end{pmatrix}, \mathbf{u}_2 = \begin{pmatrix} -4 \\ 0 \\ -2 \\ 1 \\ 0 \\ 0 \end{pmatrix}, \mathbf{u}_3 = \begin{pmatrix} -2 \\ 0 \\ 0 \\ 0 \\ 1 \\ 0 \end{pmatrix}.$$

We have expressed the general solution as the sum of $\mathbf{v}$ and *any* linear combination of $\{\mathbf{u}_1, \mathbf{u}_2, \mathbf{u}_3\}$.

Applying the same method generally to a consistent system of rank r with n unknowns, we can express the general solution of a consistent system $A\mathbf{x} = \mathbf{b}$ in the form

$$\mathbf{x} = \mathbf{v} + s_1\mathbf{u}_1 + s_2\mathbf{u}_2 + \cdots + s_{n-r}\mathbf{u}_{n-r}.$$

Note that, if we put all the s_is equal to 0, we get a solution $\mathbf{x} = \mathbf{v}$, which means that $A\mathbf{v} = \mathbf{b}$, so $\mathbf{v}$ is a *particular solution* of the system. Putting $s_1 = 1$ and the remaining s_is equal to zero, we get a solution $\mathbf{x} = \mathbf{v} + \mathbf{u}_1$, which means that $A(\mathbf{v} + \mathbf{u}_1) = \mathbf{b}$. Thus

$$\mathbf{b} = A(\mathbf{v} + \mathbf{u}_1) = A\mathbf{v} + A\mathbf{u}_1 = \mathbf{b} + A\mathbf{u}_1.$$

Comparing the first and last expressions, we see that $A\mathbf{u}_1$ is the zero vector $\mathbf{0}$. Clearly, the same equation holds for $\mathbf{u}_2, \ldots, \mathbf{u}_{n-r}$. So we have proved the following.

The general solution of $A\mathbf{x} = \mathbf{b}$ is the sum of:

- a particular solution $\mathbf{v}$ of the system $A\mathbf{x} = \mathbf{b}$ and
- a linear combination of solutions $\mathbf{u}_1, \mathbf{u}_2, \ldots, \mathbf{u}_{n-r}$ of the system $A\mathbf{x} = \mathbf{0}$.

Example We shall find, in vector notation, the general solution to the following system:

$$\begin{aligned} x_1 - x_2 + x_3 + 2x_5 &= 3, \\ x_2 - x_3 &= 4, \\ x_1 + x_2 + x_3 + x_4 + x_5 &= 6, \\ x_1 + 2x_5 &= 7. \end{aligned}$$

Since the number of equations m is less than the number of unknowns n, and trivially $r \leq m$, the system is either inconsistent, or, if it is consistent, it has infinitely many solutions depending on $n - r > 0$ free parameters. We proceed as usual by using row operations to reduce the augmented matrix to echelon form:

$$\left(\begin{array}{ccccc|c} 1 & -1 & 1 & 0 & 2 & 3 \\ 0 & 1 & -1 & 0 & 0 & 4 \\ 1 & 1 & 1 & 1 & 1 & 6 \\ 1 & 0 & 0 & 0 & 2 & 7 \end{array}\right) \rightarrow \left(\begin{array}{ccccc|c} 1 & -1 & 1 & 0 & 2 & 3 \\ 0 & 1 & -1 & 0 & 0 & 4 \\ 0 & 2 & 0 & 1 & -1 & 3 \\ 0 & 1 & -1 & 0 & 0 & 4 \end{array}\right)$$

$$\rightarrow \left(\begin{array}{ccccc|c} 1 & -1 & 1 & 0 & 2 & 3 \\ 0 & 1 & -1 & 0 & 0 & 4 \\ 0 & 0 & 2 & 1 & -1 & -5 \\ 0 & 0 & 0 & 0 & 0 & 0 \end{array}\right) \rightarrow \left(\begin{array}{ccccc|c} 1 & -1 & 1 & 0 & 2 & 3 \\ 0 & 1 & -1 & 0 & 0 & 4 \\ 0 & 0 & 1 & \frac{1}{2} & -\frac{1}{2} & -\frac{5}{2} \\ 0 & 0 & 0 & 0 & 0 & 0 \end{array}\right).$$

This echelon form represents a consistent system since it has no row of the form $(0\ 0\ \ldots\ 0 \mid a)$ with $a \neq 0$. Furthermore, the rank (number of non-zero rows) is 3, so the general solution will involve $5 - 3 = 2$ free parameters.

The leading 1s are in columns 1, 2 and 3, so we shall express the general solution in terms of the 'free' unknowns x_4 and x_5.

The echelon matrix represents the system

$$\begin{aligned} x_1 - x_2 + x_3 + 2x_5 &= 3, \\ x_2 - x_3 &= 4, \\ x_3 + \frac{1}{2}x_4 - \frac{1}{2}x_5 &= -\frac{5}{2}. \end{aligned}$$

Let $x_4 = s$ and $x_5 = t$; then, by back-substitution,

$$\begin{aligned} x_3 &= -\frac{5}{2} - \frac{1}{2}s + \frac{1}{2}t, \\ x_2 &= 4 + x_3 = \frac{3}{2} - \frac{1}{2}s + \frac{1}{2}t, \\ x_1 &= 3 + x_2 - x_3 - 2x_5 = 3 + 4 - 2t = 7 - 2t. \end{aligned}$$

So the general solution is

$$\mathbf{x} = \begin{pmatrix} x_1 \\ x_2 \\ x_3 \\ x_4 \\ x_5 \end{pmatrix} = \begin{pmatrix} 7 - 2t \\ \frac{3}{2} - \frac{1}{2}s + \frac{1}{2}t \\ -\frac{5}{2} - \frac{1}{2}s + \frac{1}{2}t \\ s \\ t \end{pmatrix}.$$

This may be written as

$$\mathbf{x} = \mathbf{v} + s\mathbf{u}_1 + t\mathbf{u}_2,$$

where

$$\mathbf{v} = \begin{pmatrix} 7 \\ \frac{3}{2} \\ -\frac{5}{2} \\ 0 \\ 0 \end{pmatrix}, \quad \mathbf{u}_1 = \begin{pmatrix} 0 \\ -\frac{1}{2} \\ -\frac{1}{2} \\ 1 \\ 0 \end{pmatrix}, \quad \mathbf{u}_2 = \begin{pmatrix} -2 \\ \frac{1}{2} \\ \frac{1}{2} \\ 0 \\ 1 \end{pmatrix}.$$

■

17.4 Arbitrage Portfolios and State Prices

In Section 15.3 we explained how matrix notation can be used to study investment decisions. If there are m assets and n states, the $m \times n$ *returns matrix* R is defined so that the element r_{ij} is the return from investing in one unit of asset i, if state j occurs. A *portfolio* $Y = (y_1, y_2, \dots, y_m)$ is a row vector (or $1 \times m$ matrix), in which y_i is the number of units of asset i held by the investor. The return on a portfolio Y in the state j is $(YR)_j$, the jth component of YR. In particular, Y is an *arbitrage portfolio* if

$$y_1 + y_2 + \cdots + y_m = 0, \quad \text{and}$$

$$YR \text{ has no negative entries and at least one positive entry.}$$

We now introduce another idea, which turns out to be closely linked to the existence of arbitrage portfolios. The positive numbers $p_1, p_2, \dots, p_n$ are said to be *state prices* for R if the column vector

$$\mathbf{p} = \begin{pmatrix} p_1 \\ p_2 \\ \vdots \\ p_n \end{pmatrix} \quad \text{satisfies} \quad R\mathbf{p} = \mathbf{u}, \quad \text{where} \quad \mathbf{u} = \begin{pmatrix} 1 \\ 1 \\ \vdots \\ 1 \end{pmatrix}.$$

Of course, the system $R\mathbf{p} = \mathbf{u}$ may be inconsistent, in which case state prices for R do not exist.

It is helpful to think of p_j as a 'weight' assigned to state j. The weights are chosen so that the weighted return on one unit of asset i, that is

$$r_{i1}p_1 + r_{i2}p_2 + \cdots + r_{in}p_n = (R\mathbf{p})_i,$$

is equal to 1 for each asset i. For this reason it is often said that, when state prices do exist, the investor is taking part in a 'fair game'.

Example Suppose there are three assets and three states, and the returns matrix is

$$R = \begin{pmatrix} 1.00 & 1.00 & 1.20 \\ 1.05 & 1.05 & 1.05 \\ 0.95 & 1.10 & 0.95 \end{pmatrix}.$$

A vector of state prices is a solution of the system whose augmented matrix is

$$(R \mid \mathbf{u}) = \left(\begin{array}{ccc|c} 1.00 & 1.00 & 1.20 & 1 \\ 1.05 & 1.05 & 1.05 & 1 \\ 0.95 & 1.10 & 0.95 & 1 \end{array}\right).$$

Reducing to echelon form in the usual way, we find that the system is consistent and that there is a unique set of state prices, $p_1 = 5/63, p_2 = 40/63, p_3 = 5/21$.

We can use this result to show that there is no arbitrage portfolio for the given R. Suppose Y is a portfolio for which $(YR)_j$, the return on Y in state j, is non-negative for $j = 1, 2, 3$ and positive for at least one j. Then, since the state prices p_j are positive, the expression

$$YR\mathbf{p} = (YR)_1 p_1 + (YR)_2 p_2 + (YR)_3 p_3$$

must be positive. However, since the state prices satisfy $R\mathbf{p} = \mathbf{u}$, this expression reduces to $Y\mathbf{u}$, which equals $y_1 + y_2 + y_3$, the net cost of Y. So the net cost cannot be zero, and there is no arbitrage portfolio. ■

In the preceding example, the fact that we have a 'fair game' (state prices exist) implies that an investor cannot get 'something for nothing' (there is no arbitrage portfolio). Indeed, the argument given in the last paragraph shows that this result holds quite generally. Equally important is the fact that the converse result is also true: if state prices do not exist, then there must be an arbitrage portfolio. The proof of this fact is rather more difficult and we shall be content to give an example (Example 17.6).

Worked Examples

Example 17.1 Find the rank and general solution of the following system of linear equations:

$$\begin{aligned} 3x_1 + x_2 + x_3 &= 3, \\ x_1 - x_2 - x_3 &= 1, \\ x_1 + 2x_2 + 2x_3 &= 1. \end{aligned}$$

Express your answer in the form $\mathbf{x} = \mathbf{v} + s\mathbf{u}$, where $\mathbf{v}$ and $\mathbf{u}$ are 3-vectors.

Solution

The augmented matrix is

$$\left(\begin{array}{ccc|c} 3 & 1 & 1 & 3 \\ 1 & -1 & -1 & 1 \\ 1 & 2 & 2 & 1 \end{array}\right),$$

which we reduce as follows:

$$\left(\begin{array}{ccc|c} 3 & 1 & 1 & 3 \\ 1 & -1 & -1 & 1 \\ 1 & 2 & 2 & 1 \end{array}\right) \to \left(\begin{array}{ccc|c} 1 & -1 & -1 & 1 \\ 3 & 1 & 1 & 3 \\ 1 & 2 & 2 & 1 \end{array}\right) \to \left(\begin{array}{ccc|c} 1 & -1 & -1 & 1 \\ 0 & 4 & 4 & 0 \\ 0 & 3 & 3 & 0 \end{array}\right)$$

$$\to \left(\begin{array}{ccc|c} 1 & -1 & -1 & 1 \\ 0 & 1 & 1 & 0 \\ 0 & 1 & 1 & 0 \end{array}\right) \to \left(\begin{array}{ccc|c} 1 & -1 & -1 & 1 \\ 0 & 1 & 1 & 0 \\ 0 & 0 & 0 & 0 \end{array}\right).$$

The system is therefore equivalent to

$$\begin{aligned} x_1 - x_2 - x_3 &= 1, \\ x_2 + x_3 &= 0. \end{aligned}$$

Since there are two non-zero rows, the rank is 2.

To find the general solution, let $x_3 = s$. Then $x_2 = -s$ and

$$x_1 = 1 + x_2 + x_3 = 1.$$

The general solution is therefore

$$\mathbf{x} = \begin{pmatrix} x_1 \\ x_2 \\ x_3 \end{pmatrix} = \begin{pmatrix} 1 \\ -s \\ s \end{pmatrix} = \begin{pmatrix} 1 \\ 0 \\ 0 \end{pmatrix} + \begin{pmatrix} 0 \\ -s \\ s \end{pmatrix} = \begin{pmatrix} 1 \\ 0 \\ 0 \end{pmatrix} + s \begin{pmatrix} 0 \\ -1 \\ 1 \end{pmatrix},$$

which, taking

$$\mathbf{v} = \begin{pmatrix} 1 \\ 0 \\ 0 \end{pmatrix}, \quad \mathbf{u} = \begin{pmatrix} 0 \\ -1 \\ 1 \end{pmatrix},$$

is of the required form.

Example 17.2 Reduce the following system to echelon form, determine its rank and find the general solution in vector notation:

$$\begin{aligned} x_1 - x_2 + 2x_3 - x_4 &= -1, \\ 2x_1 + x_2 - 2x_3 - 2x_4 &= -2, \\ -x_1 + 2x_2 - 4x_3 + x_4 &= 1, \\ 3x_1 - 3x_4 &= -3. \end{aligned}$$

Solution

The augmented matrix is

$$\left(\begin{array}{cccc|c} 1 & -1 & 2 & -1 & -1 \\ 2 & 1 & -2 & -2 & -2 \\ -1 & 2 & -4 & 1 & 1 \\ 3 & 0 & 0 & -3 & -3 \end{array}\right).$$

Reducing this to echelon form by elementary row operations,

$$\left(\begin{array}{cccc|c} 1 & -1 & 2 & -1 & -1 \\ 2 & 1 & -2 & -2 & -2 \\ -1 & 2 & -4 & 1 & 1 \\ 3 & 0 & 0 & -3 & -3 \end{array}\right) \to \left(\begin{array}{cccc|c} 1 & -1 & 2 & -1 & -1 \\ 0 & 3 & -6 & 0 & 0 \\ 0 & 1 & -2 & 0 & 0 \\ 0 & 3 & -6 & 0 & 0 \end{array}\right)$$

$$\to \left(\begin{array}{cccc|c} 1 & -1 & 2 & -1 & -1 \\ 0 & 1 & -2 & 0 & 0 \\ 0 & 1 & -2 & 0 & 0 \\ 0 & 1 & -2 & 0 & 0 \end{array}\right) \to \left(\begin{array}{cccc|c} 1 & -1 & 2 & -1 & -1 \\ 0 & 1 & -2 & 0 & 0 \\ 0 & 0 & 0 & 0 & 0 \\ 0 & 0 & 0 & 0 & 0 \end{array}\right).$$

So, the system is equivalent to

$$\begin{aligned} x_1 - x_2 + 2x_3 - x_4 &= -1, \\ x_2 - 2x_3 &= 0. \end{aligned}$$

The rank is 2. We let $x_3 = s$ and $x_4 = t$. Then $x_2 = 2x_3 = 2s$ and $x_1 = -1 + x_2 - 2x_3 + x_4 = t - 1$. In vector notation, the general solution is

$$\mathbf{x} = \begin{pmatrix} x_1 \\ x_2 \\ x_3 \\ x_4 \end{pmatrix} = \begin{pmatrix} t-1 \\ 2s \\ s \\ t \end{pmatrix} = \begin{pmatrix} -1 \\ 0 \\ 0 \\ 0 \end{pmatrix} + \begin{pmatrix} 0 \\ 2s \\ s \\ 0 \end{pmatrix} + \begin{pmatrix} t \\ 0 \\ 0 \\ t \end{pmatrix} = \mathbf{v} + s\mathbf{u}_1 + t\mathbf{u}_2,$$

where

$$\mathbf{v} = \begin{pmatrix} -1 \\ 0 \\ 0 \\ 0 \end{pmatrix}, \ \mathbf{u}_1 = \begin{pmatrix} 0 \\ 2 \\ 1 \\ 0 \end{pmatrix}, \ \mathbf{u}_2 = \begin{pmatrix} 1 \\ 0 \\ 0 \\ 1 \end{pmatrix}.$$

Example 17.3 Investigate the set of solutions of the following system:

$$\begin{aligned} x_1 + 2x_2 + 4x_3 &= 7, \\ x_1 + 2x_3 &= -2, \\ 2x_1 + 3x_2 + 7x_3 &= 9. \end{aligned}$$

Solution

The augmented matrix is

$$\left(\begin{array}{ccc|c} 1 & 2 & 4 & 7 \\ 1 & 0 & 2 & -2 \\ 2 & 3 & 7 & 9 \end{array}\right).$$

Performing row operations,

$$\left(\begin{array}{ccc|c} 1 & 2 & 4 & 7 \\ 1 & 0 & 2 & -2 \\ 2 & 3 & 7 & 9 \end{array}\right) \to \left(\begin{array}{ccc|c} 1 & 2 & 4 & 7 \\ 0 & -2 & -2 & -9 \\ 0 & -1 & -1 & -5 \end{array}\right) \to \left(\begin{array}{ccc|c} 1 & 2 & 4 & 7 \\ 0 & 1 & 1 & \frac{9}{2} \\ 0 & -1 & -1 & -5 \end{array}\right) \to \left(\begin{array}{ccc|c} 1 & 2 & 4 & 7 \\ 0 & 1 & 1 & \frac{9}{2} \\ 0 & 0 & 0 & -\frac{1}{2} \end{array}\right).$$

This echelon matrix has a row of the form $(0\ 0\ 0 \mid a)$, where $a \neq 0$, so the system is inconsistent and there are no solutions.

Example 17.4 For which value of c is the following system of equations consistent? Find all the solutions when c has this value.

$$\begin{aligned} x_1 + x_2 + x_3 &= 2, \\ 2x_1 + x_2 + 2x_3 &= 5, \\ 4x_1 + 3x_2 + 4x_3 &= c. \end{aligned}$$

Solution

The augmented matrix is

$$\left(\begin{array}{ccc|c} 1 & 1 & 1 & 2 \\ 2 & 1 & 2 & 5 \\ 4 & 3 & 4 & c \end{array}\right).$$

Performing row operations to reduce this to echelon form,

$$\left(\begin{array}{ccc|c} 1 & 1 & 1 & 2 \\ 2 & 1 & 2 & 5 \\ 4 & 3 & 4 & c \end{array}\right) \to \left(\begin{array}{ccc|c} 1 & 1 & 1 & 2 \\ 0 & -1 & 0 & 1 \\ 0 & -1 & 0 & c-8 \end{array}\right) \to \left(\begin{array}{ccc|c} 1 & 1 & 1 & 2 \\ 0 & 1 & 0 & -1 \\ 0 & 1 & 0 & 8-c \end{array}\right) \to \left(\begin{array}{ccc|c} 1 & 1 & 1 & 2 \\ 0 & 1 & 0 & -1 \\ 0 & 0 & 0 & 9-c \end{array}\right).$$

Looking at the bottom row of the last matrix, we see that the system of equations is inconsistent unless $9 - c = 0$; that is, it is consistent only when $c = 9$.

When $c = 9$, the system is equivalent to

$$\begin{aligned} x_1 + x_2 + x_3 &= 2, \\ x_2 &= -1, \end{aligned}$$

which has rank 2. Setting $x_3 = s$, we have $x_2 = -1$ and

$$x_1 = 2 - x_2 - x_3 = 3 - s.$$

The general solution when $c = 9$ is therefore $\mathbf{x} = \mathbf{v} + s\mathbf{u}$, where

$$\mathbf{v} = \begin{pmatrix} 3 \\ -1 \\ 0 \end{pmatrix}, \quad \mathbf{u} = \begin{pmatrix} -1 \\ 0 \\ 1 \end{pmatrix}.$$

Example 17.5 A firm manufactures three different types of chocolate bar, 'Abracadabra', 'Break' and 'Choca-mocha'. The main ingredients in each are cocoa, milk and coffee. To produce 1000 Abracadabra bars requires 5 units of cocoa, 3 units of milk and 2 units of coffee. To produce 1000 Break bars requires 5 units of cocoa, 4 of milk and 1 of coffee, and the production of 1000 Choca-mocha bars requires 5 units of cocoa, 2 of milk and 3 of coffee. The firm has supplies of 250 units of cocoa, 150 of milk and 100 of coffee each week (and as much as it wants of the other ingredients, such as sugar). Show that if the firm uses up its supply of cocoa, milk and coffee, then the number of Break bars produced each week equals the number of Choca-mocha bars produced. How does the number of Abracadabra bars produced relate to the production level of the other two bars? Find the maximum possible weekly production of Choca-mocha bars.

Solution

Let us denote the weekly production levels of Abracadabra, Break and Choca-mocha bars by a, b, c (respectively), measured in thousands of bars. Then, since 5 units of cocoa are needed to produce 1000 bars of each type, and since 250 units of cocoa are used each week, we must have

$$5a + 5b + 5c = 250.$$

By considering the distribution of the 150 available units of milk,

$$3a + 4b + 2c = 150.$$

Similarly, since 100 units of coffee are used, it must be the case that

$$2a + b + 3c = 100.$$

In other words, we have the system of equations

$$\begin{aligned} 5a + 5b + 5c &= 250, \\ 3a + 4b + 2c &= 150, \\ 2a + b + 3c &= 100. \end{aligned}$$

We solve this using elementary row operations, starting with the augmented matrix, as follows:

$$\left(\begin{array}{ccc|c} 5 & 5 & 5 & 250 \\ 3 & 4 & 2 & 150 \\ 2 & 1 & 3 & 100 \end{array}\right) \rightarrow \left(\begin{array}{ccc|c} 1 & 1 & 1 & 50 \\ 3 & 4 & 2 & 150 \\ 2 & 1 & 3 & 100 \end{array}\right)$$
$$\rightarrow \left(\begin{array}{ccc|c} 1 & 1 & 1 & 50 \\ 0 & 1 & -1 & 0 \\ 0 & -1 & 1 & 0 \end{array}\right)$$
$$\rightarrow \left(\begin{array}{ccc|c} 1 & 1 & 1 & 50 \\ 0 & 1 & -1 & 0 \\ 0 & 0 & 0 & 0 \end{array}\right).$$

Therefore, the system is equivalent to

$$\begin{aligned} a + b + c &= 50, \\ b - c &= 0, \end{aligned}$$

from which we obtain $b = c$ and $a = 50 - b - c = 50 - 2c$.

In other words, the weekly production levels of Break and Choca-mocha bars are equal, and the production of Abracadabra is (in thousands) $50 - 2c$, where c is the production of Choca-mocha (and Break) bars. Clearly, none of a, b, c can be negative, so the production level, c, of Choca-mocha bars must be such that $a = 50 - 2c \geq 0$; that is $c \leq 25$. Therefore, the maximum number of Choca-mocha bars which it is possible to manufacture in a week is 25 000 (in which case the same number of Break bars are produced and no Abracadabra bars will be manufactured).

Example 17.6 Suppose that an investor invests her money in three different assets and that three possible states can occur. Show that if the returns matrix is

$$R = \begin{pmatrix} 0.95 & 0.9 & 1.0 \\ 1.1 & 1.1 & 1.1 \\ 1.2 & 1.15 & 1.25 \end{pmatrix}$$

then there is no vector of state prices. Show also that

$$Y = \begin{pmatrix} 1000 & -4000 & 5000 \end{pmatrix} \quad \text{and} \quad Z = \begin{pmatrix} 0 & -5000 & 5000 \end{pmatrix}$$

are arbitrage portfolios. Which of the two would you choose, given the choice?

Solution

Recall that $\mathbf{p}$ is a state-price vector for R if

$$\mathbf{p} = \begin{pmatrix} p_1 \\ p_2 \\ p_3 \end{pmatrix} \quad \text{and} \quad R\mathbf{p} = \begin{pmatrix} 1 \\ 1 \\ 1 \end{pmatrix},$$

where p_1, p_2, p_3 are positive. Then (as in Section 17.4), the augmented matrix for the linear system which determines $\mathbf{p}$ is

$$\left(\begin{array}{ccc|c} 0.95 & 0.9 & 1.0 & 1 \\ 1.1 & 1.1 & 1.1 & 1 \\ 1.2 & 1.15 & 1.25 & 1 \end{array}\right).$$

Using elementary row operations, we proceed as follows:

$$\left(\begin{array}{ccc|c} 0.95 & 0.9 & 1 & 1 \\ 1.1 & 1.1 & 1.1 & 1 \\ 1.2 & 1.15 & 1.25 & 1 \end{array}\right) \to \left(\begin{array}{ccc|c} 1 & 1 & 1 & \frac{10}{11} \\ 0.95 & 0.9 & 1 & 1 \\ 1.2 & 1.15 & 1.25 & 1 \end{array}\right)$$

$$\to \left(\begin{array}{ccc|c} 1 & 1 & 1 & \frac{10}{11} \\ 0 & -0.05 & -0.05 & \frac{3}{22} \\ 0 & -0.05 & 0.05 & -\frac{1}{11} \end{array}\right)$$

$$\to \left(\begin{array}{ccc|c} 1 & 1 & 1 & \frac{10}{11} \\ 0 & 1 & -1 & -\frac{30}{11} \\ 0 & 0 & 0 & -\frac{5}{22} \end{array}\right).$$

The form of the bottom row implies that there are no solutions to the system, and hence there is no state-price vector.

To check that Y and Z as given are arbitrage portfolios, we need to verify that each costs nothing, cannot lose and in at least one of the states has a positive return. It is clear that each of the portfolios costs nothing, since the sum of the three entries in each is zero. The product YR is

$$YR = \begin{pmatrix} 1000 & -5000 & 4000 \end{pmatrix} \begin{pmatrix} 0.95 & 0.9 & 1.0 \\ 1.1 & 1.1 & 1.1 \\ 1.2 & 1.15 & 1.25 \end{pmatrix} = \begin{pmatrix} 250 & 0 & 500 \end{pmatrix}.$$

This column vector has no negative entries and two positive entries. In other words, in none of the possible states does the value of the portfolio diminish and in two of them its value increases. Therefore, Y is an arbitrage portfolio. The returns from portfolio Z are described by the product

$$ZR = \begin{pmatrix} 500 & 250 & 750 \end{pmatrix}.$$

Thus an investor with portfolio Z experiences an increase in the value of her portfolio no matter which state results. Furthermore, since each entry of the column ZR is greater than the corresponding entry in YR, an investor with portfolio Z makes a greater return than one with portfolio Y, regardless of which of the three states occurs. In this sense, Z is 'better' than Y.

This example illustrates the general result that if there are no state prices – that is, investment is not a 'fair game' – then arbitrage portfolios do exist.

Main Topics

- inconsistent systems and systems with infinitely many solutions
- echelon matrix and the rank of a system
- general solution of consistent system, involving free parameters
- general solution of consistent system in vector notation
- state prices and arbitrage portfolios

Key Terms, Notations and Formulae

- consistent and inconsistent systems
- rank, r
- if echelon matrix has row $(0\ 0\ \dots\ 0 \mid a)$, $a \neq 0$, system is inconsistent
- if consistent system has rank r, general solution has $n - r$ parameters
- general solution to a consistent system of rank r, in vector notation:

$$\mathbf{x} = \mathbf{v} + s_1\mathbf{u}_1 + s_2\mathbf{u}_2 + \cdots + s_{n-r}\mathbf{u}_{n-r}$$

- returns matrix, R; arbitrage portfolio
- state-price vector

$$\mathbf{p} = \begin{pmatrix} p_1 \\ p_2 \\ \vdots \\ p_n \end{pmatrix},$$

such that $p_1, p_2, \dots, p_n > 0$ and

$$R\mathbf{p} = \begin{pmatrix} 1 \\ 1 \\ \vdots \\ 1 \end{pmatrix}.$$

Exercises

17.1 Find the general solution of the following system of linear equations:

$$\begin{aligned} x_2 + 2x_3 &= 7, \\ 3x_1 - x_2 - x_3 &= 5, \\ 6x_1 + 2x_2 + 6x_3 &= 38. \end{aligned}$$

Express your answer in the form $\mathbf{x} = \mathbf{v} + s\mathbf{u}$, where $\mathbf{v}$ and $\mathbf{u}$ are 3-vectors.

17.2 Find the values of the numbers a and b if the following system of equations has infinitely many solutions:

$$\begin{aligned} x + y + z + w &= 4, \\ x + 3y - z + 5w &= 8, \\ 2x - y + 3z - w &= 3, \\ 2x - 3y + az - 5w &= b. \end{aligned}$$

Find also, when a and b take these values, all the solutions of the system.

17.3 Suppose that x, y, z satisfy the equations

$$\begin{aligned} x + y + z &= 4 + a, \\ 2x + y &= 3 + a, \\ x - y - 2z &= -4, \end{aligned}$$

where a is some number.

Use row operations to determine the values of x, y and z, in terms of a.

For what values of a will x, y, z all be positive?

17.4 Suppose that an investor invests her money in three different assets and that three possible states can occur. Show that if the returns matrix is

$$R = \begin{pmatrix} 1.05 & 0.95 & 1.0 \\ 1.05 & 1.05 & 1.05 \\ 1.2 & 1.26 & 1.23 \end{pmatrix},$$

then there is no vector of state prices.

An investor obtains a loan of \$2000 from the bank. Bank interest is given by row 2 of R. Using this loan, find, if possible, an arbitrage portfolio which guarantees a return of at least \$200 in each state.

Problems

17.1 What is the rank of the system discussed in the first example of Section 17.1? Write down the general solution in vector notation.

17.2 Find the general solution of the following system of linear equations:

$$\begin{aligned} x_1 - x_2 + 5x_3 &= 80, \\ 3x_1 + 2x_2 - x_3 &= 25, \\ 5x_1 + 9x_3 &= 185. \end{aligned}$$

Express your answer in the form $\mathbf{x} = \mathbf{v} + s\mathbf{u}$, where $\mathbf{v}$ and $\mathbf{u}$ are 3-vectors.

17.3 Reduce the following system to echelon form, determine its rank and express the general solution in vector notation:

$$\begin{aligned} 5x_1 + 3x_2 + 2x_3 + x_4 &= 66, \\ x_1 + 3x_2 + 4x_3 + 2x_4 &= 48, \\ -x_1 + x_2 + 2x_3 + x_4 &= 10, \\ 2x_1 + 2x_2 + 2x_3 + x_4 &= 38. \end{aligned}$$

17.4 Use elementary row operations to show that the following system of equations is inconsistent:

$$\begin{aligned} -x_1 + 2x_2 + 5x_3 &= 3, \\ 2x_1 + 3x_2 - x_3 &= 17, \\ x_1 + 5x_2 + 4x_3 &= 10, \\ x_2 - x_3 &= 5. \end{aligned}$$

17.5 Determine all solutions (if any) of the following system of equations:

$$\begin{aligned} 2x_1 + 10x_2 + 9x_3 + 14x_4 &= 14, \\ x_1 + 3x_2 - 2x_3 + x_4 &= 1, \\ x_1 + 5x_2 + 3x_3 + 7x_4 &= 8, \\ x_1 + 7x_2 + 5x_3 + 13x_4 &= 12. \end{aligned}$$

17.6 For what value of k is the following system of equations consistent?

$$\begin{aligned} x_1 + x_2 + x_3 &= 3, \\ -x_1 + x_2 + 2x_3 &= 2, \\ 2x_1 - 3x_2 + x_3 &= 0, \\ x_2 - 4x_3 &= k. \end{aligned}$$

Determine all solutions when k has this value.

17.7 Suppose that the following system of equations has a solution, where a, b, c are constants:

$$\begin{aligned} x_1 + x_2 + x_3 &= a, \\ 2x_1 + 3x_2 + 2x_3 &= b, \\ 4x_1 + 3x_3 &= 15a, \\ x_1 - 3x_2 + 4x_3 &= c. \end{aligned}$$

Show that it must be the case that $c = -16b$. When $c = -16b$, show that there is just one solution, regardless of the values of a and b. Find this solution in terms of a and b.

17.8 Suppose an investor invests her money in three different assets – land, bonds and stocks – and that three possible states can occur. Show that there is no state-price vector for the returns matrix

$$R = \begin{pmatrix} 1.05 & 1.20 & 1.10 \\ 1.05 & 1.05 & 1.05 \\ 0.90 & 1.05 & 0.95 \end{pmatrix}.$$

Find an arbitrage portfolio.

18 Inverse Matrices

18.1 The Square Linear System

In this chapter we study the 'square' linear system, in which the number of equations is equal to the number of unknowns. In this case it is possible that there is a *unique* solution, and we shall concentrate on establishing conditions for this to be so.

We use the matrix notation $A\mathbf{x} = \mathbf{b}$ for the system under consideration. So A is a square matrix (size $n \times n$), and $\mathbf{x}$ and $\mathbf{b}$ are column vectors (size $n \times 1$). As usual we presume that A and $\mathbf{b}$ are given, and the 'solution' $\mathbf{x}$ is required.

Consider what can happen when we solve the system $A\mathbf{x} = \mathbf{b}$ using the method of elementary row operations. It is possible that we reduce the augmented matrix $(A \mid \mathbf{b})$ to the form

$$\left(\begin{array}{ccccc|c} 1 & * & * & \cdots & * & * \\ 0 & 1 & * & \cdots & * & * \\ 0 & 0 & 1 & \cdots & * & * \\ \vdots & \vdots & \vdots & \ddots & \vdots & \vdots \\ 0 & 0 & 0 & \cdots & 1 & * \end{array}\right).$$

Here the leading 1s are placed on the main 'diagonal', which would not always be the case. As noted in Section 16.4, this is precisely the situation in which the method of back-substitution will yield a unique solution.

Example If the system reduces to the echelon form

$$\left(\begin{array}{ccc|c} 1 & 4 & 3 & 6 \\ 0 & 1 & 2 & 5 \\ 0 & 0 & 1 & 2 \end{array}\right),$$

then we get the equations

$$x_3 = 2,\ x_2 = 5 - 2x_3,\ x_1 = 6 - 4x_2 - 3x_3,$$

and back-substitution gives the solution $x_3 = 2, x_2 = 1, x_1 = -4$.

For future reference we note that an alternative method is to continue using row operations to get 0s above as well as below the leading 1s. In other words, we reduce to the form $(I \mid \mathbf{c})$, where I is the identity matrix:

$$\left(\begin{array}{ccc|c} 1 & 4 & 3 & 6 \\ 0 & 1 & 2 & 5 \\ 0 & 0 & 1 & 2 \end{array}\right) \to \left(\begin{array}{ccc|c} 1 & 4 & 3 & 6 \\ 0 & 1 & 0 & 1 \\ 0 & 0 & 1 & 2 \end{array}\right) \to \left(\begin{array}{ccc|c} 1 & 0 & 3 & 2 \\ 0 & 1 & 0 & 1 \\ 0 & 0 & 1 & 2 \end{array}\right) \to \left(\begin{array}{ccc|c} 1 & 0 & 0 & -4 \\ 0 & 1 & 0 & 1 \\ 0 & 0 & 1 & 2 \end{array}\right).$$

In this form the equations are

$$x_1 = -4,\ x_2 = 1,\ x_3 = 2,$$

and this is precisely the solution we are seeking. What we have done is to transform the augmented matrix $(A \mid \mathbf{b})$ to the form $(I \mid \mathbf{c})$, giving the solution to $A\mathbf{x} = \mathbf{b}$ explicitly as $\mathbf{x} = \mathbf{c}$. ■

It is important to notice that, when we reduce the augmented matrix $(A \mid \mathbf{b})$, what happens to A does not depend on $\mathbf{b}$. Thus, if there is a unique solution of $A\mathbf{x} = \mathbf{b}$ for one $\mathbf{b}$, then there is a unique solution for every $\mathbf{b}$. In particular, there is a unique solution to the system $A\mathbf{x} = \mathbf{0}$. Since it is obvious that $\mathbf{x} = \mathbf{0}$ is a solution to $A\mathbf{x} = \mathbf{0}$, this is the only solution.

To summarise: $A\mathbf{x} = \mathbf{b}$ has a unique solution for one $\mathbf{b}$ if and only if it has a unique solution for all $\mathbf{b}$; and equivalently, if and only if $A\mathbf{x} = \mathbf{0}$ has the unique solution $\mathbf{x} = \mathbf{0}$.

18.2 The Inverse of a Square Matrix

We say that the square matrix A has an *inverse* Z if $AZ = ZA = I$. There are two points which must be stressed at once:

- if a matrix has an inverse, it has only one;
- a matrix may have no inverse.

The first point can be established by a simple argument. Suppose Y and Z are both inverses for a given matrix A. Then we have the equations

$$AY = YA = I \quad \text{and} \quad AZ = ZA = I.$$

Using these relations, and the elementary properties of matrix multiplication, we have

$$Y = YI = Y(AZ) = (YA)Z = IZ = Z.$$

We have shown that the two supposed inverses, Y and Z, are in fact equal. So there is only one inverse and we are justified in speaking of *the* inverse of A.

A useful fact, but one which is rather difficult to prove by elementary means, is that, in order to be sure that the square matrix Z is the inverse of A, one need only verify *either* that AZ is the identity matrix *or* that ZA is the identity matrix. That is, there is no need to check that *both* matrix products are the identity matrix.

It is easy to find examples of matrices which have no inverse.

Example Any 2×2 matrix whose first row is $(0 \quad 0)$ does not have an inverse. To see this, let A be any such matrix, and suppose Z were its inverse. Then we should have $AZ = I$, that is

$$\begin{pmatrix} 0 & 0 \\ a_{21} & a_{22} \end{pmatrix} \begin{pmatrix} z_{11} & z_{12} \\ z_{21} & z_{22} \end{pmatrix} = \begin{pmatrix} 1 & 0 \\ 0 & 1 \end{pmatrix}.$$

Working out the dot product of the first row of A with the first column of Z, we get $0z_{11} + 0z_{21} = 1$. This cannot be true for any values of z_{11} and z_{21}. Hence there is no inverse Z. ■

When A has an inverse we denote it by A^{-1}. We have

$$AA^{-1} = A^{-1}A = I,$$

and we say that A is *invertible*. (The term 'non-singular' is also used.)

There is a fundamental link between the existence of an inverse matrix and the unique solution of a system of equations.

- *The square matrix A is invertible if and only if, for any $\mathbf{b}$, the system $A\mathbf{x} = \mathbf{b}$ has a unique solution.*

This assertion has two parts: the 'if' part and the 'only if' part. The 'if' part is investigated in the next section, but the 'only if' part is easy. If A is invertible then it has inverse A^{-1}. Multiplying the equation $A\mathbf{x} = \mathbf{b}$ on the left by A^{-1}, we get

$$A^{-1}A\mathbf{x} = A^{-1}\mathbf{b}, \quad \text{that is} \quad \mathbf{x} = A^{-1}\mathbf{b}.$$

In other words, the system $A\mathbf{x} = \mathbf{b}$ has the unique solution $\mathbf{x} = A^{-1}\mathbf{b}$. This means that if we know A^{-1}, then we can find the solution for any given $\mathbf{b}$.

Example Verify that, if

$$A = \begin{pmatrix} 1 & 4 & 11 \\ 2 & 8 & 16 \\ 1 & 6 & 17 \end{pmatrix}, \quad \text{then} \quad A^{-1} = \frac{1}{6}\begin{pmatrix} 20 & -1 & -12 \\ -9 & 3 & 3 \\ 2 & -1 & 0 \end{pmatrix},$$

and hence write down the solution to the system

$$\begin{aligned} x_1 + 4x_2 + 11x_3 &= b_1, \\ 2x_1 + 8x_2 + 16x_3 &= b_2, \\ x_1 + 6x_2 + 17x_3 &= b_3. \end{aligned}$$

The fact that A^{-1} is the inverse of A can be checked by working out the products AA^{-1} and $A^{-1}A$: both turn out to be the identity matrix. (As mentioned above, it would suffice to check that this is the case for either one of these two products.)

Now, the solution to the given system $A\mathbf{x} = \mathbf{b}$ is $\mathbf{x} = A^{-1}\mathbf{b}$, that is explicitly

$$x_1 = \frac{1}{6}(20b_1 - b_2 - 12b_3),$$
$$x_2 = \frac{1}{6}(-9b_1 + 3b_2 + 3b_3),$$
$$x_3 = \frac{1}{6}(2b_1 - b_2).$$

■

18.3 Calculation of the Inverse

In this section we justify the first part of the claim made above: if $A\mathbf{x} = \mathbf{b}$ has a unique solution, then A^{-1} exists. The argument will also show us how to calculate A^{-1}.

Denote by $\mathbf{e}_i$ the column vector with 0s in every component except the ith, which is 1. Note that $\mathbf{e}_i$ is the ith column of the identity matrix I. We claim that the way to find A^{-1} is to solve the n systems $A\mathbf{x} = \mathbf{e}_i$, for $i = 1, 2, \ldots, n$. To see this, suppose the solution to $A\mathbf{x} = \mathbf{e}_i$ is $\mathbf{x} = \mathbf{z}_i$, and let Z be the matrix whose columns are $\mathbf{z}_1, \mathbf{z}_2, \ldots, \mathbf{z}_n$. We write

$$Z = (\mathbf{z}_1 \mid \mathbf{z}_2 \mid \ldots \mid \mathbf{z}_n).$$

Then we have

$$AZ = A(\mathbf{z}_1 \mid \mathbf{z}_2 \mid \ldots \mid \mathbf{z}_n) = (A\mathbf{z}_1 \mid A\mathbf{z}_2 \mid \ldots \mid A\mathbf{z}_n) = (\mathbf{e}_1 \mid \mathbf{e}_2 \mid \ldots \mid \mathbf{e}_n),$$

which is just another way of writing $AZ = I$. In fact, it always turns out that $ZA = I$ also. Hence Z is the required inverse A^{-1}.

The importance of the preceding argument is that it provides us with a practical method of calculating A^{-1}, using elementary row operations. At first sight it looks as if we may have to solve each of the n systems $A\mathbf{x} = \mathbf{e}_i$ separately, but this is not so, since the same sequence of row operations is used in each case. In fact we can set out the calculation in a very simple way.

Example In order to calculate the inverse of

$$A = \begin{pmatrix} -2 & 1 & 3 \\ 0 & -1 & 1 \\ 1 & 2 & 0 \end{pmatrix},$$

we set out the given matrix and each of the 'right-hand sides' $\mathbf{e}_1, \mathbf{e}_2, \mathbf{e}_3$ as follows:

$$\left(\begin{array}{ccc|ccc} -2 & 1 & 3 & 1 & 0 & 0 \\ 0 & -1 & 1 & 0 & 1 & 0 \\ 1 & 2 & 0 & 0 & 0 & 1 \end{array}\right).$$

Then we carry out row operations as usual, operating on all the right-hand sides simultaneously:

$$\left(\begin{array}{ccc|ccc} -2 & 1 & 3 & 1 & 0 & 0 \\ 0 & -1 & 1 & 0 & 1 & 0 \\ 1 & 2 & 0 & 0 & 0 & 1 \end{array}\right) \to \left(\begin{array}{ccc|ccc} -2 & 1 & 3 & 1 & 0 & 0 \\ 0 & -1 & 1 & 0 & 1 & 0 \\ 0 & \frac{5}{2} & \frac{3}{2} & \frac{1}{2} & 0 & 1 \end{array}\right)$$

$$\to \left(\begin{array}{ccc|ccc} -2 & 1 & 3 & 1 & 0 & 0 \\ 0 & -1 & 1 & 0 & 1 & 0 \\ 0 & 0 & 4 & \frac{1}{2} & \frac{5}{2} & 1 \end{array}\right)$$

$$\to \left(\begin{array}{ccc|ccc} 1 & -\frac{1}{2} & -\frac{3}{2} & -\frac{1}{2} & 0 & 0 \\ 0 & 1 & -1 & 0 & -1 & 0 \\ 0 & 0 & 1 & \frac{1}{8} & \frac{5}{8} & \frac{1}{4} \end{array}\right)$$

$$\to \left(\begin{array}{ccc|ccc} 1 & -\frac{1}{2} & -\frac{3}{2} & -\frac{1}{2} & 0 & 0 \\ 0 & 1 & 0 & \frac{1}{8} & -\frac{3}{8} & \frac{1}{4} \\ 0 & 0 & 1 & \frac{1}{8} & \frac{5}{8} & \frac{1}{4} \end{array}\right)$$

$$\to \left(\begin{array}{ccc|ccc} 1 & 0 & 0 & -\frac{1}{4} & \frac{3}{4} & \frac{1}{2} \\ 0 & 1 & 0 & \frac{1}{8} & -\frac{3}{8} & \frac{1}{4} \\ 0 & 0 & 1 & \frac{1}{8} & \frac{5}{8} & \frac{1}{4} \end{array}\right).$$

As we noted in Section 18.1, the columns on the right are the solutions $\mathbf{z}_i$ to the systems $A\mathbf{x} = \mathbf{e}_i$, that is the columns of $Z = A^{-1}$. In other words, we have transformed $(A \mid I)$ into $(I \mid A^{-1})$. We conclude that

$$A^{-1} = \frac{1}{8}\begin{pmatrix} -2 & 6 & 4 \\ 1 & -3 & 2 \\ 1 & 5 & 2 \end{pmatrix}.$$

■

As in the example, the general procedure for finding the inverse of an invertible matrix A is to start with the $n \times 2n$ matrix $(A \mid I)$ and transform it by elementary row operations to the form $(I \mid B)$. Then the matrix B is the inverse of A.

Not only is this a simple method of calculating A^{-1}, when it exists, it is also a *test* to determine whether or not A has an inverse. Given any square matrix A, we can apply the procedure. If at any stage we create a row of the form

$$(0\ 0\ \cdots\ 0 \mid *\ *\ \cdots\ *),$$

then A is *not* invertible. Otherwise we transform $(A \mid I)$ to the form $(I \mid B)$, in which case A is invertible and $B = A^{-1}$.

Example We use elementary row operations to show that the matrix

$$A = \begin{pmatrix} 1 & 1 & 5 \\ 2 & 3 & 1 \\ 1 & 0 & 14 \end{pmatrix}$$

is not invertible. Starting with the 3×6 matrix

$$(A \mid I) = \left(\begin{array}{ccc|ccc} 1 & 1 & 5 & 1 & 0 & 0 \\ 2 & 3 & 1 & 0 & 1 & 0 \\ 1 & 0 & 14 & 0 & 0 & 1 \end{array}\right),$$

we reduce this as follows:

$$\left(\begin{array}{ccc|ccc} 1 & 1 & 5 & 1 & 0 & 0 \\ 2 & 3 & 1 & 0 & 1 & 0 \\ 1 & 0 & 14 & 0 & 0 & 1 \end{array}\right) \to \left(\begin{array}{ccc|ccc} 1 & 1 & 5 & 1 & 0 & 0 \\ 0 & 1 & -9 & -2 & 1 & 0 \\ 0 & -1 & 9 & -1 & 0 & 1 \end{array}\right)$$

$$\to \left(\begin{array}{ccc|ccc} 1 & 1 & 5 & 1 & 0 & 0 \\ 0 & 1 & -9 & -2 & 1 & 0 \\ 0 & 0 & 0 & -3 & 1 & 1 \end{array}\right).$$

There is no need to proceed further. Since this last matrix has a row whose first half is all zero, we deduce that A is not invertible. ■

18.4 The Inverse of a 2 × 2 Matrix

Some calculations are used so often that it is useful to remember the result. The formula for the inverse of a general 2×2 matrix is a case in point.

Let

$$A = \begin{pmatrix} a & b \\ c & d \end{pmatrix}.$$

Then we claim that

$$A^{-1} = \frac{1}{ad - bc} \begin{pmatrix} d & -b \\ -c & a \end{pmatrix}, \quad \text{provided that} \quad ad - bc \neq 0.$$

In order to check our claim, we have simply to verify explicitly that $AA^{-1} = I$ and $A^{-1}A = I$, which is easily done.

Note that in A^{-1} the 'diagonal' terms a and d are switched, and the 'off-diagonal' terms b and c have their sign changed; finally everything is divided by $ad - bc$. If $ad - bc = 0$, then A is not invertible (see Chapter 20 for more about this).

Example In Section 1.3 we had to find the market equilibrium when the supply and demand sets are given by

$$2q - 15p = -20 \quad \text{and} \quad q + 5p = 40.$$

We can do this using the rule for inverting a 2×2 matrix, as follows. In matrix form the equations are

$$\begin{pmatrix} 2 & -15 \\ 1 & 5 \end{pmatrix} \begin{pmatrix} q \\ p \end{pmatrix} = \begin{pmatrix} -20 \\ 40 \end{pmatrix}.$$

So the solution is given by

$$\begin{pmatrix} q^* \\ p^* \end{pmatrix} = \begin{pmatrix} 2 & -15 \\ 1 & 5 \end{pmatrix}^{-1} \begin{pmatrix} -20 \\ 40 \end{pmatrix}$$
$$= \frac{1}{25} \begin{pmatrix} 5 & 15 \\ -1 & 2 \end{pmatrix} \begin{pmatrix} -20 \\ 40 \end{pmatrix}$$
$$= \begin{pmatrix} 20 \\ 4 \end{pmatrix},$$

which agrees with the result found in Section 1.3. ∎

18.5 IS–LM Analysis

IS–LM analysis is the part of elementary macroeconomics which studies the equilibrium values of the national income Y and the interest rate r, in terms of certain *policy parameters* and *behavioural parameters*. IS stands for 'investment savings' and LM for 'liquidity money'. We shall present the analysis as a problem of inverting a 2×2 matrix.

First we consider equilibrium in the 'goods sector'. This can be represented by the equation $Y = C + I + G$, which simply says that the national income Y is equal to the national expenditure. Expenditure can be split into consumption C, investment I and government spending G, the last being considered as a policy parameter. Behavioural parameters enter when we consider how consumption is related to the national income, and how investment is affected by the interest rate. In elementary work it is usual to assume that both relationships are linear, that is

$$C = C_0 + bY \quad \text{and} \quad I = I_0 - ar \quad (a, b, C_0, I_0 \text{ positive constants}).$$

Thus the equilibrium condition becomes

$$Y = (C_0 + bY) + (I_0 - ar) + G, \quad \text{or} \quad (1 - b)Y + ar = C_0 + I_0 + G.$$

Equilibrium in the 'money sector' is represented by the statement that the supply of money is equal to the demand for money: $M_s = M_d$. The money supply M_s is assumed to be given; it is another policy parameter. The money demand M_d is assumed to depend on Y and r, and again we assume a linear relationship:

$$M_d = M_0 + fY - gr \quad (f, g, M_0 \text{ positive constants}).$$

Thus the equilibrium condition becomes

$$M_s = M_0 + fY - gr, \quad \text{or} \quad fY - gr = M_s - M_0.$$

The two equilibrium conditions can be written as a system of two linear equations in the two unknowns Y and r:

$$fY - gr = M_s - M_0,$$
$$(1 - b)Y + ar = C_0 + I_0 + G.$$

In matrix terms this system is

$$\begin{pmatrix} f & -g \\ 1-b & a \end{pmatrix} \begin{pmatrix} Y \\ r \end{pmatrix} = \begin{pmatrix} M_s - M_0 \\ C_0 + I_0 + G \end{pmatrix}$$

and the solution is

$$\begin{pmatrix} Y^* \\ r^* \end{pmatrix} = \begin{pmatrix} f & -g \\ 1-b & a \end{pmatrix}^{-1} \begin{pmatrix} M_s - M_0 \\ C_0 + I_0 + G \end{pmatrix}$$
$$= \frac{1}{fa + g(1-b)} \begin{pmatrix} a & g \\ -(1-b) & f \end{pmatrix} \begin{pmatrix} M_s - M_0 \\ C_0 + I_0 + G \end{pmatrix}.$$

Thus we have explicit formulae for the equilibrium values Y^* and r^*:

$$Y^* = \frac{a(M_s - M_0) + g(C_0 + I_0 + G)}{fa + g(1-b)},$$
$$r^* = \frac{-(1-b)(M_s - M_0) + f(C_0 + I_0 + G)}{fa + g(1-b)}.$$

These expressions enable us to answer simple questions about what happens when the policy parameters or the behavioural parameters change.

Example Suppose that the government decides to allow an increase in the money supply M_s. What will be the effect on the equilibrium value of the national income?

From the formula for Y^* we see that

$$\frac{\partial Y^*}{\partial M_s} = \frac{a}{fa + g(1-b)}.$$

Thus an increase in money supply will result in an increase in Y^* if

$$\frac{a}{fa + g(1-b)}$$

is positive. Remembering that the behavioural parameters a, b, f, g are assumed to be positive, we see that this condition will certainly be satisfied if $1 - b > 0$, that is if $b < 1$.

In fact, this condition will certainly hold in any realistic model. We have $Y = C + I + G$, so $C < Y$, and $C = C_0 + bY$, so $bY < C$. Hence $bY < Y$, or $b < 1$. ■

Worked Examples

Example 18.1 (a) Use the method of elementary row operations to find the inverse of

$$B = \begin{pmatrix} 3 & 2 & -1 \\ 1 & 7 & 5 \\ -1 & 0 & 1 \end{pmatrix}.$$

(b) Check your answer by working out BB^{-1}.

(c) Write down the solution of the system $B\mathbf{x} = \mathbf{c}$ for a general 3-vector $\mathbf{c}$.

Solution

(a) We start with the matrix

$$(B \mid I) = \left(\begin{array}{ccc|ccc} 3 & 2 & -1 & 1 & 0 & 0 \\ 1 & 7 & 5 & 0 & 1 & 0 \\ -1 & 0 & 1 & 0 & 0 & 1 \end{array}\right)$$

and reduce this, using row operations, to $(I \mid B^{-1})$.

We have

$$\left(\begin{array}{ccc|ccc} 3 & 2 & -1 & 1 & 0 & 0 \\ 1 & 7 & 5 & 0 & 1 & 0 \\ -1 & 0 & 1 & 0 & 0 & 1 \end{array}\right) \to \left(\begin{array}{ccc|ccc} -1 & 0 & 1 & 0 & 0 & 1 \\ 1 & 7 & 5 & 0 & 1 & 0 \\ 3 & 2 & -1 & 1 & 0 & 0 \end{array}\right)$$

$$\to \left(\begin{array}{ccc|ccc} 1 & 0 & -1 & 0 & 0 & -1 \\ 1 & 7 & 5 & 0 & 1 & 0 \\ 3 & 2 & -1 & 1 & 0 & 0 \end{array}\right) \to \left(\begin{array}{ccc|ccc} 1 & 0 & -1 & 0 & 0 & -1 \\ 0 & 7 & 6 & 0 & 1 & 1 \\ 0 & 2 & 2 & 1 & 0 & 3 \end{array}\right)$$

$$\to \left(\begin{array}{ccc|ccc} 1 & 0 & -1 & 0 & 0 & -1 \\ 0 & 2 & 2 & 1 & 0 & 3 \\ 0 & 7 & 6 & 0 & 1 & 1 \end{array}\right) \to \left(\begin{array}{ccc|ccc} 1 & 0 & -1 & 0 & 0 & -1 \\ 0 & 1 & 1 & \frac{1}{2} & 0 & \frac{3}{2} \\ 0 & 7 & 6 & 0 & 1 & 1 \end{array}\right)$$

$$\to \left(\begin{array}{ccc|ccc} 1 & 0 & -1 & 0 & 0 & -1 \\ 0 & 1 & 1 & \frac{1}{2} & 0 & \frac{3}{2} \\ 0 & 0 & -1 & -\frac{7}{2} & 1 & -\frac{19}{2} \end{array}\right) \to \left(\begin{array}{ccc|ccc} 1 & 0 & -1 & 0 & 0 & -1 \\ 0 & 1 & 1 & \frac{1}{2} & 0 & \frac{3}{2} \\ 0 & 0 & 1 & \frac{7}{2} & -1 & \frac{19}{2} \end{array}\right)$$

$$\to \left(\begin{array}{ccc|ccc} 1 & 0 & 0 & \frac{7}{2} & -1 & \frac{17}{2} \\ 0 & 1 & 0 & -3 & 1 & -8 \\ 0 & 0 & 1 & \frac{7}{2} & -1 & \frac{19}{2} \end{array}\right).$$

So

$$B^{-1} = \begin{pmatrix} \frac{7}{2} & -1 & \frac{17}{2} \\ -3 & 1 & -8 \\ \frac{7}{2} & -1 & \frac{19}{2} \end{pmatrix}.$$

(Of course, there is nothing special about the sequence of row operations used here: there are many possible routes to the required final matrix.)

(b) The product BB^{-1} is

$$\begin{pmatrix} 3 & 2 & -1 \\ 1 & 7 & 5 \\ -1 & 0 & 1 \end{pmatrix} \begin{pmatrix} \frac{7}{2} & -1 & \frac{17}{2} \\ -3 & 1 & -8 \\ \frac{7}{2} & -1 & \frac{19}{2} \end{pmatrix} = \begin{pmatrix} 1 & 0 & 0 \\ 0 & 1 & 0 \\ 0 & 0 & 1 \end{pmatrix},$$

the identity matrix I, as it should be.

(c) The system $B\mathbf{x} = \mathbf{c}$ has exactly one solution for each $\mathbf{c}$. That solution is $\mathbf{x} = B^{-1}\mathbf{c}$; that is

$$\mathbf{x} = \begin{pmatrix} \frac{7}{2} & -1 & \frac{17}{2} \\ -3 & 1 & -8 \\ \frac{7}{2} & -1 & \frac{19}{2} \end{pmatrix} \begin{pmatrix} c_1 \\ c_2 \\ c_3 \end{pmatrix} = \begin{pmatrix} \frac{7}{2}c_1 - c_2 + \frac{17}{2}c_3 \\ -3c_1 + c_2 - 8c_3 \\ \frac{7}{2}c_1 - c_2 + \frac{19}{2}c_3 \end{pmatrix}.$$

Example 18.2 Determine whether or not the following matrix is invertible:

$$A = \begin{pmatrix} 1 & 2 & 1 \\ 3 & 4 & 1 \\ 2 & 2 & 0 \end{pmatrix}.$$

Solution

We reduce the matrix

$$(A \mid I) = \left(\begin{array}{ccc|ccc} 1 & 2 & 1 & 1 & 0 & 0 \\ 3 & 4 & 1 & 0 & 1 & 0 \\ 2 & 2 & 0 & 0 & 0 & 1 \end{array}\right)$$

using elementary row operations, as follows:

$$\left(\begin{array}{ccc|ccc} 1 & 2 & 1 & 1 & 0 & 0 \\ 3 & 4 & 1 & 0 & 1 & 0 \\ 2 & 2 & 0 & 0 & 0 & 1 \end{array}\right) \to \left(\begin{array}{ccc|ccc} 1 & 2 & 1 & 1 & 0 & 0 \\ 0 & -2 & -2 & -3 & 1 & 0 \\ 0 & -2 & -2 & -2 & 0 & 1 \end{array}\right)$$

$$\to \left(\begin{array}{ccc|ccc} 1 & 2 & 1 & 1 & 0 & 0 \\ 0 & -2 & -2 & -3 & 1 & 0 \\ 0 & 0 & 0 & 1 & -1 & 1 \end{array}\right).$$

The last row, which has zero entries in the first half, indicates that A is not invertible.

Example 18.3 Let A be the matrix in the previous question. What is the general solution of the system $A\mathbf{x} = \mathbf{0}$? Verify that the zero vector is a linear combination of the columns of A, with non-zero coefficients.

Solution

In general, to solve $A\mathbf{x} = \mathbf{b}$, we would reduce the augmented matrix $(A \mid \mathbf{b})$. However, if $\mathbf{b} = 0$, then no matter what elementary row operations are carried out, the last column of this 3×4 matrix remains 0. Referring to the solution of the previous worked example, we see that we can reduce A to the matrix

$$B = \begin{pmatrix} 1 & 2 & 1 \\ 0 & 1 & 1 \\ 0 & 0 & 0 \end{pmatrix}.$$

It follows that $(A \mid 0)$ would reduce to $(B \mid 0)$ and hence $A\mathbf{x} = 0$ if and only if $B\mathbf{x} = 0$; that is if and only if

$$\begin{aligned} x_1 + 2x_2 + x_3 &= 0, \\ x_2 + x_3 &= 0. \end{aligned}$$

This has infinitely many solutions. If we set $x_3 = s$, then $x_2 = -s$ and $x_1 = -x_3 - 2x_2 = s$. The general solution is therefore

$$\mathbf{x} = \begin{pmatrix} x_1 \\ x_2 \\ x_3 \end{pmatrix} = \begin{pmatrix} s \\ -s \\ s \end{pmatrix} = s \begin{pmatrix} 1 \\ -1 \\ 1 \end{pmatrix}.$$

Taking $s = 1$, a solution to $A\mathbf{x} = 0$ is

$$\mathbf{x} = \begin{pmatrix} 1 \\ -1 \\ 1 \end{pmatrix}.$$

We therefore have

$$\begin{pmatrix} 1 & 2 & 1 \\ 3 & 4 & 1 \\ 2 & 2 & 0 \end{pmatrix} \begin{pmatrix} 1 \\ -1 \\ 1 \end{pmatrix} = \begin{pmatrix} 0 \\ 0 \\ 0 \end{pmatrix};$$

that is

$$\begin{pmatrix} 1(1) - 1(2) + 1(1) \\ 1(3) - 1(4) + 1(1) \\ 1(2) - 1(2) + 1(0) \end{pmatrix} = \begin{pmatrix} 0 \\ 0 \\ 0 \end{pmatrix}$$

or

$$1 \begin{pmatrix} 1 \\ 3 \\ 2 \end{pmatrix} - 1 \begin{pmatrix} 2 \\ 4 \\ 2 \end{pmatrix} + 1 \begin{pmatrix} 1 \\ 1 \\ 0 \end{pmatrix} = \begin{pmatrix} 0 \\ 0 \\ 0 \end{pmatrix}.$$

In other words, the zero vector is a linear combination of the columns of A.

Main Topics

- the inverse of a matrix
- square systems of linear equations and invertibility
- determining if a matrix is invertible and calculating the inverse if it is
- the inverse of a general 2×2 matrix
- IS–LM analysis

Key Terms, Notations and Formulae

- inverse, A^{-1}, of A: $AA^{-1} = A^{-1}A = I$
- invertible
- if A square and invertible, then $A\mathbf{x} = \mathbf{b}$ has unique solution $A^{-1}\mathbf{b}$
- if $(A \mid I) \to (I \mid B)$ by row operations, then A is invertible and $B = A^{-1}$
- if $A = \begin{pmatrix} a & b \\ c & d \end{pmatrix}$ and $ad - bc \neq 0$, then A is invertible and

$$A^{-1} = \frac{1}{ad - bc}\begin{pmatrix} d & -b \\ -c & a \end{pmatrix}$$

- national income, Y; interest rate, r; consumption, C; investment, I
- $Y = C + I + G$
- government spending, G; demand for money, M_d; money supply, M_s
- IS–LM equations: $C = C_0 + bY$, $I = I_0 - ar$, $M_d = M_0 + fY - gr$
- equilibrium values of national income and interest rate, Y^*, r^*

Exercises

18.1 Find the inverse of the matrix

$$A = \begin{pmatrix} 1 & 2 & 4 \\ 1 & 3 & 6 \\ -1 & 0 & 1 \end{pmatrix}.$$

18.2 Use elementary row operations to find any inverses of the matrices

$$A = \begin{pmatrix} 1 & 2 & -1 \\ 0 & 1 & 2 \\ 3 & 8 & 1 \end{pmatrix} \quad \text{and} \quad B = \begin{pmatrix} -1 & 2 & 1 \\ 0 & 1 & 2 \\ 3 & 1 & 4 \end{pmatrix}.$$

18.3 Write down the inverse of the matrix

$$A = \begin{pmatrix} 1 & 2 \\ 1 & 3 \end{pmatrix}.$$

Problems

18.1 (a) Use the method of elementary row operations to find the inverse of

$$B = \begin{pmatrix} 1 & 4 & 3 \\ 1 & 5 & 5 \\ 2 & 11 & 13 \end{pmatrix}.$$

(b) Check your answer by working out BB^{-1}.

(c) Write down the solution of the system $B\mathbf{x} = \mathbf{c}$ for a general 3-vector $\mathbf{c}$.

18.2 Find the inverses of the matrices

$$\begin{pmatrix} 1 & -4 & 7 \\ 2 & -5 & 5 \\ 3 & -18 & 3 \end{pmatrix} \quad \text{and} \quad \begin{pmatrix} 5 & 7 & -2 \\ 10 & 15 & -2 \\ 5 & 8 & -1 \end{pmatrix},$$

using elementary row operations.

18.3 Use elementary row operations to find the inverse of the following matrix. (It may help to multiply by 10 first and then use a calculator!)

$$\begin{pmatrix} 0.8 & -0.3 & -0.2 \\ -0.4 & 0.9 & -0.2 \\ -0.1 & -0.3 & 0.8 \end{pmatrix}.$$

18.4 Using elementary row operations, show that the following matrix is not invertible:

$$\begin{pmatrix} 1 & -1 & 2 \\ 3 & 5 & -7 \\ 1 & 7 & -11 \end{pmatrix}.$$

18.5 Use elementary row operations to find the inverse of the following matrix:

$$\begin{pmatrix} 1 & 1 & 1 & 2 \\ 1 & 3 & 7 & -2 \\ 2 & 6 & 16 & -6 \\ 2 & 9 & 24 & -10 \end{pmatrix}.$$

18.6 Using elementary row operations, show that the following matrix is not invertible:

$$\begin{pmatrix} 1 & -2 & 3 & 1 \\ 2 & -3 & 7 & 2 \\ -1 & 4 & -1 & 1 \\ 2 & -3 & 7 & 3 \end{pmatrix}.$$

18.7 Using elementary row operations, determine whether the matrix

$$\begin{pmatrix} 1 & 3 & 5 \\ 2 & 1 & 5 \\ 4 & 0 & 8 \end{pmatrix}$$

is invertible and, if it is, find its inverse.

18.8 The supply and demand sets for a good are

$$S = \{(q,p) \mid 3q - 10p = 10\} \quad \text{and} \quad D = \{(q,p) \mid 2q + p = 22\}.$$

Use matrix inversion to determine the equilibrium quantity and price.

18.9 Suppose that the national economy is described by the following IS–LM equations (where the symbols have their usual meanings, and where C, I, M_d, M_s and G are measured in billions of dollars):

$$\begin{aligned} C &= 30 + 0.2Y, \\ I &= 10 - 20r, \\ M_d &= 5 + 0.1Y - 10r, \\ M_s &= 10, \\ G &= 10. \end{aligned}$$

Find the equilibrium values Y^* and r^*.

18.10 Let Y^* be the equilibrium national income in the IS–LM analysis of Section 18.5. By using partial differentiation, discuss how Y^* changes if government spending G is increased slightly.

19 The Input–Output Model

19.1 An Economy with Many Industries

In this chapter we shall investigate a more general version of the 'grommets-and-widgets' economy discussed in Section 16.1. Specifically, we postulate an economy with n interdependent production processes, manufacturing commodities C_1, C_2, ..., C_n. The production process for any one of the commodities requires an input, which uses part of the output of some of the others. In addition, there is an external demand for each commodity.

The problem is to determine the production schedule which enables each process to meet all the demands for its product. At first sight it is not at all obvious that it is possible to satisfy all the interlinked requirements, but we shall see that, under very reasonable conditions, there is indeed a unique solution.

Example The production processes for three goods, C_1, C_2, C_3, are interlinked (Figure 19.1). To produce one dollar's worth of C_1 requires the input of \$0.2 worth of C_1, \$0.4 of C_2 and \$0.1 of C_3. To produce one dollar's worth of C_2 requires \$0.3 worth of C_1, \$0.1 worth of C_2 and \$0.3 worth of C_3, and to produce one dollar's worth of C_3 requires \$0.2 worth of each of C_1, C_2 and C_3.

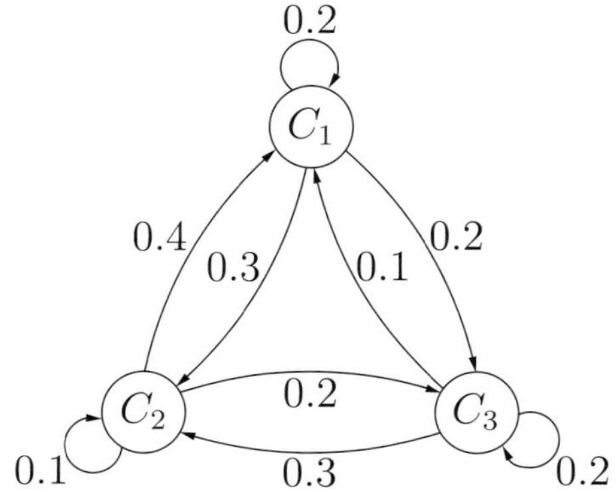

Figure 19.1 An input–output model with three industries.

Suppose that, in a given time period, there is an external demand for d_1 dollars' worth of C_1, d_2 of C_2 and d_3 of C_3. We wish to know the production levels x_1, x_2, x_3 of

C_1, C_2, C_3 required to satisfy all demands in the given period. (From now on we shall suppress references to the unit of measurement (a dollar's worth) and the time period.)

Consider first the total demand for C_1. This is d_1, the external demand, plus the quantity required to produce C_1, C_2 and C_3. Each unit of C_1 requires 0.2 units of C_1, each unit of C_2 requires 0.3 of C_1 and each unit of C_3 requires 0.2 of C_1. Since the quantities of C_1, C_2, C_3 being produced are x_1, x_2, x_3, the total demand for C_1 is therefore

$$x_1 = d_1 + 0.2x_1 + 0.3x_2 + 0.2x_3.$$

Similarly, considering the total demands for C_2 and C_3 shows that

$$x_2 = d_2 + 0.4x_1 + 0.1x_2 + 0.2x_3,$$
$$x_3 = d_3 + 0.1x_1 + 0.3x_2 + 0.2x_3.$$

■

19.2 The Technology Matrix

As in the example above, it is convenient to measure the amounts of each commodity in terms of a special commodity, such as dollars, which we take for granted. In general, some of commodity C_i will be needed to produce each unit of C_j. We define

- a_{ij} = amount of C_i needed to produce one unit of C_j

and consider these quantities as the entries of a matrix $A = (a_{ij})$. This is known as the *technology matrix*.

We suppose that in a given time period:

- d_i = amount of C_i required to satisfy the external demand;
- x_i = amount of C_i required to satisfy all needs.

Our problem is: given the technology matrix A and the external demands d_i, can we find the production schedules x_i?

As in the example, for any one of the commodities C_i, the external demand is d_i and the demand from the C_j process is $a_{ij}x_j$. So we have n equations

$$x_i = d_i + a_{i1}x_1 + a_{i2}x_2 + \cdots + a_{in}x_n,$$

one equation for each value of $i = 1, 2, \ldots, n$. We can write this system of linear equations very compactly in matrix form. Let

$$\mathbf{x} = \begin{pmatrix} x_1 \\ x_2 \\ \vdots \\ x_n \end{pmatrix}, \quad \mathbf{d} = \begin{pmatrix} d_1 \\ d_2 \\ \vdots \\ d_n \end{pmatrix} \quad \text{and} \quad A = \begin{pmatrix} a_{11} & a_{12} & \ldots & a_{1n} \\ a_{21} & a_{22} & \ldots & a_{2n} \\ \vdots & \vdots & \ddots & \vdots \\ a_{n1} & a_{n2} & \ldots & a_{nn} \end{pmatrix}.$$

The system of equations is simply

$$\mathbf{x} = \mathbf{d} + A\mathbf{x}.$$

Rearranging we get $\mathbf{x} - A\mathbf{x} = \mathbf{d}$, and using the fact that $I\mathbf{x} = \mathbf{x}$, where I is the identity matrix,

$$(I - A)\mathbf{x} = \mathbf{d}.$$

Thus, provided the matrix $(I - A)$ is invertible, there is a unique solution

$$\mathbf{x} = (I - A)^{-1}\mathbf{d}.$$

The matrix $I - A$ is often known as the *Leontief matrix*, after the Russian-American economist Wassily Leontief (who won the Nobel Prize in 1973).

Example (continued) The technology matrix for the example described in Figure 19.1 is

$$A = \begin{pmatrix} 0.2 & 0.3 & 0.2 \\ 0.4 & 0.1 & 0.2 \\ 0.1 & 0.3 & 0.2 \end{pmatrix}.$$

The input–output equations in matrix form are

$$\mathbf{x} = \begin{pmatrix} x_1 \\ x_2 \\ x_3 \end{pmatrix} = \begin{pmatrix} d_1 \\ d_2 \\ d_3 \end{pmatrix} + \begin{pmatrix} 0.2 & 0.3 & 0.2 \\ 0.4 & 0.1 & 0.2 \\ 0.1 & 0.3 & 0.2 \end{pmatrix} \begin{pmatrix} x_1 \\ x_2 \\ x_3 \end{pmatrix} = \mathbf{d} + A\mathbf{x}.$$

(*Note*: a common mistake is to write down a matrix which is the technology matrix with its rows and columns interchanged. Care is always needed at this point.)

By the general theory, $\mathbf{x} = (I - A)^{-1}\mathbf{d}$. Now

$$I - A = \begin{pmatrix} 0.8 & -0.3 & -0.2 \\ -0.4 & 0.9 & -0.2 \\ -0.1 & -0.3 & 0.8 \end{pmatrix}$$

and (see Problem 18.3)

$$(I - A)^{-1} = \begin{pmatrix} 1.72 & 0.78 & 0.63 \\ 0.89 & 1.61 & 0.63 \\ 0.55 & 0.70 & 1.56 \end{pmatrix},$$

where the terms are correct to two decimal places. This enables us to compute the required production schedule $\mathbf{x}$ for any demand $\mathbf{d}$, from the equation $\mathbf{x} = (I - A)^{-1}\mathbf{d}$. Explicitly, we have

$$\begin{aligned} x_1 &= 1.72d_1 + 0.78d_2 + 0.63d_3, \\ x_2 &= 0.89d_1 + 1.61d_2 + 0.63d_3, \\ x_3 &= 0.55d_1 + 0.70d_2 + 1.56d_3. \end{aligned}$$

Notice that a single matrix inversion has answered the problem for any $\mathbf{d}$. ■

19.3 Why Is There a Solution?

In the example we were able to find the inverse matrix $(I - A)^{-1}$, thus guaranteeing a unique solution for any external demand $\mathbf{d}$. Furthermore, the entries of $(I - A)^{-1}$ turned out to be non-negative. Thus, making the reasonable assumption that the external demands d_1, d_2, d_3 are non-negative, the resulting values of x_1, x_2, x_3 are also non-negative. It is important to discover whether these desirable events are merely fortuitous.

In fact, it can be shown that a simple mathematical condition, based on a realistic economic hypothesis, is sufficient to ensure that all is well. Consider the production of commodity C_j. Each dollar's worth produced requires a_{1j} dollars' worth of C_1, a_{2j} dollars' worth of C_2, and so on. Thus the total cost of producing a dollar's worth of C_j is

$$a_{1j} + a_{2j} + \cdots + a_{nj}.$$

It is very reasonable to assume that this total is less than one dollar, because otherwise the production of C_j would make a loss. In other words, it is economically realistic to assume that the condition

$$\sum_{i=1}^{n} a_{ij} < 1$$

holds for each $j = 1, 2, \ldots, n$.

In order to use this fact mathematically, we have to look at an alternative way of calculating $(I - A)^{-1}$. This depends on the formula

$$(I - A)(I + A + A^2 + A^3 + \cdots + A^m) = I - A^{m+1},$$

which can be proved by simple algebra. The significance of the formula is that, if we ignore the term A^{m+1} on the right-hand side, it says that

$$Z = I + A + A^2 + A^3 + \cdots + A^m$$

satisfies $(I - A)Z = I$. In other words, Z is almost an inverse for $I - A$. More precisely, if all the entries of A^{m+1} tend to zero as $m \to \infty$, then the sum

$$I + A + A^2 + A^3 + \cdots$$

will approach $(I - A)^{-1}$. This argument also explains why $(I - A)^{-1}$ has non-negative entries: it is because it is the sum of the matrices $I, A, A^2, \ldots$, and all these have non-negative entries.

The one remaining step required to solve our problem is a mathematical theorem (which we shall not prove here):

- the conditions

$$\sum_{i=1}^{n} a_{ij} < 1$$

imply that the entries of A^{m+1} tend to zero as $m \to \infty$.

So a realistic economic hypothesis, that each production process is profitable, can be shown mathematically to lead to a desirable conclusion, that all the requirements can be satisfied simultaneously.

Example Suppose the A matrix is

$$A = \begin{pmatrix} 0.3 & 0.2 \\ 0.1 & 0.2 \end{pmatrix} = \frac{1}{10}\begin{pmatrix} 3 & 2 \\ 1 & 2 \end{pmatrix}.$$

The sum of each column of A is strictly less than 1, so the theorem tells us that the entries of the powers of A tend to zero. By simple calculation we find

$$A^2 = \frac{1}{100}\begin{pmatrix} 11 & 10 \\ 5 & 6 \end{pmatrix}, \qquad A^3 = \frac{1}{1000}\begin{pmatrix} 43 & 42 \\ 21 & 22 \end{pmatrix},$$
$$A^4 = \frac{1}{10000}\begin{pmatrix} 171 & 170 \\ 85 & 86 \end{pmatrix}, \quad A^5 = \frac{1}{100000}\begin{pmatrix} 683 & 682 \\ 341 & 342 \end{pmatrix}.$$

Using the approximation $(I - A)^{-1} = I + A + A^2 + A^3 + A^4 + A^5$ we get, for the term in the first row and column, $1 + 0.3 + 0.11 + 0.043 + 0.0171 + 0.00683$, which is approximately 1.48. Working out the other terms we find

$$(I - A)^{-1} = \begin{pmatrix} 1.48 & 0.37 \\ 0.18 & 1.29 \end{pmatrix},$$

approximately. This may be compared with the result obtained by other methods (which may, of course, be more efficient in this case). ■

Worked Examples

Example 19.1 Suppose we are given an input–output model with two industries, for which the matrix of coefficients is

$$A = \begin{pmatrix} 0.3 & 0.1 \\ 0.2 & 0.4 \end{pmatrix}.$$

Write down the equation which determines the production schedule $\mathbf{x}$ in terms of the external demand $\mathbf{d}$ and solve it.

Solution

The equation required is $\mathbf{x} = \mathbf{d} + A\mathbf{x}$ or, equivalently, $(I - A)\mathbf{x} = \mathbf{d}$, so that $\mathbf{x} = (I - A)^{-1}\mathbf{d}$. Now

$$I - A = \begin{pmatrix} 1 & 0 \\ 0 & 1 \end{pmatrix} - \begin{pmatrix} 0.3 & 0.1 \\ 0.2 & 0.4 \end{pmatrix} = \begin{pmatrix} 0.7 & -0.1 \\ -0.2 & 0.6 \end{pmatrix}.$$

Using the formula of Section 18.4 for the inverse of a 2×2 matrix, we find

$$(I - A)^{-1} = \begin{pmatrix} 1.50 & 0.25 \\ 0.50 & 1.75 \end{pmatrix}.$$

Then

$$\mathbf{x} = (I - A)^{-1}\mathbf{d} = \begin{pmatrix} 1.50 & 0.25 \\ 0.50 & 1.75 \end{pmatrix} \begin{pmatrix} d_1 \\ d_2 \end{pmatrix} = \begin{pmatrix} 1.50d_1 + 0.25d_2 \\ 0.50d_1 + 1.75d_2 \end{pmatrix}.$$

Example 19.2 A factory makes two goods, grommets and widgets. To make \$1 worth of grommets requires \$0.1 worth of widgets, and to make \$1 worth of widgets requires \$0.15 worth of grommets and \$0.05 worth of widgets. There is an external demand for \$500 worth of grommets and \$1000 worth of widgets. What should be the total production of each commodity?

Solution

Let x_1 be the amount of grommets produced, in dollars, and x_2 the amount of widgets. The total demand for grommets is $500 + 0.15x_2$ and the total demand for widgets is $1000 + 0.1x_1 + 0.05x_2$. Therefore

$$x_1 = 500 + 0.15x_2,$$
$$x_2 = 1000 + 0.1x_1 + 0.05x_2.$$

In matrix form, the system is

$$\mathbf{x} = \begin{pmatrix} x_1 \\ x_2 \end{pmatrix} = \begin{pmatrix} 500 \\ 1000 \end{pmatrix} + \begin{pmatrix} 0 & 0.15 \\ 0.1 & 0.05 \end{pmatrix} \begin{pmatrix} x_1 \\ x_2 \end{pmatrix} = \mathbf{d} + A\mathbf{x},$$

and so $(I - A)\mathbf{x} = \mathbf{d}$ and

$$\mathbf{x} = (I - A)^{-1}\mathbf{d} = \begin{pmatrix} 1 & -0.15 \\ -0.1 & 0.95 \end{pmatrix}^{-1} \begin{pmatrix} 500 \\ 1000 \end{pmatrix}.$$

As in the previous example, the inverse matrix can be found using the formula of Section 18.4:

$$(I - A)^{-1} = \begin{pmatrix} 1.016043 & 0.160428 \\ 0.106952 & 1.069519 \end{pmatrix}.$$

Thus

$$\mathbf{x} = (I - A)^{-1}\mathbf{d} = \begin{pmatrix} 668.45 \\ 1122.99 \end{pmatrix}.$$

(Of course, it is also quite easy to solve the simultaneous equations directly in this case.)

Example 19.3 Let A be the matrix of the previous example. Use the method of Section 19.3 to find an approximation to $(I - A)^{-1}$, using enough terms to ensure that your answer is correct to two decimal places.

Solution

The matrix A is

$$A = \begin{pmatrix} 0 & 0.15 \\ 0.1 & 0.05 \end{pmatrix}.$$

The sum of each column of A is strictly less than 1, so the theorem tells us that the entries of the powers of A tend to zero. By simple calculation we find

$$A^2 = \begin{pmatrix} 0.0150 & 0.0075 \\ 0.0050 & 0.0175 \end{pmatrix} \quad \text{and} \quad A^3 = \begin{pmatrix} 0.00075 & 0.00263 \\ 0.00175 & 0.00163 \end{pmatrix}.$$

Using the approximation $(I - A)^{-1} = I + A + A^2 + A^3$ we then get

$$(I - A)^{-1} = \begin{pmatrix} 1.01575 & 0.16013 \\ 0.10675 & 1.06913 \end{pmatrix},$$

which, compared to the answer above, is correct to two decimal places.

Main Topics

- interdependent production processes
- deriving equations for required total production of each commodity
- approximating the inverse of $I - A$ using powers of A

Key Terms, Notations and Formulae

- technology matrix, A: a_{ij} = amount of C_i needed for one unit of C_j
- d_i, external demand for C_i; x_i required total production of C_i
- $\mathbf{x} = A\mathbf{x} + \mathbf{d}$, or $(I - A)\mathbf{x} = \mathbf{d}$
- if $a_{1j} + a_{2j} + \cdots + a_{nj} < 1$, then $(I - A)$ is invertible and $(I - A)^{-1} = I + A + A^2 + A^3 + \cdots$

Exercises

19.1 Consider an economy with three industries: coal, electricity and railways. To produce \$1 of coal requires \$0.25 worth of electricity and \$0.25 rail costs for transportation. To produce \$1 of electricity requires \$0.65 worth of coal for fuel, \$0.05 of electricity for the auxiliary equipment and \$0.05 for transport. To provide \$1 worth of transport, the railway requires \$0.55 worth of coal for fuel and \$0.10 electricity. Each week the external demand for coal is \$50 000 and the external demand for electricity is \$25 000. There is no external demand for the railway. What should be the weekly production schedule for each industry?

19.2 An economy has four industries: coal, iron, electricity and railways. To produce \$1 of coal requires \$0.10 worth of iron, \$0.20 of electricity and \$0.50 rail costs for transportation. To produce \$1 of iron requires \$0.60 worth of coal for fuel, \$0.01 of electricity and \$0.20 for transport. To produce \$1 of electricity requires \$0.50 worth of coal for fuel, \$0.02 of electricity for the auxiliary equipment and \$0.01 for transport. To provide \$1 worth of transport, the railway requires \$0.50 coal for fuel, \$0.20 iron and \$0.25 electricity. Each week the external demand for coal is \$8000, the external demand for iron is \$1000, the external demand for electricity is \$4000 and the external demand for the railway is \$6000. Write down the equation which determines the production schedule for these industries.

Problems

19.1 Suppose we are given an input–output model with two industries, for which the matrix of coefficients is

$$A = \begin{pmatrix} 0.2 & 0.2 \\ 0.1 & 0.3 \end{pmatrix}.$$

Write down the equation which determines the production schedule $\mathbf{x}$ in terms of the external demand $\mathbf{d}$ and solve it.

19.2 A factory makes two goods, grommets and widgets. To make \$1 worth of grommets requires \$0.2 worth of grommets and \$0.1 worth of widgets, and to make \$1 worth of widgets requires \$0.05 worth of grommets and \$0.1 worth of widgets. There is a market demand for \$750 worth of grommets and \$500 worth of widgets. What should the total production of each be to meet the market demand?

19.3 A factory makes three products, X, Y and Z. The production processes for these products are interrelated. To produce \$1 of X requires 0.05 units (in dollars) of X, 0.1 of Y and 0.1 of Z. To produce \$1 of Y requires 0.4 of X and 0.1 of Z. To produce \$1 worth of Z requires 0.1 of X and 0.2 of Y. Each week the external demands for X, Y and Z are 200, 500 and 1500 units, respectively. What should be the weekly production level of each good?

19.4 Let A be the matrix of Problem 19.1. Use the method of Section 19.3 to find an approximation to $(I - A)^{-1}$, using enough terms to ensure that your answer is correct to one decimal place.

20 Determinants

20.1 Determinants

Determinants can be used to solve a system of linear equations, and to find the inverse of a matrix. In the first half of the twentieth century determinants played a very prominent part in every mathematics course. Many people who were later to become famous economists took such courses, and consequently they sometimes used determinants in their work. However, it must be stressed that anything in this book which can be done with determinants can also be achieved using the easier and more efficient techniques developed in the previous chapters. For that reason our discussion will be restricted to the main points.

The *determinant* of a square matrix A is a particular number associated with A, written $\det A$ or $|A|$. When A is a 2×2 matrix, the determinant is given by the formula

$$\det \begin{pmatrix} a & b \\ c & d \end{pmatrix} = \begin{vmatrix} a & b \\ c & d \end{vmatrix} = ad - bc.$$

For a 3×3 matrix, the determinant is given as follows:

$$\begin{aligned} \det \begin{pmatrix} a & b & c \\ d & e & f \\ g & h & i \end{pmatrix} &= \begin{vmatrix} a & b & c \\ d & e & f \\ g & h & i \end{vmatrix} \\ &= a \begin{vmatrix} e & f \\ h & i \end{vmatrix} - b \begin{vmatrix} d & f \\ g & i \end{vmatrix} + c \begin{vmatrix} d & e \\ g & h \end{vmatrix} \\ &= aei - afh + bfg - bdi + cdh - ceg. \end{aligned}$$

Example Suppose

$$A = \begin{pmatrix} 1 & 2 & 3 \\ 4 & 1 & 1 \\ -1 & 3 & 0 \end{pmatrix}.$$

Then, according to the formula just given:

$$\begin{aligned}
\det A &= \begin{vmatrix} 1 & 2 & 3 \\ 4 & 1 & 1 \\ -1 & 3 & 0 \end{vmatrix} \\
&= 1\begin{vmatrix} 1 & 1 \\ 3 & 0 \end{vmatrix} - 2\begin{vmatrix} 4 & 1 \\ -1 & 0 \end{vmatrix} + 3\begin{vmatrix} 4 & 1 \\ -1 & 3 \end{vmatrix} \\
&= 1(-3) - 2(1) + 3(13) \\
&= 34\,.
\end{aligned}$$

■

There is a general formula for the determinant of an $n \times n$ matrix, which (just as the determinant of a 3×3 matrix is defined via the determinants of three related 2×2 matrices) expresses it in terms of the determinants of n certain $(n-1) \times (n-1)$ submatrices and, by recursion, this leads to a method of calculation. But this method for calculating the determinant is already quite complicated even for a 3×3 matrix and gets worse very rapidly indeed for larger matrices: for an $n \times n$ matrix the resulting number of terms in the final expression turns out to be $n! = n(n-1)(n-2)\ldots 2.1$. This is a large number (e.g., $10! = 3\,628\,800$) and so the formula is useless for practical purposes as a means of calculation.

If we have to calculate a determinant in practice, we use a method based on row operations. This depends on the following facts.

(i) If any row of a matrix is multiplied by a constant c, its determinant is also multiplied by c.
(ii) If a multiple of one row is added to another, the determinant is unchanged.
(iii) If two rows are interchanged, the determinant is multiplied by -1.
(iv) $\det I = 1$.

Although we have given explicit formulae for the determinants of a 2×2 and a 3×3 matrix, we have not yet defined explicitly what we mean by the determinant of a larger matrix. However, as we shall see, rules (i) to (iv), together with elementary row operations, provide a method for computing the determinant of a matrix. In fact, we can regard these rules as *defining* the determinant of a matrix. (This may seem a little abstract, but all it means is that these rules give rise to a method for computing a single number associated with a given matrix, and that number is called the determinant.)

We observe that the first three rules tell us how the determinant behaves when we apply elementary row operations, as defined in Section 16.3. We shall now explain how the rules enable us to work out any determinant.

First, we note that the determinant of any matrix which has non-zero entries only on the main diagonal is determined by rules (i) and (iv): Explicitly, the matrix

$$C = \begin{pmatrix} c_1 & 0 & 0 & \dots & 0 \\ 0 & c_2 & 0 & \dots & 0 \\ 0 & 0 & c_3 & \dots & 0 \\ \vdots & \vdots & \vdots & \ddots & \vdots \\ 0 & 0 & 0 & \dots & c_n \end{pmatrix}$$

is obtained from I by multiplying the first row by c_1, the second row by c_2, and so on. Hence, using rules (i) and (iv):

$$\det C = c_1 c_2 \dots c_n \det I = c_1 c_2 \dots c_n.$$

Next, we consider any matrix in which every entry below the main diagonal is zero, and those on the main diagonal are not zero – this is known as an *upper-triangular* matrix. Such a matrix can be reduced to the form C by adding multiples of one row to another, *without altering the diagonal entries*. Hence, by rule (ii):

$$\begin{vmatrix} c_1 & * & * & \dots & * \\ 0 & c_2 & * & \dots & * \\ 0 & 0 & c_3 & \dots & * \\ \vdots & \vdots & \vdots & \ddots & \vdots \\ 0 & 0 & 0 & \dots & c_n \end{vmatrix} = \begin{vmatrix} c_1 & 0 & 0 & \dots & 0 \\ 0 & c_2 & 0 & \dots & 0 \\ 0 & 0 & c_3 & \dots & 0 \\ \vdots & \vdots & \vdots & \ddots & \vdots \\ 0 & 0 & 0 & \dots & c_n \end{vmatrix} = c_1 c_2 \dots c_n.$$

In other words, we have proved that

- the determinant of an upper-triangular matrix is equal to the product of its diagonal entries.

It follows that we can calculate the determinant of a matrix by reducing it to upper-triangular form, using row operations of types (i), (ii) and (iii), bearing in mind the effects each type of operation has on the value of the determinant.

Example The matrix

$$A = \begin{pmatrix} 1 & 3 & 2 & 5 \\ 0 & 0 & 3 & 2 \\ 1 & 5 & 4 & 0 \\ 1 & 2 & 1 & 1 \end{pmatrix}$$

is reduced to an upper-triangular matrix T as follows:

$$A = \begin{pmatrix} 1 & 3 & 2 & 5 \\ 0 & 0 & 3 & 2 \\ 1 & 5 & 4 & 0 \\ 1 & 2 & 1 & 1 \end{pmatrix} \to \begin{pmatrix} 1 & 3 & 2 & 5 \\ 0 & 0 & 3 & 2 \\ 0 & 2 & 2 & -5 \\ 0 & -1 & -1 & -4 \end{pmatrix}$$

$$\to \begin{pmatrix} 1 & 3 & 2 & 5 \\ 0 & 2 & 2 & -5 \\ 0 & 0 & 3 & 2 \\ 0 & -1 & -1 & -4 \end{pmatrix} \to \begin{pmatrix} 1 & 3 & 2 & 5 \\ 0 & 2 & 2 & -5 \\ 0 & 0 & 3 & 2 \\ 0 & 0 & 0 & -\frac{13}{2} \end{pmatrix} = T.$$

Here, the second step is an interchange of rows (rule (iii)) and the other steps are covered by rule (ii). It follows that

$$\det A = -\det T = -\left(1 \times 2 \times 3 \times \left(-\frac{13}{2}\right)\right) = 39.$$

■

Example We use row operations to evaluate

$$|A| = \begin{vmatrix} 1 & 2 & -1 & 4 \\ -1 & 3 & 0 & 2 \\ 2 & 1 & 1 & 2 \\ 1 & 4 & 1 & 3 \end{vmatrix}$$

by reducing A to an upper-triangular matrix. First we obtain zeros below the leading 1 by adding multiples of row 1 to the rows below. The new matrix will have the same determinant as A. So

$$|A| = \begin{vmatrix} 1 & 2 & -1 & 4 \\ 0 & 5 & -1 & 6 \\ 0 & -3 & 3 & -6 \\ 0 & 2 & 2 & -1 \end{vmatrix}.$$

Next, we observe that

$$\begin{vmatrix} 1 & 2 & -1 & 4 \\ 0 & 5 & -1 & 6 \\ 0 & -3 & 3 & -6 \\ 0 & 2 & 2 & -1 \end{vmatrix} = -3 \begin{vmatrix} 1 & 2 & -1 & 4 \\ 0 & 5 & -1 & 6 \\ 0 & 1 & -1 & 2 \\ 0 & 2 & 2 & -1 \end{vmatrix},$$

where we factored -3 from the third row. (We would need to multiply the resulting determinant on the right by -3 in order to put the -3 back into the third row, and get back a matrix with the same determinant as A.) Next we switch row 2 and row 3, with the effect of changing the sign of the determinant:

$$|A| = 3 \begin{vmatrix} 1 & 2 & -1 & 4 \\ 0 & 1 & -1 & 2 \\ 0 & 5 & -1 & 6 \\ 0 & 2 & 2 & -1 \end{vmatrix}.$$

We use operations of type (ii) to achieve upper-triangular form. These operations result in no change in the value of the determinant. So we have

$$|A| = 3 \begin{vmatrix} 1 & 2 & -1 & 4 \\ 0 & 1 & -1 & 2 \\ 0 & 0 & 4 & -4 \\ 0 & 0 & 4 & -5 \end{vmatrix} = 3 \begin{vmatrix} 1 & 2 & -1 & 4 \\ 0 & 1 & -1 & 2 \\ 0 & 0 & 4 & -4 \\ 0 & 0 & 0 & -1 \end{vmatrix}.$$

Finally we evaluate the determinant of the upper-triangular matrix, obtaining

$$|A| = 3\begin{vmatrix} 1 & 2 & -1 & 4 \\ 0 & 1 & -1 & 2 \\ 0 & 0 & 4 & -4 \\ 0 & 0 & 0 & -1 \end{vmatrix} = 3(1 \times 1 \times 4 \times (-1)) = -12.$$

■

20.2 The Determinant as a Test for Invertibility

In Section 18.4 we showed that a general 2×2 matrix is invertible if and only if $ad - bc$ is not zero. Since this expression is the determinant of the matrix, we can now say that a 2×2 matrix A is invertible if and only if $\det A \neq 0$.

In fact the same result is true in general. In Chapter 18 we observed that a square matrix A is invertible if and only if, when we reduce it to echelon form, the leading 1s are on the main diagonal. The upper-triangular form discussed in the previous section corresponds to this echelon form, except that we allow a diagonal term to take any non-zero value. It follows that A is invertible if and only if $\det A \neq 0$.

So we have another condition for there to be a unique solution of a system of n linear equations in n unknowns:

- $A\mathbf{x} = \mathbf{b}$ *has a unique solution if and only if* $\det A \neq 0$.

As we mentioned at the beginning of this chapter, determinants are of historical importance. However, we now see the determinant as a device which may occasionally be useful in theory and in more advanced work, but which has little significance in actual computations of the type needed for the topics of this book.

Consider, for example, the statement that a matrix is invertible if and only if its determinant is not zero. It might be argued that this is a very useful result, because 'just by working out the determinant' we can decide if a matrix has an inverse, and thus if the corresponding system of linear equations has a unique solution. However, if we work out the determinant through row reduction, this process more or less answers our question directly in any case, without the need for any mention of the determinant. For, if the aim is to determine whether a system of linear equations has a (unique) solution, we need only row reduce the augmented matrix, and if it is to determine invertibility and produce an inverse, we could just implement the method of Section 18.3. Neither of these is much more work than calculating the determinant.

Some readers may have met other methods of working out determinants, such as the one called 'expansion in terms of a row'. (The formula we gave in Section 20.1 for the determinant of a 3×3 matrix is an explicit case of this.) As we noted earlier, such methods are computationally inefficient – that is, although they work for small matrices, the amount of computation involved for any matrix of realistic size is enormous. Thus the existence of such methods has no bearing on the usefulness of determinants.

20.3 Cramer's Rule

Another once-popular technique for solving $A\mathbf{x} = \mathbf{b}$, where A is an invertible $n \times n$ matrix, is *Cramer's rule* (named after the mathematician Gabriel Cramer (1704–1752)). In general, Cramer's rule is not much used nowadays, but the rule for $n = 2$ is perhaps worth remembering. It can be derived very easily from the formula for the inverse of a 2×2 matrix obtained in Section 18.4.

Let us write the general system of two equations in two unknowns in the matrix form $A\mathbf{x} = \mathbf{b}$, that is

$$\begin{pmatrix} a_{11} & a_{12} \\ a_{21} & a_{22} \end{pmatrix} \begin{pmatrix} x_1 \\ x_2 \end{pmatrix} = \begin{pmatrix} b_1 \\ b_2 \end{pmatrix}.$$

Using the formula for the inverse, we get

$$\begin{aligned} \begin{pmatrix} x_1 \\ x_2 \end{pmatrix} &= \begin{pmatrix} a_{11} & a_{12} \\ a_{21} & a_{22} \end{pmatrix}^{-1} \begin{pmatrix} b_1 \\ b_2 \end{pmatrix} \\ &= \frac{1}{a_{11}a_{22} - a_{12}a_{21}} \begin{pmatrix} a_{22} & -a_{12} \\ -a_{21} & a_{11} \end{pmatrix} \begin{pmatrix} b_1 \\ b_2 \end{pmatrix} \\ &= \frac{1}{a_{11}a_{22} - a_{12}a_{21}} \begin{pmatrix} a_{22}b_1 - a_{12}b_2 \\ -a_{21}b_1 + a_{11}b_2 \end{pmatrix}. \end{aligned}$$

After a minor rearrangement, this tells us that the solutions x_1 and x_2 can each be written in terms of determinants:

$$x_1 = \frac{\begin{vmatrix} b_1 & a_{12} \\ b_2 & a_{22} \end{vmatrix}}{\begin{vmatrix} a_{11} & a_{12} \\ a_{21} & a_{22} \end{vmatrix}} \quad \text{and} \quad x_2 = \frac{\begin{vmatrix} a_{11} & b_1 \\ a_{21} & b_2 \end{vmatrix}}{\begin{vmatrix} a_{11} & a_{12} \\ a_{21} & a_{22} \end{vmatrix}}.$$

Cramer's rule in the 2×2 case can be stated in the following way: $x_i = \Delta_i/\Delta$, where Δ is the determinant of A and Δ_i is the determinant of the matrix obtained by replacing the ith column of A with $\mathbf{b}$.

The result in the $n \times n$ case is just the same, so it is a neat way of writing down the solution. But we must remember that it requires the computation of $n+1$ determinants of size $n \times n$. For this reason Cramer's rule is not to be recommended as a practical technique.

20.4 The Inverse of a 3 × 3 Matrix via Determinants

We have, in Section 18.3, discussed a general method to determine whether a matrix is invertible and, if it is, to produce the inverse. We have also seen an explicit formula for the inverse of a 2×2 matrix, in Section 18.4. When applicable (i.e., when the matrix is invertible), this involves the quantity $ad - bc$, which we now recognise as the determinant of the matrix. Explicitly:

$$\text{if } A = \begin{pmatrix} a & b \\ c & d \end{pmatrix} \text{ and } \det A = ad - bc \neq 0, \text{ then } A^{-1} = \frac{1}{\det A} \begin{pmatrix} d & -b \\ -c & a \end{pmatrix}.$$

It is natural to ask if there is an explicit formula, involving determinants, for the inverse of a 3×3 matrix, and it turns out that there is. Although we think the row operations approach is easier and, as noted, extends to larger matrices too, for the sake of completeness we present this alternative approach here.

Suppose we have the 3×3 matrix

$$A = \begin{pmatrix} a_{11} & a_{12} & a_{13} \\ a_{21} & a_{22} & a_{23} \\ a_{31} & a_{32} & a_{33} \end{pmatrix}.$$

Then A is invertible precisely when $\det A \neq 0$. For each i and j between 1 and 3, let A_{ij} be the determinant of the 2×2 matrix obtained from A by deleting row i and column j. For example

$$A_{23} = \begin{vmatrix} a_{11} & a_{12} \\ a_{31} & a_{32} \end{vmatrix}.$$

It turns out that if $\det A \neq 0$, then A is invertible and

$$A^{-1} = \frac{1}{\det A} \begin{pmatrix} +A_{11} & -A_{21} & +A_{31} \\ -A_{12} & +A_{22} & -A_{32} \\ +A_{13} & -A_{23} & +A_{33} \end{pmatrix}.$$

Note carefully the 'alternating' pattern of signs attached to the entries of the matrix. The matrix

$$\begin{pmatrix} +A_{11} & -A_{12} & +A_{13} \\ -A_{21} & +A_{22} & -A_{23} \\ +A_{31} & -A_{32} & +A_{33} \end{pmatrix}$$

is often called the *co-factor* matrix, and the matrix that appears in the above formula for the inverse is

$$\begin{pmatrix} +A_{11} & -A_{21} & +A_{31} \\ -A_{12} & +A_{22} & -A_{32} \\ +A_{13} & -A_{23} & +A_{33} \end{pmatrix}.$$

Note that this is the *transpose* of the co-factor matrix; that is the matrix obtained by interchanging the rows and columns. It is known as the *adjugate* of the matrix A, denoted $\operatorname{adj}(A)$. For this reason you might sometimes see the formula

$$A^{-1} = \frac{1}{\det A} \operatorname{adj}(A).$$

Calculating the inverse in this way requires care, since we have to compute a 3×3 determinant ($\det A$), then nine different 2×2 determinants, then we have to attach the correct signs to these, and place them in the right positions.

Example Let

$$A = \begin{pmatrix} -2 & 1 & 3 \\ 0 & -1 & 1 \\ 1 & 2 & 0 \end{pmatrix}.$$

We calculate its inverse using the method just described. First, we find its determinant. We can use the formula from Section 20.1 (we could equally well use row operations):

$$\begin{aligned} \det A &= -2\begin{vmatrix} -1 & 1 \\ 2 & 0 \end{vmatrix} - 1\begin{vmatrix} 0 & 1 \\ 1 & 0 \end{vmatrix} + 3\begin{vmatrix} 0 & -1 \\ 1 & 2 \end{vmatrix} \\ &= -2(-2) - 1(-1) + 3(1) = 8. \end{aligned}$$

Since this is non-zero, we know that the inverse exists. We now calculate the relevant 2×2 determinants:

$$A_{11} = \begin{vmatrix} -1 & 1 \\ 2 & 0 \end{vmatrix} = -2, \quad A_{12} = \begin{vmatrix} 0 & 1 \\ 1 & 0 \end{vmatrix} = -1, \quad A_{13} = \begin{vmatrix} 0 & -1 \\ 1 & 2 \end{vmatrix} = 1,$$

$$A_{21} = \begin{vmatrix} 1 & 3 \\ 2 & 0 \end{vmatrix} = -6, \quad A_{22} = \begin{vmatrix} -2 & 3 \\ 1 & 0 \end{vmatrix} = -3, \quad A_{23} = \begin{vmatrix} -2 & 1 \\ 1 & 2 \end{vmatrix} = -5,$$

$$A_{31} = \begin{vmatrix} 1 & 3 \\ -1 & 1 \end{vmatrix} = 4, \quad A_{32} = \begin{vmatrix} -2 & 3 \\ 0 & 1 \end{vmatrix} = -2, \quad A_{33} = \begin{vmatrix} -2 & 1 \\ 0 & -1 \end{vmatrix} = 2.$$

The inverse is then

$$A^{-1} = \frac{1}{\det A}\operatorname{adj}(A) = \frac{1}{8}\begin{pmatrix} +A_{11} & -A_{21} & +A_{31} \\ -A_{12} & +A_{22} & -A_{32} \\ +A_{13} & -A_{23} & +A_{33} \end{pmatrix} = \frac{1}{8}\begin{pmatrix} -2 & 6 & 4 \\ 1 & -3 & 2 \\ 1 & 5 & 2 \end{pmatrix}.$$

You can easily check that this is correct by checking that the product of A with this matrix is the identity matrix. ■

Worked Examples

Example 20.1 Evaluate the determinants of the matrices

$$A = \begin{pmatrix} 1 & 2 & 1 \\ 3 & 1 & 1 \\ 4 & 2 & 3 \end{pmatrix} \quad \text{and} \quad B = \begin{pmatrix} 5 & 2 & 1 \\ 1 & -2 & 3 \\ 5 & 1 & -1 \end{pmatrix}.$$

Solution

We reduce each matrix to upper-triangular form by interchanging rows and adding a multiple of one row to another. (Remember that we do *not* multiply any row by a constant, because this changes the numerical value of the determinant.) For A, we have

$$\begin{pmatrix} 1 & 2 & 1 \\ 3 & 1 & 1 \\ 4 & 2 & 3 \end{pmatrix} \to \begin{pmatrix} 1 & 2 & 1 \\ 0 & -5 & -2 \\ 0 & -6 & -1 \end{pmatrix} \to \begin{pmatrix} 1 & 2 & 1 \\ 0 & -5 & -2 \\ 0 & 0 & \frac{7}{5} \end{pmatrix} = T.$$

Thus, since we did not interchange any rows:

$$\det A = \det T = 1(-5)\left(\frac{7}{5}\right) = -7.$$

For B, we have

$$\begin{pmatrix} 5 & 2 & 1 \\ 1 & -2 & 3 \\ 5 & 1 & -1 \end{pmatrix} \to \begin{pmatrix} 1 & -2 & 3 \\ 5 & 2 & 1 \\ 5 & 1 & -1 \end{pmatrix} \to \begin{pmatrix} 1 & -2 & 3 \\ 0 & 12 & -14 \\ 0 & 11 & -16 \end{pmatrix} \to \begin{pmatrix} 1 & -2 & 3 \\ 0 & 12 & -14 \\ 0 & 0 & -\frac{38}{12} \end{pmatrix} = U.$$

In transforming B to U, we performed exactly one row interchange; thus

$$\det B = (-1)\det U = (-1)(1)(12)\left(-\frac{38}{12}\right) = 38.$$

Example 20.2 Show that the following determinant is zero:

$$\begin{vmatrix} 1 & -2 & -3 & 4 \\ -2 & 3 & 4 & -5 \\ 3 & -4 & -5 & 6 \\ -4 & 5 & 6 & -7 \end{vmatrix}.$$

Solution

We have the following calculation:

$$\begin{aligned} \begin{vmatrix} 1 & -2 & -3 & 4 \\ -2 & 3 & 4 & -5 \\ 3 & -4 & -5 & 6 \\ -4 & 5 & 6 & -7 \end{vmatrix} &= \begin{vmatrix} 1 & -2 & -3 & 4 \\ 0 & -1 & -2 & 3 \\ 0 & 2 & 4 & -6 \\ 0 & -3 & -6 & 9 \end{vmatrix} \\ &= \begin{vmatrix} 1 & -2 & -3 & 4 \\ 0 & -1 & -2 & 3 \\ 0 & 0 & 0 & 0 \\ 0 & 0 & 0 & 0 \end{vmatrix} \\ &= 1(-1)(0)(0) = 0. \end{aligned}$$

Example 20.3 Evaluate the determinant of the matrix

$$\begin{pmatrix} 1 & 0 & r \\ 0 & 1 & s \\ x & y & t \end{pmatrix}.$$

Solution

We proceed by elementary row operations:

$$\begin{pmatrix} 1 & 0 & r \\ 0 & 1 & s \\ x & y & t \end{pmatrix} \to \begin{pmatrix} 1 & 0 & r \\ 0 & 1 & s \\ 0 & y & t - xr \end{pmatrix} \to \begin{pmatrix} 1 & 0 & r \\ 0 & 1 & s \\ 0 & 0 & t - xr - ys \end{pmatrix}.$$

Therefore the determinant is $1 \times 1 \times (t - xr - ys)$; that is $t - xr - ys$.

Example 20.4 Suppose that the numbers x_1 and x_2 satisfy the equations

$$\begin{aligned} x_1 - 2x_2 &= 3, \\ 3x_1 + 5x_2 &= 20. \end{aligned}$$

Find x_1 and x_2 using Cramer's rule.

Solution

(Note that this is the same problem as Example 1.1, with x and y replaced by x_1 and x_2.) We have to solve a 2×2 system of linear equations which, in matrix form, may be written as $A\mathbf{x} = \mathbf{b}$, where

$$A = \begin{pmatrix} 1 & -2 \\ 3 & 5 \end{pmatrix}, \quad \mathbf{x} = \begin{pmatrix} x_1 \\ x_2 \end{pmatrix} \quad \text{and} \quad \mathbf{b} = \begin{pmatrix} 3 \\ 20 \end{pmatrix}.$$

Cramer's rule tells us that

$$x_1 = \frac{\begin{vmatrix} 3 & -2 \\ 20 & 5 \end{vmatrix}}{\begin{vmatrix} 1 & -2 \\ 3 & 5 \end{vmatrix}} \quad \text{and} \quad x_2 = \frac{\begin{vmatrix} 1 & 3 \\ 3 & 20 \end{vmatrix}}{\begin{vmatrix} 1 & -2 \\ 3 & 5 \end{vmatrix}}.$$

Recalling that

$$\begin{vmatrix} a & b \\ c & d \end{vmatrix} = ad - bc,$$

we have

$$x_1 = \frac{3(5) - (-2)(20)}{1(5) - (-2)(3)} = \frac{55}{11} = 5 \quad \text{and} \quad x_2 = \frac{1(20) - 3(3)}{11} = \frac{11}{11} = 1.$$

Main Topics

- the determinant of a square matrix
- calculating determinants using row operations
- determinants and invertibility
- Cramer's rule
- formula for the inverse of a 3×3 matrix in terms of determinants

Key Terms, Notations and Formulae

- determinant of A, $\det A$ or $|A|$
- for a 2×2 matrix, $\begin{vmatrix} a & b \\ c & d \end{vmatrix} = ad - bc$
- for $n \times n$ matrices, $\det A$ can be defined in terms of row operations
- $\det A$ calculated by reducing to upper-triangular matrix T
- using row operations, *except* multiplying a row by a constant
- A is invertible if and only if $\det A \neq 0$
- $A\mathbf{x} = \mathbf{b}$ has a unique solution if and only if $\det A \neq 0$
- Cramer's rule: if A is a 2×2 matrix where $\det A \neq 0$, then the solution of $A\mathbf{x} = \mathbf{b}$ is

$$x_1 = \frac{\begin{vmatrix} b_1 & a_{12} \\ b_2 & a_{22} \end{vmatrix}}{|A|} = \frac{\Delta_1}{\Delta} \quad \text{and} \quad x_2 = \frac{\begin{vmatrix} a_{11} & b_1 \\ a_{21} & b_2 \end{vmatrix}}{|A|} = \frac{\Delta_2}{\Delta}$$

- the inverse of a 3×3 matrix in terms of determinants:

$$A^{-1} = \frac{1}{\det A} \begin{pmatrix} +A_{11} & -A_{21} & +A_{31} \\ -A_{12} & +A_{22} & -A_{32} \\ +A_{13} & -A_{23} & +A_{33} \end{pmatrix}.$$

Exercises

20.1 Calculate the determinant of the matrix

$$\begin{pmatrix} 5 & 7 & -2 \\ 10 & 15 & -2 \\ 5 & 8 & -1 \end{pmatrix}.$$

20.2 Calculate the determinant of the matrix

$$\begin{pmatrix} 1 & 2 & -1 & 4 \\ -1 & 3 & 0 & 2 \\ 2 & 1 & 1 & 2 \\ 1 & 4 & 1 & 3 \end{pmatrix}.$$

20.3 Use the determinants-based method of Section 20.4 to find the inverse of the matrix

$$A = \begin{pmatrix} -1 & 2 & 1 \\ 0 & 1 & 2 \\ 3 & 1 & 4 \end{pmatrix}.$$

Problems

20.1 Calculate the determinant of the matrix

$$\begin{pmatrix} 1 & -4 & 7 \\ 2 & -5 & 5 \\ 3 & -18 & 3 \end{pmatrix}.$$

20.2 By computing the determinant, show that the following matrix is not invertible:

$$\begin{pmatrix} 1 & -2 & 3 & 1 \\ 2 & -3 & 7 & 2 \\ -1 & 4 & -1 & 1 \\ 2 & -3 & 7 & 3 \end{pmatrix}.$$

20.3 Calculate the determinant of the matrix

$$\begin{pmatrix} 1 & 1 & 1 & 2 \\ 1 & 3 & 7 & -2 \\ 2 & 6 & 16 & -6 \\ 2 & 9 & 24 & -10 \end{pmatrix}.$$

20.4 Let a, b, c be any three real numbers. Show that the determinant of the matrix

$$\begin{pmatrix} 1 & 1 & 1 \\ a & b & c \\ a^2 & b^2 & c^2 \end{pmatrix}$$

is $(a-b)(b-c)(c-a)$. In what circumstances is the matrix invertible?

20.5 Solve the following system of equations using Cramer's rule:

$$2x + y = 9,$$
$$x - 3y = 1.$$

20.6 Write down the IS–LM equations (see Section 18.5) and obtain the solution by Cramer's rule.

20.7 Let

$$A = \begin{pmatrix} a & b \\ c & d \end{pmatrix} \quad \text{and} \quad B = \begin{pmatrix} e & f \\ g & h \end{pmatrix}$$

be two general 2×2 matrices. Show that

$$\det(AB) = \det A \times \det B.$$

Hence show that if a 2×2 matrix is invertible, then

$$\det(A^{-1}) = \frac{1}{\det A}.$$

(These results are true for general $n \times n$ matrices and can be shown to follow from the rules (i) to (iv) of Section 20.1.)

21 Constrained Optimisation

21.1 The Elementary Theory of the Firm

You will recall that a *firm* can be described in terms of a function whose inputs are amounts k (capital) and l (labour), and whose output is a quantity q (production). The value of q depends upon the values of k and l, and we call the function $q(k, l)$ the *production function* for the firm under consideration.

We are now going to consider the question of how the cost C of producing an amount q is determined. Clearly, it depends on the cost of using the amounts of capital and labour required to produce q. Suppose that the cost of a unit of capital is v and the cost of a unit of labour (the *wage*) is w. Then the cost of k units of capital and l units of labour is

$$vk + wl.$$

In order to produce a given quantity q^* of its product, the firm must select a combination (k, l) of capital and labour for which $q(k, l) = q^*$. As we noted in Chapter 12, there are many combinations which result in a given output, and they all lie on the *isoquant* $q(k, l) = q^*$. The cost of any particular combination (k, l) is given by the expression displayed above, and it is reasonable to suppose that the firm will choose that combination for which the cost is least. In other words, the cost is found by solving the following problem:

$$\text{minimise} \quad vk + wl \quad \text{subject to} \quad q(k, l) = q^*.$$

This is known as a *constrained optimisation* problem. The function to be minimised (or maximised) is called the *objective function* and the condition satisfied by the variables is the *constraint*.

Example Suppose $v = 10$ and $w = 8$, the production function is $5kl$ and the firm wishes to produce $q^* = 1600$ units. We have to solve the problem

$$\text{minimise} \quad 10k + 8l \quad \text{subject to} \quad 5kl = 1600.$$

In this case we can simplify the problem by using the constraint to eliminate one variable, say l. Because $5kl = 1600$, we have $l = 320/k$. Thus the problem is reduced to finding the minimum value of $10k + 8(320/k)$, and we can do this by the standard method – putting the derivative equal to zero. We have

$$10 - \frac{2560}{k^2} = 0, \quad \text{which gives} \quad 10k^2 = 2560.$$

Thus we get the solution

$$k = 16 \quad \text{and} \quad l = \frac{320}{k} = 20,$$

and the minimum cost is $C = 10k + 8l = 160 + 160 = 320$. Note that in this case it is easy to check, by calculating the second derivative, that $k = 16$ is a minimum, rather than any other kind of critical point. ■

It is helpful to look at this problem geometrically, using a method similar to that discussed in Section 14.5. In the first diagram of Figure 21.1, the isoquant $5kl = 1600$ (i.e., $kl = 320$) is sketched. The problem is to find the point on this isoquant for which the cost $10k + 8l$ is least. The lines in the second diagram are typical contours $10k + 8l =$ constant for this function. Note that, as the constant decreases, the contours move towards the origin. The point on the isoquant $kl = 320$ for which $10k + 8l$ is least is determined by the contour nearest the origin which meets the isoquant. It is clear from the third diagram that this contour is a *tangent* to the isoquant.

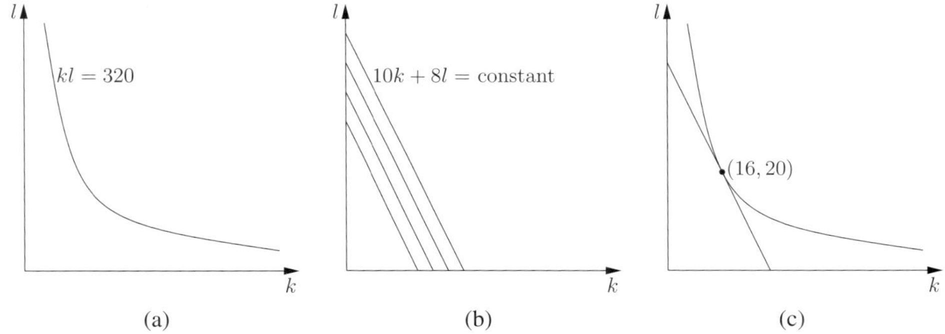

Figure 21.1 Geometrical illustration of the example.

21.2 The Method of Lagrange Multipliers

The analytical method used in the previous section relied on the fact that we could 'solve' the constraint $5kl = 1600$ and obtain l in terms of k. In this way we reduced the problem to the problem of finding the minimum of a function of k only, which we solved in the standard way. In general, given a constrained optimisation problem

$$\text{maximise or minimise} \quad F(x, y) \quad \text{subject to} \quad G(x, y) = 0,$$

it may be difficult or impossible to eliminate one variable. In order to deal with such situations, we use the following method. We form the *Lagrangean* (or 'Lagrangian')

$$L(x, y, \lambda) = F(x, y) - \lambda G(x, y),$$

where λ is a new variable which we call the *Lagrange multiplier*. Then the solution of the constrained optimisation problem can be found by solving the equations

$$\frac{\partial L}{\partial x} = 0, \quad \frac{\partial L}{\partial y} = 0 \quad \text{and} \quad G(x, y) = 0.$$

(Note that the last equation can also be written as $\partial L/\partial \lambda = 0$.) The method is named after the Italian mathematician Joseph-Louis Lagrange, born Giuseppe Lodovico Lagrangia (1736–1813).

To see why this works, observe that the constraint $G(x, y) = 0$ defines y implicitly as a function of x, say $y = y(x)$. As in the discussion of implicit functions in Chapter 12, we cannot necessarily write down a formula for $y(x)$. We have to minimise $F(x, y(x))$, considered as a function $\Phi(x) = F(x, y(x))$ of x. Using the condition $\Phi'(x) = 0$ and the chain rule, we have

$$0 = \frac{d\Phi}{dx} = \frac{\partial F}{\partial x} + \frac{\partial F}{\partial y}\frac{dy}{dx}.$$

But we also have the constraint $G(x, y(x)) = 0$, and differentiating this we get

$$0 = \frac{\partial G}{\partial x} + \frac{\partial G}{\partial y}\frac{dy}{dx}.$$

Eliminating dy/dx from the two equations gives

$$\frac{\partial F}{\partial x}\bigg/\frac{\partial G}{\partial x} = \frac{\partial F}{\partial y}\bigg/\frac{\partial G}{\partial y},$$

and defining λ to be the value of both ratios we obtain

$$\frac{\partial F}{\partial x} - \lambda\frac{\partial G}{\partial x} = 0 \quad \text{and} \quad \frac{\partial F}{\partial y} - \lambda\frac{\partial G}{\partial y} = 0, \quad \text{that is} \quad \frac{\partial L}{\partial x} = 0 \quad \text{and} \quad \frac{\partial L}{\partial y} = 0.$$

These are precisely the first two conditions stated above. Together with the constraint equation, they determine the optimum values of x and y (and the corresponding value of λ).

Example The problem studied in Section 21.1 is to minimise the function $F(k, l) = 10k + 8l$ subject to the constraint $5kl = 1600$, which may be written as $G(k, l) = 5kl - 1600 = 0$. The Lagrangean is

$$L = 10k + 8l - \lambda(5kl - 1600),$$

and the optimal values of k and l are determined by the equations

$$\frac{\partial L}{\partial k} = 10 - 5\lambda l = 0,$$
$$\frac{\partial L}{\partial l} = 8 - 5\lambda k = 0,$$
$$5kl - 1600 = 0.$$

We can eliminate λ using the first two equations. The first equation implies $5\lambda = 10/l$ and the second implies $5\lambda = 8/k$. Therefore, $10/l = 8/k$, or $l = 5k/4$. Using this information in the third equation, we obtain

$$5k\left(\frac{5k}{4}\right) = 1600, \quad \text{that is} \quad 25k^2 = 6400.$$

Thus $k = 16$ and $l = 5k/4 = 20$, as we found earlier. ■

The reader might ask how we know that we have indeed found the minimum rather than a maximum. When we studied 'unconstrained optimisation' in Chapter 13, we made use of second-order conditions to indicate the nature of a critical point. There is also a set of second-order conditions for constrained optimisation problems of the type discussed here. However, we shall not discuss these explicitly as they are substantially more complex than those of Chapter 13. In practice, particularly in economics problems, it often suffices to use common sense.

21.3 The Cost Function

In Chapters 8 to 10 we began to investigate the problem of profit maximisation. In that discussion, the cost function $C(q)$ – which at that stage was simply some *given* function – played a crucial role. But we did not explain how the cost function might be obtained. In this section we shall obtain the cost function as the solution to a constrained optimisation problem.

We consider the general problem of minimising total capital and labour costs, given that a prescribed output must be produced. This means that the resulting cost function will depend on the firm's production function and on the costs per unit for capital and labour.

In Section 21.1 we considered an example in which the required production level was 1600 units. The answer to that problem, the minimum cost of producing 1600 units, is just $C(1600)$, the cost function $C(q)$ evaluated at $q = 1600$. In this light it is clear how we can determine a *general* expression for the cost function using the technique of Lagrange multipliers: we simply replace the constraint $5kl = 1600$ by $5kl = q^*$, where q^* is an arbitrary level of desired production.

Working through the calculation with q^* instead of 1600, the optimal values of k and l turn out to be

$$k = \frac{2}{5}\sqrt{q^*} \quad \text{and} \quad l = \frac{1}{2}\sqrt{q^*},$$

and so the minimum cost of producing q^* units is

$$C(q^*) = 10k + 8l = 8\sqrt{q^*}.$$

Since q^* here represents any given quantity, it follows that the cost function is $C(q) = 8\sqrt{q}$.

The following example is superficially similar, but in Section 21.5 we shall see that it leads to a supply set with quite different characteristics.

Example We derive the cost function $C(q)$ for a firm with production function $8k^{1/2}l^{1/4}$, given that capital costs 4 per unit and labour costs 5 per unit.

The cost-minimisation problem for a given output q^* is

$$\text{minimise} \quad 4k + 5l \quad \text{subject to} \quad 8k^{1/2}l^{1/4} = q^*,$$

so the Lagrangean is $L = 4k + 5l - \lambda(8k^{1/2}l^{1/4} - q^*)$. The solution is determined by the three equations

$$\frac{\partial L}{\partial k} = 4 - \lambda\left(\frac{8}{2}k^{-1/2}l^{1/4}\right) = 0,$$
$$\frac{\partial L}{\partial l} = 5 - \lambda\left(\frac{8}{4}k^{1/2}l^{-3/4}\right) = 0,$$
$$8k^{1/2}l^{1/4} - q^* = 0.$$

From the first two equations we get

$$2\lambda = 2k^{1/2}l^{-1/4} = 5k^{-1/2}l^{3/4},$$

so that $l = 2k/5$. Substituting this into the last equation gives

$$8k^{1/2}\left(\frac{2}{5}\right)^{1/4}k^{1/4} = q^*,$$

so

$$k = \frac{1}{16}\left(\frac{5}{2}\right)^{1/3}(q^*)^{4/3} \quad \text{and} \quad l = \frac{2}{5}k = \frac{1}{16}\left(\frac{2}{5}\right)^{2/3}(q^*)^{4/3}.$$

It follows that the minimum cost of producing q^* units is

$$4\left(\frac{1}{16}\left(\frac{5}{2}\right)^{1/3}(q^*)^{4/3}\right) + 5\left(\frac{1}{16}\left(\frac{2}{5}\right)^{2/3}(q^*)^{4/3}\right) = \frac{15}{16}\left(\frac{2}{5}\right)^{2/3}(q^*)^{4/3}.$$

Thus the firm's cost function is $C(q) = Zq^{4/3}$, where Z is the constant $\frac{15}{16}\left(\frac{2}{5}\right)^{2/3}$, which is approximately 0.509. ■

21.4 The Efficient Small Firm Again

In Chapter 10 we developed a procedure for finding the supply set for an efficient small firm, where 'efficient' is interpreted to mean 'profit-maximising'. Roughly speaking, the result obtained there was that the supply set may be determined in terms of the derivative of the cost function.

Example (continued) Suppose that the firm whose cost function was found to be $C(q) = Zq^{4/3}$ is an efficient small firm. When the market price is fixed at p, the firm's profit from producing q is

$$\Pi_p(q) = pq - C(q) = pq - Zq^{4/3}.$$

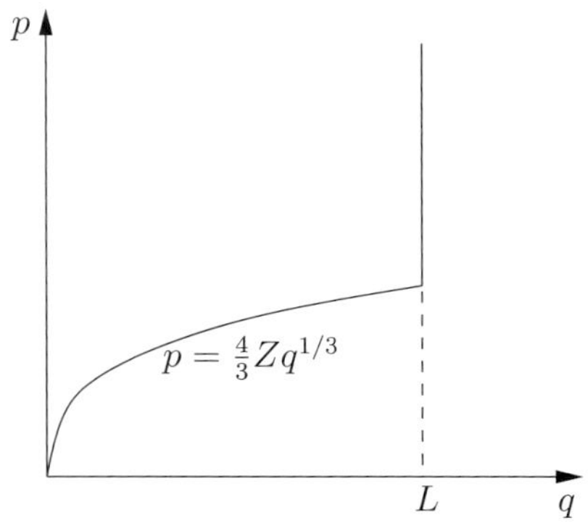

Figure 21.2 A supply set.

For each p, the profit-maximising value of q is determined by the first-order condition

$$\Pi_p'(q) = p - C'(q) = p - \frac{4}{3}Zq^{1/3} = 0.$$

So p and q are related by $p = (4Z/3)q^{1/3}$; in other words, the inverse supply function $p^S(q)$ is $(4Z/3)q^{1/3}$.

The supply function q^S is obtained by inverting the relation $p = (4Z/3)q^{1/3}$, which yields

$$q = \frac{27p^3}{64Z^3} = \frac{16}{5}p^3,$$

since $Z = \frac{15}{16}\left(\frac{2}{5}\right)^{2/3}$.

Note that when the firm is producing at its profit-maximising level q, the actual profit is

$$\Pi(q) = qC'(q) - C(q) = q\left(\frac{4}{3}Zq^{1/3}\right) - Zq^{4/3} = \frac{Z}{3}q^{4/3}.$$

This is zero when $q = 0$ and positive for $q > 0$. So the startup point is $q_s = 0$.

If L is the upper limit on the firm's production capacity, and $0 \leq q \leq L$, then a point (q,p) is in the supply set if $q = (16/5)p^3$. For larger values of p, when $L < (16/5)p^3$ or $p > (5L/16)^{1/3}$, the firm should produce L. So for these values of p the points (L,p) are in the supply set. See Figure 21.2. ■

It might be hoped that we can carry out this procedure more generally. There is, however, one subtle consideration. The first step in our analysis is to say that, when the going price is p, the profit obtained from producing q is $pq - C(q)$, and so the maximum profit occurs at the critical point, given by $p = C'(q)$. However, we must remember that *a critical point is not necessarily a maximum*. In the next section we shall see that it can happen that the equation $p = C'(q)$ determines a *minimum* value of $pq - C(q)$. In that case our analysis of the efficient small firm must proceed along different lines.

21.5 The Cobb–Douglas Firm

Suppose that a firm has a Cobb–Douglas production function

$$q(k,l) = Ak^{\alpha}l^{\beta} \quad (A, \alpha, \beta > 0).$$

We begin by deriving a general expression for the cost function, given the production function and the unit costs. If the unit capital and labour costs are, respectively, v and w, then the minimum cost of producing q units may be found by minimising $vk + wl$ subject to the constraint $Ak^{\alpha}l^{\beta} = q$. The Lagrangean for this problem is

$$L = vk + wl - \lambda\left(Ak^{\alpha}l^{\beta} - q\right).$$

The solution is given by the three equations

$$\begin{aligned}\frac{\partial L}{\partial k} &= v - \alpha\lambda Ak^{\alpha-1}l^{\beta} = 0,\\ \frac{\partial L}{\partial l} &= w - \beta\lambda Ak^{\alpha}l^{\beta-1} = 0,\\ Ak^{\alpha}l^{\beta} &- q = 0.\end{aligned}$$

From the first two equations, we obtain two expressions for $A\lambda$:

$$A\lambda = \frac{v}{\alpha k^{\alpha-1}l^{\beta}} = \frac{w}{\beta k^{\alpha}l^{\beta-1}},$$

so that

$$l = \left(\frac{v}{w}\frac{\beta}{\alpha}\right)k.$$

Observe that the ratio of the optimal values of l and k varies inversely with the relative expense w/v, the relative cost of labour over capital, while it varies directly with β/α, which is a measure of the 'productivity' of labour compared with that of capital.

Substituting for l in the constraint equation, we have

$$Ak^{\alpha}\left(\frac{v\beta}{w\alpha}k\right)^{\beta} = q,$$

which is of the form $Uk^{\alpha+\beta} = q$, where $U = A\left(v\beta/w\alpha\right)^{\beta}$ is a constant depending only on the given values of A, α, β, v and w. Thus we have

$$k = Vq^{1/(\alpha+\beta)} \quad \text{and} \quad l = Wq^{1/(\alpha+\beta)},$$

where $V = (1/U)^{1/(\alpha+\beta)}$ and $W = \left(v\beta/w\alpha\right)V$. Substituting these values in the total cost $vk + wl$, we see that the minimum cost is

$$vk + wl = (vV + wW)q^{1/(\alpha+\beta)}.$$

So the dependence on q is really very simple:

$$C(q) = Zq^{1/(\alpha+\beta)},$$

where Z depends only on the given values of A, α, β, v and w. Note that the examples considered in Section 21.3 are both of this form.

Now let us try to work out the firm's supply set. For convenience, we write

$$C(q) = Zq^f, \quad \text{where} \quad f = \frac{1}{\alpha + \beta}.$$

So the profit from producing q when the going price is p is $\Pi_p(q) = pq - Zq^f$. This has a critical point where $\Pi'_p = 0$, that is at the point q_0 satisfying

$$p - Zfq_0^{f-1} = 0.$$

The nature of the critical point at q_0 is determined by the sign of Π''_p at that point. We have

$$\Pi''_p(q_0) = -Zf(f-1)q_0^{f-2} = (1-f)\frac{p}{q_0}.$$

Since both p and q_0 are positive, the sign is simply that of $1 - f$.

We consider two cases: $f < 1$ and $f > 1$.

Case 1. Suppose first that $f < 1$. Since $f = 1/(\alpha+\beta)$, this is the case where $\alpha+\beta > 1$; in other words, we have *increasing returns to scale* (see Section 12.4). Here the sign of Π''_p is positive, and so q_0 is a *minimum*. In this case the graph of a typical Π_p is as shown in Figure 21.3a. It will be seen that Π_p is negative for small values of q, has a minimum at q_0 and then increases without bound.

Since we are assuming that the firm is 'small', there will be a limit L on its production, which is independent of the going price p. The profit-maximisation problem is

$$\text{maximise} \quad \Pi_p(q), \quad \text{where} \quad q \in [0, L].$$

If the firm is very small, $\Pi_p(L)$ will be negative and in the interval $[0, L]$ the maximum value of Π_p will occur at 0. This means that the firm will produce 0 when $\Pi_p(L) < 0$, that is when

$$pL - ZL^f < 0, \quad \text{or} \quad p < ZL^{f-1}.$$

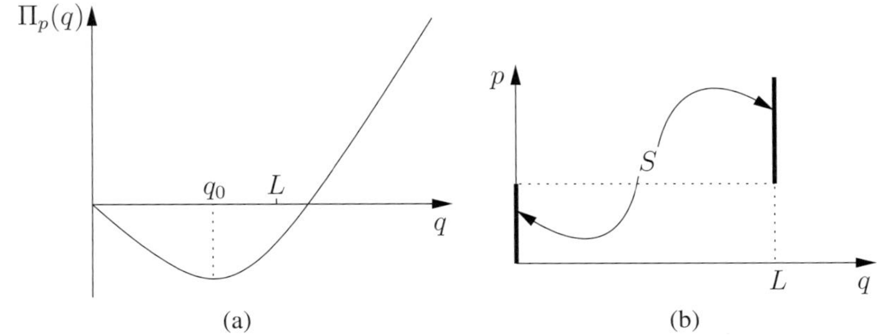

Figure 21.3 The case of increasing returns to scale.

On the other hand, if $p > ZL^{f-1}$, $\Pi_p(L) > 0$ and the firm will produce L. Thus the supply set consists of the points $(0,p)$ for $p < ZL^{f-1}$ and the points (L,p) for $p > ZL^{f-1}$ (see Figure 21.3b).

Case 2. The second case is when $f > 1$. Here we have $\alpha + \beta < 1$, which corresponds to *decreasing returns to scale*. In this case the sign of Π_p'' at q_0 is negative and so q_0 is a *maximum*, and the graph of a typical Π_p is as shown in Figure 21.4a. The profit-maximising strategy is to produce q_0, provided it is within the firm's capability to do so, in other words provided $q_0 \leq L$. So the supply set consists of the points on the curve $p = Zfq^{f-1}$ for $p \leq ZfL^{f-1}$ together with the points (L,p) for $p > ZfL^{f-1}$ (see Figure 21.4b).

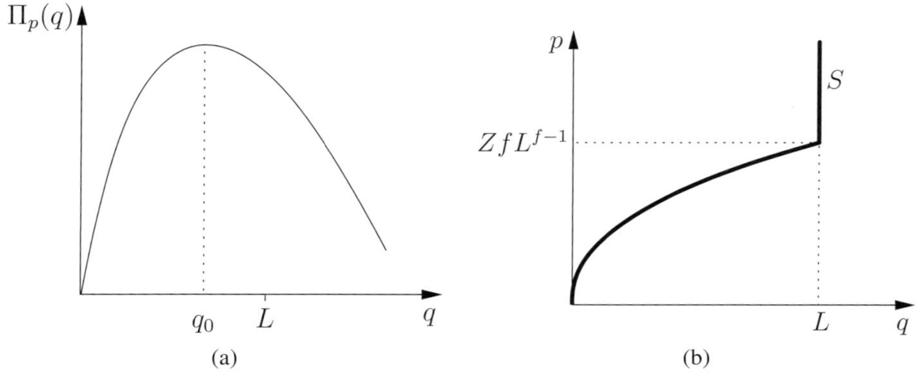

Figure 21.4 The case of decreasing returns to scale.

Worked Examples

Example 21.1 A firm's weekly output is given by the production function $q(k,l) = k^{3/4}l^{1/4}$, and the unit costs for capital and labour are $v = 1$ and $w = 5$ per week. Find the minimum cost of producing a weekly output of 5000 and the corresponding values of k and l.

Solution

The problem to be solved is the constrained optimisation problem

$$\text{minimise} \quad k + 5l \quad \text{subject to} \quad k^{3/4}l^{1/4} = 5000.$$

The Lagrangean for the problem is

$$L = k + 5l - \lambda(k^{3/4}l^{1/4} - 5000),$$

and the optimal values of k and l are the solutions to the three equations

$$\frac{\partial L}{\partial k} = 1 - \frac{3}{4}\lambda k^{-1/4}l^{1/4} = 0,$$
$$\frac{\partial L}{\partial l} = 5 - \frac{1}{4}\lambda k^{3/4}l^{-3/4} = 0,$$
$$k^{3/4}l^{1/4} - 5000 = 0.$$

The first two equations imply that

$$\frac{1}{3}k^{1/4}l^{-1/4} = 5k^{-3/4}l^{3/4},$$

so $l = k/15$ at the optimal values. Substituting this into the third equation gives

$$k^{3/4}\left(\frac{1}{15}\right)^{1/4}k^{1/4} = 5000,$$

so

$$k = 5000(15)^{1/4} \quad \text{and} \quad l = \frac{k}{15} = 5000(15)^{-3/4}.$$

Then, the minimum cost is

$$k + 5l = 5000(15)^{1/4} + 5(5000)(15)^{-3/4} = 100\,000(15)^{-3/4},$$

which is approximately 13120.

Example 21.2 A firm has production function $q(k, l) = k^{1/4}l^{1/3}$, and the unit costs for capital and labour are $v = 60$ and $w = 10$. Assuming that it is an efficient small firm, with an upper limit L on its production level, find its supply set.

Solution

We first find the firm's cost function. This, as a function of q, is the solution to the constrained optimisation problem

$$\text{minimise} \quad 60k + 10l \quad \text{subject to} \quad k^{1/4}l^{1/3} = q.$$

The Lagrangean is

$$L = 60k + 10l - \lambda(k^{1/4}l^{1/3} - q),$$

and the optimal k and l are the solutions to

$$\frac{\partial L}{\partial k} = 60 - \frac{1}{4}\lambda k^{-3/4}l^{1/3} = 0,$$
$$\frac{\partial L}{\partial l} = 10 - \frac{1}{3}\lambda k^{1/4}l^{-2/3} = 0,$$
$$k^{1/4}l^{1/3} - q = 0.$$

The first two equations show that $l = 8k$. Substituting this into the third equation, we obtain

$$k^{1/4}(8k)^{1/3} = q, \quad \text{so} \quad 2k^{7/12} = q.$$

It follows that

$$k = \left(\frac{q}{2}\right)^{12/7} \quad \text{and} \quad l = 8\left(\frac{q}{2}\right)^{12/7}.$$

The cost function, the minimum cost of producing q units, is

$$C(q) = 60k + 10l = 140\left(\frac{q}{2}\right)^{12/7} = Zq^{12/7},$$

where $Z = 140/2^{12/7}$. For any fixed price p, the profit function is

$$\Pi(q) = pq - Zq^{12/7}.$$

The condition $\Pi'(q) = 0$ yields

$$p = \frac{12}{7}Zq^{5/7}$$

so, for $q > 0$, the profit is

$$\Pi(q) = \left(\frac{12}{7}Zq^{5/7}\right)q - Zq^{12/7} = \frac{5}{7}Zq^{12/7}.$$

Since $\Pi(q) > \Pi(0) = 0$ for all $q > 0$, the startup point is 0.

Thus, for $0 \leq q \leq L$, the points (q, p) in the supply set are those for which $p = 12Zq^{5/7}/7$. The corresponding range for p is $0 \leq p \leq 12ZL^{5/7}/7$. If the going price p is greater than $12ZL^{5/7}/7$, the firm should produce L, so for these values of p the points (L, p) belong to the supply set.

Example 21.3 A firm has production function $q(k,l) = 50k^{2/3}l^{1/3}$, and unit capital and labour costs of 6 and 4, respectively. What is the *maximum output* achievable in a week if the firm spends no more than 1000 each week?

Solution

This is a rather different type of problem from those we have concentrated on in this chapter, but the technique of Lagrange multipliers will yield the result once we have framed the problem as a constrained optimisation problem. Here, we seek to maximise the output $50k^{2/3}l^{1/3}$, and the constraint is that the firm spends no more than 1000. The amount the firm spends is $6k + 4l$ and, clearly, the optimal values of k and l will satisfy $6k + 4l = 1000$. This is the constraint then, and the Lagrangean is

$$L = 50k^{2/3}l^{1/3} - \lambda(6k + 4l - 1000).$$

The equations to solve are

$$\begin{aligned}
\frac{\partial L}{\partial k} &= \frac{100}{3}k^{-1/3}l^{1/3} - 6\lambda = 0, \\
\frac{\partial L}{\partial l} &= \frac{50}{3}k^{2/3}l^{-2/3} - 4\lambda = 0, \\
& 6k + 4l - 1000 = 0.
\end{aligned}$$

From the first two equations:

$$\lambda = \frac{100}{18}k^{-1/3}l^{1/3} = \frac{50}{12}k^{2/3}l^{-2/3},$$

so that $l = 3k/4$. The third equation then gives

$$6k + 4\left(\frac{3}{4}k\right) = 1000, \quad \text{so} \quad k = \frac{1000}{9},$$

from which it follows that $l = 1000/12$ and the maximal output is

$$50\left(\frac{1000}{9}\right)^{2/3}\left(\frac{1000}{12}\right)^{1/3} = \frac{50\,000}{12^{1/3}9^{2/3}},$$

which is approximately 5047.6.

Example 21.4 The quantity of its good which is produced from x units of X and y of Y is given by $Q(x, y) = x^{1/4}y^{3/4}$. If the firm spends no more than \$1280 each week on the raw materials, what is its maximum possible weekly production, given that one unit of X costs \$16 and one unit of Y costs \$1?

Solution

The problem is to maximise $Q(x, y)$ subject to the constraint that the amount spent on raw materials is at most \$1280. Clearly, the optimal values of x and y will satisfy the constraint $16x + y = 1280$. The Lagrangean is

$$L = x^{1/4}y^{3/4} - \lambda(16x + y - 1280),$$

and the equations to solve are

$$\begin{aligned}
\frac{\partial L}{\partial x} &= \frac{1}{4}x^{-3/4}y^{3/4} - 16\lambda = 0, \\
\frac{\partial L}{\partial y} &= \frac{3}{4}x^{1/4}y^{-1/4} - \lambda = 0, \\
16x + y - 1280 &= 0.
\end{aligned}$$

The first two equations, on eliminating λ, yield $y = 48x$. Then the third equation implies $64x = 1280$, so that $x = 20$ and $y = 48x = 960$. The maximum quantity is therefore $Q(20, 960) = 20^{1/4}960^{3/4}$.

Example 21.5 A firm manufactures two products, X and Y, and sells these in related markets. Suppose that the firm is the only producer of X and Y and that the inverse demand functions for X and Y are

$$p^X = 100 - y - 4x \quad \text{and} \quad p^Y = 50 - y - x.$$

Determine the production levels that maximise weekly profit, given that the cost function is $C(x, y) = 10x + 5y$ and that *the firm must spend no more than* 100 *each week*.

Solution

This is like the problems we studied in Chapter 13, where we looked at what we may now call 'unconstrained optimisation'. However, there is a constraint in this problem, namely $10x + 5y = 100$. The function to be maximised is the profit function

$$\Pi(x, y) = xp^{\mathrm{X}} + yp^{\mathrm{Y}} - C(x, y) = 90x - 4x^2 + 45y - y^2 - 2xy.$$

The Lagrangean for the problem is

$$L = 90x - 4x^2 + 45y - y^2 - 2xy - \lambda(10x + 5y - 100),$$

and the equations to solve are

$$\begin{aligned}\frac{\partial L}{\partial x} &= 90 - 8x - 2y - 10\lambda = 0,\\ \frac{\partial L}{\partial y} &= 45 - 2y - 2x - 5\lambda = 0,\\ 10x + 5y - 100 &= 0.\end{aligned}$$

The first two equations, on eliminating λ, yield $y = 2x$. Then the third equation gives $20x = 100$, so $x = 5$ and $y = 10$.

Main Topics

- theory of the firm; minimising total cost of capital and labour
- the cost function as the solution to a constrained optimisation problem
- general constrained optimisation and Lagrange multiplier method
- deriving the cost function and supply set of an efficient small firm given the production function and unit capital and labour costs

Key Terms, Notations and Formulae

- capital, k; labour, l
- unit cost of capital, v; unit cost of labour (wage), w
- total production costs $vk + wl$
- constrained optimisation: optimise $F(x, y)$ subject to $G(x, y) = 0$
- objective function; constraint
- Lagrange multiplier, λ
- Lagrangean $L(x, y, \lambda) = F(x, y) - \lambda G(x, y)$
- solve equations $\dfrac{\partial L}{\partial x} = 0$, $\dfrac{\partial L}{\partial y} = 0$ and $G = 0$
- cost function $C(q)$; $C(q^*)$ is solution to minimise $vk + wl$ subject to $q(k, l) = q^*$
- efficient small firm with Cobb–Douglas production function $q(k, l) = Ak^{\alpha}l^{\beta}$ has $C(q) = Zq^{1/(\alpha+\beta)} = Zq^f$
- form of supply curve of the efficient small firm depends on f

Exercises

21.1 Find the positive values of x and y which maximise

$$\frac{2y}{y+2} + \frac{x}{x+1}$$

subject to the constraint $x + y = 120$.

21.2 A firm has a production function given by

$$q(k, l) = k^{1/4} l^{1/12}$$

where k and l are the amounts of capital and labour used.

(i) If the prices of capital and labour are v and w per unit, respectively, find the values of k and l that will minimise the firm's costs when they produce a quantity Q.

(ii) The firm sells its product at a fixed price of p per unit. What is the firm's profit function?

(iii) Find the quantity that maximises the firm's profit.

Problems

21.1 A firm has weekly production function $q(k, l) = k^{1/4} l^{1/2}$ and the unit weekly costs for capital and labour are $v = 20$ and $w = 10$. The firm wishes to produce 200 units a week of its good. Find the minimum cost of doing so.

21.2 Find the values of x and y which maximise the function $3\sqrt{x} + 4\sqrt{y}$ subject to the constraint $x + y = 100$. Give a geometrical interpretation of the result, using new variables $X = \sqrt{x}$ and $Y = \sqrt{y}$.

21.3 A firm has production function $q(k, l) = 10k^{1/4} l^{1/4}$ and the unit costs for capital and labour are $v = 300$ and $w = 100$. Find the firm's cost function.

If the manufactured good can be sold for 2000 per unit, what amount should be produced in order to maximise profit?

21.4 An efficient small firm has production function $q(k, l) = k^{1/5} l^{1/5}$ and the unit costs for capital and labour are $v = 2$ and $w = 3$. Find its supply set assuming that there is a limit L on its production level.

21.5 Suppose that a small efficient firm has cost function $C(q) = Zq$ for some constant Z, and that its maximum production level is L. Determine its supply set.

21.6 A firm manufactures a good from two raw materials, X and Y. The quantity of the good which is produced from x units of X and y of Y is given by

$$Q(x, y) = (\sqrt{x} + 2\sqrt{y})^2.$$

Each unit of X costs the firm \$2 and each unit of Y costs \$1. Find the minimum cost of producing 100 units of the manufactured good.

22 Lagrangeans and the Consumer

22.1 Lagrangeans: A More General Formulation

The method of Lagrange multipliers is useful in many areas of economic theory. In addition to the theory of the firm, discussed in the previous chapter, there are clearly applications to the consumer choice problem (Section 14.5). For these purposes a more general formulation is often needed.

In Section 21.2 we introduced a simple version, with two variables and one constraint. This can be extended to the more general case where there are n variables and m constraints in the following way. Suppose the problem is

$$\text{maximise or minimise} \quad F(x_1, \ldots, x_n)$$

$$\text{subject to} \quad G_1(x_1, \ldots, x_n) = 0, \quad \ldots, \quad G_m(x_1, \ldots, x_n) = 0.$$

We define the Lagrangean

$$L = F(x_1, \ldots, x_n) - \lambda_1 G_1(x_1, \ldots, x_n) - \cdots - \lambda_m G_m(x_1, \ldots, x_n),$$

where there are now m Lagrange multipliers $\lambda_1, \lambda_2, \ldots, \lambda_m$. The solution is obtained by solving the n first-order equations

$$\frac{\partial L}{\partial x_1} = 0, \quad \frac{\partial L}{\partial x_2} = 0, \quad \ldots, \quad \frac{\partial L}{\partial x_n} = 0,$$

together with the m constraint equations

$$G_1 = 0, \quad G_2 = 0, \quad \ldots, \quad G_m = 0.$$

The proof is similar to that for the two-variable case, given in Section 21.2. Notice that in the general case there are $n + m$ equations (the n first-order conditions and the m constraints) and $n + m$ unknowns (the n variables and the m Lagrange multipliers). So, in principle at least, there is some hope of finding a solution.

22.2 The Elementary Theory of the Consumer

We now apply the Lagrangean method to a more general form of the consumer choice problem introduced in Section 14.5. Specifically, we consider a consumer (also known as a 'household') who may purchase quantities $x_1, x_2, \ldots, x_n$ of n goods, at given prices $p_1, p_2, \ldots, p_n$, subject to a budget constraint. The behaviour of such a consumer can be formulated as a constrained optimisation problem.

Suppose that the consumer wishes to maximise a utility function $u(\mathbf{x})$, where $\mathbf{x} = (x_1, x_2, \ldots, x_n)$, and that the budget available is M. We shall assume that an increase in any of the n goods results in an increase in utility, so that the maximal utility will be achieved when the total expenditure is actually equal to M. That is, we shall take the budget constraint to be

$$p_1x_1 + p_2x_2 + \cdots + p_nx_n = M.$$

In plain language, since each of the n goods is 'good', the consumer will spend all his or her money to attain maximum utility.

In vector notation, the consumer's problem is

$$\text{maximise} \quad u(\mathbf{x}) \quad \text{subject to} \quad \mathbf{p}.\mathbf{x} = M.$$

The Lagrangean for this problem is

$$L = u(x_1, x_2, \ldots, x_n) - \lambda(p_1x_1 + p_2x_2 + \cdots + p_nx_n - M).$$

We have $n+1$ unknowns (the n quantities x_i and the Lagrange multiplier λ), and there are $n+1$ equations to be solved: the first-order conditions $\partial L/\partial x_i = 0$, $i = 1, 2, \ldots, n$, and the constraint equation. Explicitly, these equations are

$$\frac{\partial u}{\partial x_1} - \lambda p_1 = 0, \quad \frac{\partial u}{\partial x_2} - \lambda p_2 = 0, \quad \ldots, \quad \frac{\partial u}{\partial x_n} - \lambda p_n = 0,$$

$$p_1x_1 + p_2x_2 + \cdots + p_nx_n = M.$$

In any given case where the utility function u, the prices p_i and the budget M are given, it may be possible to solve these equations to find the quantities of each good which the consumer should purchase in order to maximise utility.

Example Suppose there are two goods with prices $p_1 = 2$ and $p_2 = 5$, the income is $M = 40$ and the utility function is

$$u(x_1, x_2) = x_1^{1/3}x_2^{1/2}.$$

The budget constraint is

$$2x_1 + 5x_2 = 40$$

and the Lagrangean is

$$L = x_1^{1/3}x_2^{1/2} - \lambda(2x_1 + 5x_2 - 40).$$

We have to solve the three equations

$$\frac{1}{3}x_1^{-2/3}x_2^{1/2} - 2\lambda = 0, \quad \frac{1}{2}x_1^{1/3}x_2^{-1/2} - 5\lambda = 0, \quad 2x_1 + 5x_2 = 40$$

for the three unknowns x_1, x_2, λ. From the first two equations we get

$$\lambda = \frac{1}{6}x_1^{-2/3}x_2^{1/2} = \frac{1}{10}x_1^{1/3}x_2^{-1/2}.$$

Solving for x_2 in terms of x_1, we get $x_2 = 3x_1/5$. Substituting this in the budget constraint gives

$$2x_1 + 5\left(\frac{3}{5}\right)x_1 = 40, \quad \text{that is} \quad 5x_1 = 40.$$

From this we get the optimum values (which we denote by asterisks to distinguish them from the symbols for the variables):

$$x_1^* = 8, \quad x_2^* = \frac{24}{5} \quad \text{and} \quad \lambda^* = \frac{1}{\sqrt{120}}.$$

■

22.3 The Price Ratio and the Tangency Condition

In microeconomics, important questions about the existence and uniqueness of solutions to the Lagrangean equations are discussed. Here we shall confine ourselves to pointing out some simple consequences, which hold under the assumptions that the solution $(x_1^*, x_2^*, \ldots, x_n^*, \lambda^*)$ is unique, that $\lambda^* \neq 0$ and that no x_i^* is zero.

We have already obtained the Lagrangean for the case of n goods:

$$L = u(x_1, x_2, \ldots, x_n) - \lambda(p_1x_1 + p_2x_2 + \cdots + p_nx_n - M).$$

The ith and jth first-order equations are

$$\frac{\partial L}{\partial x_i} = \frac{\partial u}{\partial x_i} - \lambda p_i = 0,$$
$$\frac{\partial L}{\partial x_j} = \frac{\partial u}{\partial x_j} - \lambda p_j = 0.$$

Eliminating λ, we get

$$\frac{p_i}{p_j} = \frac{\partial u}{\partial x_i} \Big/ \frac{\partial u}{\partial x_j}.$$

This says that the *price ratio* p_i/p_j is equal to another ratio, known to economists as the *marginal rate of substitution*, evaluated at the optimal point.

In the case $n = 2$, the result

$$\frac{p_1}{p_2} = \frac{\partial u}{\partial x_1} \Big/ \frac{\partial u}{\partial x_2}$$

corresponds to the fact that the budget line is tangent to the indifference curve at the optimum point (Section 14.5). To see this, we note that the gradient of the budget line

$p_1x_1 + p_2x_2 = M$ is $-p_1/p_2$, and the gradient of the tangent to the indifference curve $u(x_1,x_2) = c$ is

$$\frac{dx_2}{dx_1} = -\frac{\partial u/\partial x_1}{\partial u/\partial x_2}.$$

Therefore, at the optimal point, the gradients of the budget line and the indifference curve are equal, as claimed.

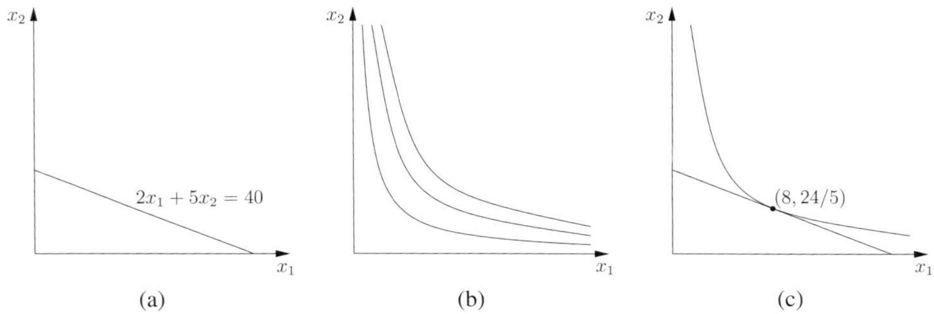

Figure 22.1 Geometrical interpretation of the consumer's problem.

Figure 22.1 illustrates the conclusion for the specific example discussed in the previous section. The budget constraint tells us that we are looking for a solution which lies on the line $2x_1 + 5x_2 = 40$, which is drawn in Figure 22.1a. Figure 22.1b shows some indifference curves $u(x_1,x_2) =$ constant. As the constant increases, this curve moves in a 'north-easterly' direction away from the origin. In order to maximise u, subject to the budget constraint, we want the curve which is furthest from the origin and which intersects the budget line. Intuitively, it is clear that this curve will be the one which touches the budget line at the optimum point $(8, 24/5)$. The value of the utility at this point is approximately

$$u(8, 24/5) = (8)^{1/3}(24/5)^{1/2} = 4.38.$$

So this particular indifference curve is tangent to the budget line, as shown in Figure 22.1c.

It is important to note that it is the convexity of U_c, the set of points for which $u(x_1,x_2) \geq c$, which ensures that the tangency condition holds and the method of the Lagrange multiplier works.

22.4 The Consumer's Demand Functions

The optimum quantities x_i^* obtained by solving the Lagrangean problem tell us how much of each good an individual consumer will demand, assuming that he behaves rationally and optimises his utility within his budget. Clearly, these quantities depend upon the prices p_i and the budget limit M. In order to make this clear, we can write

$$x_i^* = q_i(p_1, p_2, \ldots, p_n, M) = q_i(\mathbf{p}, M),$$

for each $i = 1, 2, \ldots, n$. The functions q_i are the *consumer's demand functions*. They are also known as the *Marshallian* demand functions, after the British economist Alfred Marshall (1842–1924).

Example Consider again the consumer with utility function $x_1^{1/3}x_2^{1/2}$. We know that, in general, the price ratio satisfies the 'tangency condition'

$$\frac{p_1}{p_2} = \frac{\partial u}{\partial x_1} \Big/ \frac{\partial u}{\partial x_2}.$$

Here we have $u(x_1, x_2) = x_1^{1/3}x_2^{1/2}$, so

$$\frac{\partial u}{\partial x_1} \Big/ \frac{\partial u}{\partial x_2} = \frac{(1/3)x_1^{-2/3}x_2^{1/2}}{(1/2)x_1^{1/3}x_2^{-1/2}} = \frac{2x_2}{3x_1}.$$

So the condition is that the optimal values satisfy

$$\frac{p_1}{p_2} = \frac{2x_2^*}{3x_1^*}.$$

Substituting for x_2^* in the budget constraint $p_1x_1^* + p_2x_2^* = M$ we obtain

$$p_1x_1^* + p_2\left(\frac{3}{2}\frac{p_1}{p_2}x_1^*\right) = M, \quad \text{that is} \quad x_1^* = \frac{2M}{5p_1},$$

and then $x_2^* = 3M/5p_2$. The consumer's demand functions are therefore

$$q_1(p_1, p_2, M) = \frac{2M}{5p_1} \quad \text{and} \quad q_2(p_1, p_2, M) = \frac{3M}{5p_2}.$$

Note that the optimal quantities when the budget is 40 and the prices are, respectively, 2 and 5 are $x_1^* = q_1(2, 5, 40) = 8$ and $x_2^* = q_2(2, 5, 40) = 24/5$, as we found in Section 22.2. ■

The consumer's demand functions are not independent of each other. Since the optimum values satisfy the budget constraint, $\mathbf{p}.\mathbf{x}^* = M$, and $x_i^* = q_i(\mathbf{p}, M)$, the demand functions satisfy the identity

$$p_1q_1(\mathbf{p}, M) + p_2q_2(\mathbf{p}, M) + \cdots + p_nq_n(\mathbf{p}, M) = M.$$

In the example above, this is easy to check: we have

$$p_1q_1(p_1, p_2, M) + p_2q_2(p_1, p_2, M) = p_1\left(\frac{2M}{5p_1}\right) + p_2\left(\frac{3M}{5p_2}\right) = \frac{2M}{5} + \frac{3M}{5} = M.$$

22.5 The Indirect Utility Function

The utility of the optimum bundle $\mathbf{x}^* = (x_1^*, x_2^*, \ldots, x_n^*)$, in the case of n goods, is $u(x_1^*, x_2^*, \ldots, x_n^*)$. Since each x_i^* is a function of the price vector $\mathbf{p}$ and the income M, so also is the optimal utility. Using the notation $x_i^* = q_i(\mathbf{p}, M)$ for the consumer's demand functions, we have

$$u(\mathbf{x}^*) = u(q_1(\mathbf{p}, M), q_2(\mathbf{p}, M), \ldots, q_n(\mathbf{p}, M)) = V(\mathbf{p}, M),$$

say. The function V is called the *indirect utility function*. It specifies the individual consumer's optimal utility when the price vector is $\mathbf{p}$ and the income is M. The indirect utility function provides us with a means of comparing one possible combination of income and prices with another, as the following example illustrates.

Example We continue the example of the previous section, in which the consumer has utility function $u(x_1, x_2) = x_1^{1/3} x_2^{1/2}$. We saw that the demand functions are

$$q_1(p_1, p_2, M) = \frac{2M}{5p_1} \quad \text{and} \quad q_2(p_1, p_2, M) = \frac{3M}{5p_2}.$$

The indirect utility function is therefore

$$\begin{aligned} V(p_1, p_2, M) &= u\left(q_1(p_1, p_2, M), q_2(p_1, p_2, M)\right) \\ &= u\left(\frac{2M}{5p_1}, \frac{3M}{5p_2}\right) \\ &= \left(\frac{2M}{5p_1}\right)^{1/3} \left(\frac{3M}{5p_2}\right)^{1/2} \\ &= \frac{2^{1/3}3^{1/2}}{5^{5/6}} \frac{M^{5/6}}{p_1^{1/3} p_2^{1/2}}. \end{aligned}$$

We may use V to rank, in terms of the consumer's preference, a number of income and price combinations. For example, suppose that the consumer's income is 40 and the prices are $p_1 = 2$, $p_2 = 5$ (the situation described earlier) and that he has been offered a new job in a new city. The new job pays an income of 42 and, because of regional differences in demand, in the new city, the prices of the two goods he spends his income on are $p_1 = 3$ and $p_2 = 4.5$. We may use the indirect utility function to determine which of the two options, staying or leaving and taking the new job, is preferable. The initial indirect utility $V(2, 5, 40)$ is 4.38 (approximately) and the indirect utility were he to change jobs would be $V(3, 4.5, 42) = 4.20$, approximately. Since this is less than the original indirect utility, we may conclude that if the consumer's aim is to maximise his utility, he should *not* take the new job. ■

We shall now explain the connection between V and the Lagrange multiplier λ^*. The partial derivative $\partial V / \partial M$ is the *marginal utility of income*. It tells us what change in optimal utility will result from a small change in income, given that prices remain constant.

Using the chain rule and the first-order conditions $\partial u/\partial x_i = \lambda^* p_i$, we have

$$\begin{aligned}\frac{\partial V}{\partial M} &= \frac{\partial u}{\partial x_1}\frac{\partial q_1}{\partial M} + \frac{\partial u}{\partial x_2}\frac{\partial q_2}{\partial M} + \cdots + \frac{\partial u}{\partial x_n}\frac{\partial q_n}{\partial M} \\ &= \lambda^* \left(p_1 \frac{\partial q_1}{\partial M} + p_2 \frac{\partial q_2}{\partial M} + \cdots + p_n \frac{\partial q_n}{\partial M} \right).\end{aligned}$$

Now, at the end of the previous section we obtained the equation

$$p_1 q_1(\mathbf{p}, M) + p_2 q_2(\mathbf{p}, M) + \cdots + p_n q_n(\mathbf{p}, M) = M.$$

Taking the partial derivative with respect to M, we get

$$p_1 \frac{\partial q_1}{\partial M} + p_2 \frac{\partial q_2}{\partial M} + \cdots + p_n \frac{\partial q_n}{\partial M} = 1.$$

So, comparing this with the expression for $\partial V/\partial M$ obtained above, we have the simple result

$$\frac{\partial V}{\partial M} = \lambda^*.$$

In words:

- *the value of the Lagrange multiplier which solves the equations is precisely the marginal utility of income.*

Suppose prices are fixed and the consumer's income is increased, from M to $M + \Delta M$, say. Then the increase ΔM results in an increase ΔV in the consumer's optimal utility, and we have shown that

$$\Delta V \simeq \frac{\partial V}{\partial M} \Delta M = \lambda^* \Delta M.$$

Example Suppose, as before, that there are two goods with prices $p_1 = 2, p_2 = 5$, and the utility function is $x_1^{1/3} x_2^{1/2}$. When the income is $M = 40$, we found that the maximum utility is $u(8, 24/5)$, which is about 4.38. How will the maximum utility change if the income rises to 42?

A good approximate answer is obtained by using the fact that $\lambda^* = 1/\sqrt{120}$, as we found in Section 22.2. Here we have $\Delta M = 2$, and

$$\Delta V \simeq \lambda^* \Delta M = \frac{1}{\sqrt{120}} \times 2,$$

which is approximately 0.18. So when the income increases from 40 to 42, the maximum utility increases approximately from 4.38 to 4.56.

Of course, such numerical estimates of utility are not particularly helpful in practice. One can certainly say that an income of 42 is better than an income of 40, since the indirect utility increases when the income is raised from 40 to 42. However, as emphasised in Section 14.3, we cannot make any conclusion on *how much* better it is: that is we should not conclude, for instance, that an income of 42 is better than an income of 40 'by a factor of 4.56/4.38'. ■

Worked Examples

Example 22.1 Dave's utility function for apples and bananas is $u(x_1, x_2) = x_1^2 x_2$, where x_1 is the number of apples and x_2 the number of bananas. Each apple costs \$0.15 and each banana costs \$0.10. Use the method of Lagrange multipliers to determine how many of each Dave will buy to maximise his utility if he has \$1.50 to spend. (Assume that fractions of apples and bananas can be bought. Note that this is the same Dave as in Chapter 14, and so we now have another method for determining his optimal bundle.)

Solution

We have to maximise the utility $u(x_1, x_2) = x_1^2 x_2$ subject to the budget constraint $0.15x_1 + 0.1x_2 = 1.5$. The Lagrangean is

$$L = x_1^2 x_2 - \lambda(0.15x_1 + 0.1x_2 - 1.5),$$

and we therefore solve the equations

$$\begin{aligned} \frac{\partial L}{\partial x_1} &= 2x_1x_2 - 0.15\lambda = 0, \\ \frac{\partial L}{\partial x_2} &= x_1^2 - 0.1\lambda = 0, \\ 0.15x_1 + 0.1x_2 - 1.5 &= 0. \end{aligned}$$

From the first two equations

$$\frac{2x_1x_2}{0.15} = \frac{x_1^2}{0.1}, \quad \text{and so} \quad x_2 = \frac{3}{4}x_1.$$

Then, from the third equation

$$0.15x_1 + 0.1\left(\frac{3x_1}{4}\right) = 1.5,$$

so that $x_1^* = 20/3$ and $x_2^* = 3x_1^*/4 = 5$. Dave should therefore buy $20/3$ apples and 5 bananas.

Example 22.2 A student has a part-time job in a restaurant that pays \$8 per hour. Her utility function for earning \$$I$ and spending S hours studying is

$$u(I, S) = I^{1/4} S^{3/4}.$$

The total amount of time she spends each week working in the restaurant and studying is 100 hours. How should she divide up her time in order to maximise her utility?

Solution

A little care needs to be taken in determining the constraint. Since the number of hours spent working in the restaurant is $I/8$, the income divided by the hourly rate, the constraint 'total time is 100' is $I/8 + S = 100$. The Lagrangean is therefore

$$L = I^{1/4}S^{3/4} - \lambda\left(\frac{I}{8} + S - 100\right)$$

and the equations to solve are

$$\frac{\partial L}{\partial I} = \frac{1}{4}I^{-3/4}S^{3/4} - \frac{\lambda}{8} = 0,$$
$$\frac{\partial L}{\partial S} = \frac{3}{4}I^{1/4}S^{-1/4} - \lambda = 0,$$
$$\frac{I}{8} + S - 100 = 0.$$

The first two equations, on elimination of λ, yield $I = 8S/3$, and substituting this into the third equation, we obtain $S = 75$. Thus the optimal division of time is 75 hours study and 25 hours restaurant work (generating $I = 200$ dollars income).

Example 22.3 A consumer's utility function for two goods is $u(x_1,x_2) = x_1^{1/2}x_2^{1/2}$ and their prices are $p_1 = 1$ and $p_2 = 2$. If the consumer has an income of 40, how many units of each good should he buy in order to maximise his utility?

Solution

The Lagrangean is

$$L = x_1^{1/2}x_2^{1/2} - \lambda(x_1 + 2x_2 - 40)$$

and the resulting equations are

$$\frac{\partial L}{\partial x_1} = \frac{1}{2}x_1^{-1/2}x_2^{1/2} - \lambda = 0,$$
$$\frac{\partial L}{\partial x_2} = \frac{1}{2}x_1^{1/2}x_2^{-1/2} - 2\lambda = 0,$$
$$x_1 + 2x_2 - 40 = 0.$$

The first two equations show that $x_1 = 2x_2$ and then the third implies that $x_2 = 10$. So, the optimal quantities are $x_1^* = 20$ and $x_2^* = 10$.

Example 22.4 Assume that $u(x_1,x_2) = x_1^{1/2}x_2^{1/2}$, as in the previous example. Write down the Lagrangean equations for general values of p_1,p_2 and M. Find the optimal values x_1^*,x_2^* of the quantities, and determine the corresponding value λ^* of the Lagrange multiplier. What are the consumer's demand functions $q_1(p_1,p_2,M)$ and $q_2(p_1,p_2,M)$? Find the indirect utility function V and verify that $\lambda^* = \partial V/\partial M$. Show also that for any number ΔM,

$$\Delta V = V(p_1,p_2,M+\Delta M) - V(p_1,p_2,M) = \lambda^*\Delta M.$$

Solution

The Lagrangean for the general problem is

$$L = x_1^{1/2}x_2^{1/2} - \lambda(p_1x_1 + p_2x_2 - M)$$

and the resulting equations are

$$\frac{\partial L}{\partial x_1} = \frac{1}{2}x_1^{-1/2}x_2^{1/2} - \lambda p_1 = 0,$$
$$\frac{\partial L}{\partial x_2} = \frac{1}{2}x_1^{1/2}x_2^{-1/2} - \lambda p_2 = 0,$$
$$p_1x_1 + p_2x_2 - M = 0.$$

From the first two equations, $x_2 = (p_1/p_2)x_1$. The third equation then gives

$$p_1x_1 + p_2\left(\frac{p_1}{p_2}x_1\right) = M,$$

so that $x_1^* = M/(2p_1)$ and $x_2^* = M/(2p_2)$. The value of λ^* is, from the first equation:

$$\lambda^* = \frac{1}{2p_1}(x_1^*)^{-1/2}(x_2^*)^{1/2} = \frac{1}{2p_1}\left(\frac{M}{2p_1}\right)^{-1/2}\left(\frac{M}{2p_2}\right)^{1/2} = \frac{1}{2\sqrt{p_1p_2}}.$$

The demand functions are just x_1^* and x_2^* considered as functions of p_1, p_2 and M; that is

$$q_1(p_1,p_2,M) = \frac{M}{2p_1} \quad \text{and} \quad q_2(p_1,p_2,M) = \frac{M}{2p_2}.$$

The indirect utility function $V(p_1,p_2,M)$ is the optimal utility

$$V(p_1,p_2,M) = u\left(\frac{M}{2p_1},\frac{M}{2p_2}\right) = \left(\frac{M}{2p_1}\right)^{1/2}\left(\frac{M}{2p_2}\right)^{1/2} = \frac{M}{2\sqrt{p_1p_2}}.$$

The partial derivative of V with respect to M, the marginal utility of income, is

$$\frac{\partial V}{\partial M} = \frac{1}{2\sqrt{p_1p_2}},$$

which equals λ^*, as predicted by the theory.

For any change ΔM in income, with constant prices, the new indirect utility is

$$\begin{aligned} V(p_1,p_2,M + \Delta M) &= \frac{M + \Delta M}{2\sqrt{p_1p_2}} \\ &= \frac{M}{2\sqrt{p_1p_2}} + \frac{\Delta M}{2\sqrt{p_1p_2}} \\ &= V(p_1,p_2,M) + \lambda^*\Delta M. \end{aligned}$$

So in this case the approximation $\Delta V \simeq \lambda^*\Delta M$ is *exact*. This is true only because the indirect utility is linear in M.

Example 22.5 A consumer's preferences are represented by the utility function

$$u(x_1,x_2) = 2\ln x_1 + \ln x_2.$$

If the budget constraint is $p_1x_1 + p_2x_2 = M$, determine the demand functions, that is the optimal values x_1^* and x_2^* in terms of p_1, p_2 and M.

Solution

The Lagrangean for the problem is

$$L = 2\ln x_1 + \ln x_2 - \lambda(p_1x_1 + p_2x_2 - M)$$

and the resulting equations are

$$\begin{aligned}
\frac{\partial L}{\partial x_1} &= \frac{2}{x_1} - \lambda p_1 = 0,\\
\frac{\partial L}{\partial x_2} &= \frac{1}{x_2} - \lambda p_2 = 0,\\
p_1x_1 + p_2x_2 - M &= 0.
\end{aligned}$$

From the first two equations

$$\lambda = \frac{2}{p_1x_1} = \frac{1}{p_2x_2},$$

so that $x_1 = (2p_2/p_1)x_2$. The third equation then gives

$$p_1\left(\frac{2p_2}{p_1}x_2\right) + p_2x_2 = M,$$

so that the optimal values, the solutions of these equations, are

$$x_1^* = \frac{2M}{3p_1} \quad \text{and} \quad x_2^* = \frac{M}{3p_2}.$$

An alternative approach is to note that, since $2\ln x_1 + \ln x_2 = \ln(x_1^2x_2)$, the problem is equivalent to determining the x_1, x_2 that maximise the utility function $x_1^2x_2$.

Main Topics

- general formulation of the Lagrangean method
- maximising a consumer's utility subject to the budget constraint
- price ratio, demand functions, indirect utility
- Lagrange multiplier as marginal utility of income

Key Terms, Notations and Formulae

- to optimise $F(\mathbf{x})$ subject to $G_1(\mathbf{x}) = \cdots = G_m(\mathbf{x}) = 0$, form the Lagrangean

  $$L = F - \lambda_1G_1 - \cdots - \lambda_mG_m$$

  and solve

  $$\frac{\partial L}{\partial x_1} = \cdots = \frac{\partial L}{\partial x_n} = 0, \quad G_1(\mathbf{x}) = \cdots = G_m(\mathbf{x}) = 0$$

- consumer utility maximisation: maximise $u(\mathbf{x})$ subject to $\mathbf{p}.\mathbf{x} = M$

- at optimum, price ratio = marginal rate of substitution, $\dfrac{p_i}{p_j} = \dfrac{\partial u/\partial x_i}{\partial u/\partial x_j}$
- this expresses a tangency condition
- demand function, $x_i^* = q_i(p_1, p_2, \ldots, p_n, M) = q_i(\mathbf{p}, M)$
- indirect utility function, $V(\mathbf{p}, M) = u(\mathbf{x}^*) = u(q_1(\mathbf{p}, M), \ldots, q_n(\mathbf{p}, M))$
- Lagrange multiplier is marginal utility of income, $\lambda^* = \dfrac{\partial V}{\partial M}$

Exercises

22.1 A consumer buys two goods, X and Y. The price of one unit of X is \$1 and the price of one unit of Y is \$16. The consumer's utility function is given by

$$u(x, y) = x^{3/4}y^{1/4}.$$

She has a budget of \$1280 in total each year to spend on X and Y. Find the values of x and y which will maximise the consumer's utility function $u(x, y)$ subject to the constraint on her budget.

Use the value of the Lagrange multiplier to estimate the increase in the maximum obtainable utility if the consumer's budget for the goods rises to \$1282.

22.2 A consumer has utility function

$$u(x, y) = \frac{xy}{x + y}$$

for two goods, X and Y. Here, $x > 0$ denotes the amount of X consumed and $y > 0$ the amount of Y consumed. Each unit of X costs p dollars, each unit of Y costs q dollars, and the consumer has a budget for X and Y of M dollars.

Find the quantities x^* of X and y^* of Y the consumer will consume in order to maximise his utility subject to the budget constraint.

Show that

$$x^* + y^* = \frac{M}{\sqrt{pq}}.$$

Hence find the optimal value, λ^*, of the Lagrange multiplier.

Find the indirect utility function V and verify that $\lambda^* = \partial V/\partial M$.

Problems

22.1 A consumer purchases quantities of two commodities, fruit and chocolate, each month. The consumer's utility function is

$$u(x_1, x_2) = x_1^{5/6}x_2^{1/3}$$

for a bundle (x_1, x_2) of x_1 units of fruit and x_2 units of chocolate. The consumer has a total of \$49 to spend on fruit and chocolate each month. Fruit costs \$1 per unit and chocolate costs \$2 per unit. How many units of each should the consumer buy each month in order to maximise her utility?

22.2 A consumer buys apples and bananas and has utility function

$$u(x_1, x_2) = x_1 x_2^2,$$

where x_1 is the number of apples and x_2 the number of bananas. Suppose that he has \$1.80 to spend on a bundle of apples and bananas, and that apples cost \$0.12 each, bananas cost \$0.20 each. Write down the budget equation and the Lagrangean for the problem of finding the optimal bundle. What is the optimal bundle?

22.3 George is a graduate student and he divides his working week between working on his research project and teaching classes in mathematics for economists. He estimates that his utility function for earning \$$W$ by teaching classes and spending R hours on his research is

$$u(W, R) = W^{3/4} R^{1/4}.$$

He is paid \$16 per hour for teaching and works for a *total* of 40 hours each week. How should he divide his time in order to maximise his utility?

22.4 It is thought that a consumer measures the utility $u(x, y)$ of possessing a quantity x of apples and y of bananas by the formula

$$u(x, y) = x^{\alpha} y^{1-\alpha}.$$

It is known that when the consumer's budget for apples and bananas is \$1, he will buy 1 apple and 2 bananas when they are equally priced. Use this information to find α.

The price of apples falls to half that of bananas, with the price of bananas unchanged. How many apples and bananas will the consumer then buy for \$10?

22.5 Assume, as in Problem 22.1, that a consumer has utility function

$$u(x_1, x_2) = x_1^{5/6} x_2^{1/3}$$

for fruit and chocolate. Determine the consumer's demand functions $q_1(p_1, p_2, M)$ and $q_2(p_1, p_2, M)$.

Determine also λ^* in terms of p_1, p_2 and M. Find the indirect utility function and show that $\lambda^* = \partial V / \partial M$.

Suppose, as before, that fruit costs \$1 per unit and chocolate \$2 per unit. If the income is raised from \$36 to \$36.5, determine the precise value of the resulting change in the indirect utility function. Show that this is approximately equal to $0.5\lambda^*$, where λ^* is evaluated at $p_1 = 1, p_2 = 2$ and $M = 36$.

22.6 A consumer's preferences are represented by the utility function

$$u(x_1, x_2) = 3 \ln x_1 + 2 \ln x_2.$$

If the budget constraint is $p_1 x_1 + p_2 x_2 = M$, determine the demand functions $q_1(p_1, p_2, M)$ and $q_2(p_1, p_2, M)$. Determine also the indirect utility function $V(p_1, p_2, M)$.

23 Second-Order Recurrence Equations

23.1 A Simplified National Economy

We shall consider a national economy, under the two simplifying assumptions that it is *closed* (no external trade) and that there is *no government* (no taxes, welfare benefits, and so on). In this situation we can identify four quantities which tell us something about the state of the economy:

Investment (I), Production (Q), Income (Y), Consumption (C).

Macroeconomists are concerned with the definitions of these terms and the relationships between them. We shall rely on the simple intuition that each of them affects the others, either directly or indirectly, as in Figure 23.1.

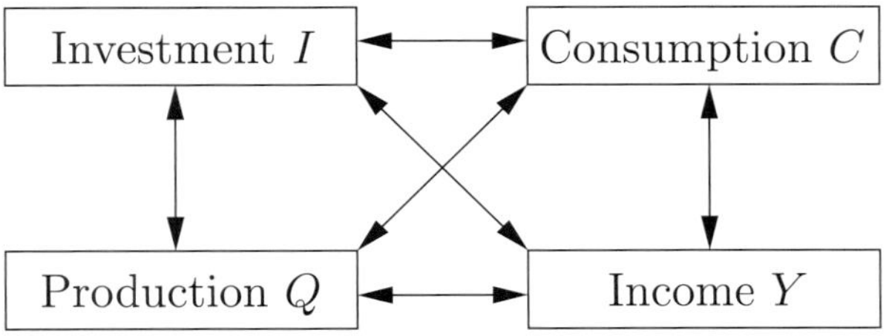

Figure 23.1 A simple model of a national economy.

In general, it is dangerous to think of the links between the variables as being one-way. However, there is a sense in which we can imagine an anticlockwise flow around the outer square: *investment* yields *production*, which is converted into *income*, some of which is required for *consumption*, leaving the rest for *investment*, which yields *production*, and so on.

There are simple relationships between the quantities I, Q, Y, C which express the fact that, in very broad terms, the 'supply' and 'demand' are equal. Specifically, production is balanced by the total income, and income is split between consumption and investment:

$$Q = Y \quad \text{and} \quad Y = C + I.$$

It is convenient to refer to these equations as *equilibrium conditions*, although it must be stressed that the word 'equilibrium' is being used rather loosely here. In particular, it does not imply that the economy is static.

23.2 Dynamics of the Economy

We look now at a very simple model of the way in which the economy evolves over time. Suppose we can measure each of the quantities in successive time periods of equal length (e.g., each year). Denote by I_t, Q_t, Y_t, C_t the values of the four key quantities in time period t so that, for instance, I_t is the total investment during the tth time period. Then we have a sequence of values $I_0, I_1, I_2, \ldots$, and similarly for the other quantities. We shall assume that the equilibrium conditions $Q_t = Y_t$ and $Y_t = C_t + I_t$ hold for each t.

In order to describe the evolution, or 'dynamics', of the economy, we must stipulate some additional relationships between the key variables. These correspond to the assumptions underlying the *behavioural parameters* introduced in our discussion of the IS equations (Section 18.5), and we shall call them *behavioural conditions*.

You may recall that in Example 3.5 we made some very simple and unrealistic assumptions: $C_t = \frac{1}{2}Y_t$ and $Y_{t+1} = kI_t$. These led to a first-order recurrence, which we could solve quite easily. More realistic assumptions form the basis of the *multiplier-accelerator* model, developed by the American economist Paul Samuelson. As we shall see, these assumptions lead to a more complex recurrence equation, but one which we can still solve, and which exhibits very interesting behaviour.

The first behavioural condition is that this year's consumption is linearly related to last year's income:

$$C_t = c + bY_{t-1} \quad (c, b \text{ positive constants}).$$

Note that this is very like the first IS equation: the only difference is that the time factor is taken into account. The second behavioural condition is that this year's investment is linearly related to last year's increase in production:

$$I_t = i + v(Q_{t-1} - Q_{t-2}) \quad (i, v \text{ positive constants}).$$

Using the equilibrium conditions $Y_t = C_t + I_t$ and $Q_t = Y_t$, we can obtain an equation involving only the Ys:

$$\begin{aligned} Y_t &= C_t + I_t \\ &= [c + bY_{t-1}] + \left[i + v\,(Q_{t-1} - Q_{t-2})\right] \\ &= c + bY_{t-1} + i + v\,(Y_{t-1} - Y_{t-2}) \\ &= (c + i) + (b + v)Y_{t-1} - vY_{t-2}. \end{aligned}$$

In other words, Y_t, Y_{t-1} and Y_{t-2} are related by the equation

$$Y_t - (b + v)Y_{t-1} + vY_{t-2} = c + i.$$

This is an example of a *second-order recurrence equation*, an equation relating each term of a sequence to the two previous terms. In the rest of this chapter we shall study such equations in detail.

23.3 Linear Homogeneous Recurrences

An equation of the form

$$y_t + a_1 y_{t-1} + a_2 y_{t-2} = k,$$

in which a_1 and a_2 are constants, is a *linear second-order* recurrence equation, with *constant coefficients*. This is precisely the type of equation we found for Y_t in the previous section. When $k = 0$, we have the *homogeneous* equation

$$y_t + a_1 y_{t-1} + a_2 y_{t-2} = 0.$$

(The use of the word 'homogeneous' here is unconnected with its use in Chapter 12.)

In the homogeneous case, two very useful principles apply:

- a constant multiple of a solution is a solution;
- the sum of two solutions is a solution.

These principles can be verified by simple algebra (Problem 23.9). It follows that if we know two solutions $y_t^{(1)}$ and $y_t^{(2)}$ of the recurrence, then

$$Ay_t^{(1)} + By_t^{(2)}$$

is also a solution for any constants A and B.

Suppose we are given a homogeneous recurrence $y_t + a_1 y_{t-1} + a_2 y_{t-2} = 0$. In order to determine the sequence of values y_t completely, we must know the *initial values* y_0 and y_1. Given these values, y_2 is determined by the equation with $t = 2$, y_3 is then determined by the equation with $t = 3$, and so on. So if we are looking for a solution $y_t = Ay_t^{(1)} + By_t^{(2)}$, we have to choose A and B so that the formula fits the initial conditions when $t = 0$ and $t = 1$. These two conditions determine appropriate values for the two arbitrary constants. This means that the *general solution* of the homogeneous recurrence is given by the formula displayed above.

We shall now describe a practical method for finding two solutions $y_t^{(1)}$ and $y_t^{(2)}$, based on the *auxiliary equation*

$$z^2 + a_1 z + a_2 = 0.$$

This is a quadratic equation. Back in Section 2.4, we observed that such an equation may have two distinct solutions, or just one solution, or no solutions, depending on the value of the quantity $a_1^2 - 4a_2$. (Recall that we are looking for solutions in the set $\mathbb{R}$ of real numbers.) We consider each case in turn.

Case 1: The Auxiliary Equation Has Two Distinct Solutions
Suppose the solutions of the auxiliary equation are α and β, where $\alpha \neq \beta$. Then it is easy to check that $y_t = \alpha^t$, the tth power of α, is a solution of the recurrence. We have

$$\begin{aligned} y_t + a_1 y_{t-1} + a_2 y_{t-2} &= \alpha^t + a_1 \alpha^{t-1} + a_2 \alpha^{t-2} \\ &= \alpha^{t-2}(\alpha^2 + a_1\alpha + a_2) \\ &= 0, \end{aligned}$$

because the expression in parentheses is zero (α satisfies the auxiliary equation). Thus, α^t is a solution of the recurrence. Similarly, β^t is a solution of the recurrence.

Now the argument given above shows that the general solution is

$$y_t = A\alpha^t + B\beta^t \quad (A, B \text{ constants}).$$

In any specific case, A and B are determined by the initial values y_0 and y_1, as in the following example.

Example We find an explicit formula for y_t when

$$y_t - 5y_{t-1} + 6y_{t-2} = 0, \quad \text{with} \quad y_0 = 0, y_1 = 1.$$

The auxiliary equation is

$$z^2 - 5z + 6 = 0, \quad \text{that is} \quad (z-2)(z-3) = 0.$$

The solutions are $\alpha = 2$ and $\beta = 3$, so the general solution is $A2^t + B3^t$. The specified values of y_0 and y_1 yield the equations

$$A + B = 0 \quad \text{and} \quad 2A + 3B = 1,$$

from which it follows that $A = -1$ and $B = 1$. Hence the formula for y_t is $y_t = 3^t - 2^t$. ■

Case 2: The Auxiliary Equation Has Just One Solution
In this case it is clear that we cannot get a solution involving two arbitrary constants by the method used above. If the (one) solution of the auxiliary equation is α, then $y_t = \alpha^t$ is a solution of the recurrence as before, but we need to find another.

The auxiliary equation has exactly one solution when $a_1^2 - 4a_2 = 0$; that is when $a_2 = a_1^2/4$. Then the equation $z^2 + a_1 z + a_2 = 0$ can be written in the form

$$z^2 + a_1 z + \frac{a_1^2}{4} = 0, \quad \text{or} \quad \left(z + \frac{a_1}{2}\right)^2 = 0,$$

and the (one) solution is $\alpha = -a_1/2$. In this case we claim that a second solution of the recurrence is $y_t = t\alpha^t$. Substituting:

$$\begin{aligned} y_t + a_1 y_{t-1} + a_2 y_{t-2} &= t\alpha^t + a_1(t-1)\alpha^{t-1} + a_2(t-2)\alpha^{t-2} \\ &= t\alpha^{t-2}(\alpha^2 + a_1\alpha + a_2) - \alpha^{t-2}(a_1\alpha + 2a_2). \end{aligned}$$

Because α satisfies the auxiliary equation, we have $\alpha^2 + a_1\alpha + a_2 = 0$. Furthermore, since $\alpha = -a_1/2$ and $a_1^2 = 4a_2$, it follows that $a_1\alpha + 2a_2 = -a_1^2/2 + 2a_2 = 0$. Hence $t\alpha^t$ is a solution, as claimed.

The general solution is therefore

$$y_t = Ct\alpha^t + D\alpha^t = (Ct + D)\alpha^t.$$

As in the previous case, the values of the constants C and D can be determined using the initial values y_0 and y_1.

Example Consider the recurrence

$$y_t - 6y_{t-1} + 9y_{t-2} = 0, \quad \text{with} \quad y_0 = 1, y_1 = 1.$$

The auxiliary equation is

$$z^2 - 6z + 9 = 0, \quad \text{that is} \quad (z - 3)^2 = 0.$$

There is therefore just one solution, $\alpha = 3$, of the auxiliary equation. The general solution to the recurrence equation is $(Ct + D)3^t$. Using the facts that $y_0 = 1$ and $y_1 = 1$, we must have

$$D = 1 \quad \text{and} \quad 3(C + D) = 1,$$

so that $C = -2/3$ and $D = 1$, giving

$$y_t = \left(-\frac{2}{3}t + 1\right) 3^t.$$

■

Case 3: The Auxiliary Equation has No Solutions

The auxiliary equation has no solutions when the quantity $a_1^2 - 4a_2$ is negative. In that case, $4a_2 - a_1^2$ is positive, and hence so is a_2. Thus there is a positive square root r of a_2; that is we can define $r = \sqrt{a_2}$. For convenience, we also define $p = -a_1/2$.

In order to write down the general solution in this case, we use the diagram in Figure 23.2. We construct a right-angled triangle with base p and hypotenuse r, and use it to define the angle θ as shown; that is $\cos\theta = p/r$.

(In the diagram, we have taken p to be positive, which is the case if a_1 is negative. If a_1 is positive, p is negative and θ is an obtuse angle.)

We claim that, with these definitions, the functions

$$r^t \cos(\theta t) \quad \text{and} \quad r^t \sin(\theta t)$$

are solutions of the recurrence. The proof that $r^t \cos(\theta t)$ is a solution is given in Example 23.5. (The reader who has studied complex numbers will probably see an easier way to carry out this proof.)

So the general solution in this case is

$$y_t = Er^t \cos(\theta t) + Fr^t \sin(\theta t).$$

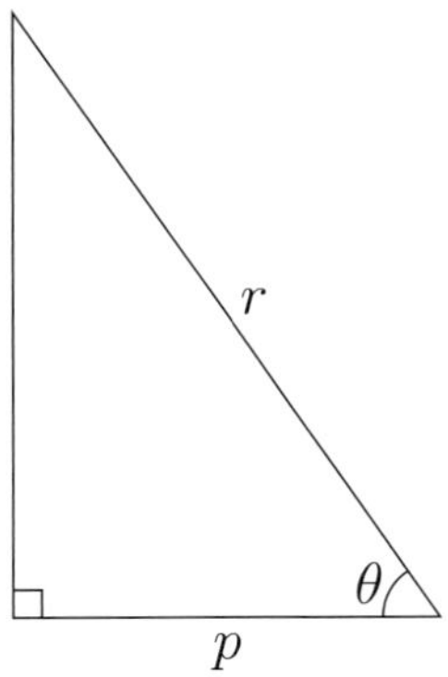

$$r = \sqrt{a_2}$$

$$p = -a_1/2$$

$$\cos\theta = p/r$$

Figure 23.2 Definition of r, p and θ.

Example Find the solution of the equation

$$y_t - 2y_{t-1} + 2y_{t-2} = 0, \quad \text{with} \quad y_0 = 0, y_1 = 1.$$

Here the auxiliary equation has no solutions, since $2^2 < 4 \times 2$. We have $a_1 = -2$ and $a_2 = 2$, so $p = 1$, $r = \sqrt{2}$ and

$$\cos\theta = 1/\sqrt{2}, \quad \text{so} \quad \theta = \pi/4.$$

Hence the general solution is

$$y_t = \left(\sqrt{2}\right)^t \left[E\cos\left(\frac{\pi t}{4}\right) + F\sin\left(\frac{\pi t}{4}\right)\right].$$

Substituting the given values for y_0 and y_1, we obtain

$$E = 0 \quad \text{and} \quad \sqrt{2}\left(\frac{E}{\sqrt{2}} + \frac{F}{\sqrt{2}}\right) = 1,$$

so that $E = 0$ and $F = 1$. Thus the required solution is

$$y_t = \left(\sqrt{2}\right)^t \sin\left(\frac{\pi t}{4}\right).$$

See Section 24.2 for more discussion of this example. ∎

23.4 Non-homogeneous Recurrences

When the right-hand side of the second-order linear equation is a constant, $k \neq 0$, it is still easy to write down the general solution. By analogy with the first-order case, we start by looking for a *time-independent* solution $y_t = y^*$ for all t. For this we require

$$y^* + a_1 y^* + a_2 y^* = k, \quad \text{or} \quad y^* = \frac{k}{1 + a_1 + a_2}.$$

We call y^* a *particular solution*.

We now put $y_t = y^* + z_t$, so that

$$\begin{aligned} y_t + a_1 y_{t-1} + a_2 y_{t-2} &= (y^* + z_t) + a_1(y^* + z_{t-1}) + a_2(y^* + z_{t-2}) \\ &= (y^* + a_1 y^* + a_2 y^*) + (z_t + a_1 z_{t-1} + a_2 z_{t-2}) \\ &= k + (z_t + a_1 z_{t-1} + a_2 z_{t-2}). \end{aligned}$$

Thus y_t is a solution of the non-homogeneous equation provided that $z_t + a_1 z_{t-1} + a_2 z_{t-2} = 0$; in other words, provided that z_t satisfies the corresponding *homogeneous* equation in which $k = 0$. So we have the following principle:

- *General solution of the non-homogeneous equation =*
  particular solution + general solution of the homogeneous equation.

This principle holds even when k is a function of t, but it is most useful when k is constant, because then it is easy to find a particular solution.

Example Find the general solution of the recurrence

$$y_t - 5y_{t-1} + 6y_{t-2} = 12,$$

and the solution which satisfies the initial conditions $y_0 = 0$, $y_1 = 1$.

Apart from the constant 12 on the right-hand side, this is the same as the first example in Section 23.3. We saw there that the general solution to the corresponding homogeneous equation is $A2^t + B3^t$.

The constant y^* is a particular solution of the non-homogeneous equation if $y^* - 5y^* + 6y^* = 12$, that is $y^* = 6$. Therefore, the general solution to the homogeneous equation is

$$y_t = 6 + A2^t + B3^t.$$

To determine A and B, given the initial values y_0 and y_1, we simply put $t = 0$ and $t = 1$. Since $y_0 = 0$ we require $6 + A + B = 0$, and since $y_1 = 1$ we require $6 + 2A + 3B = 1$. Solving these equations, we find $A = -13$ and $B = 7$. So the solution satisfying the given initial conditions is

$$y_t = 6 - 13(2^t) + 7(3^t).$$

■

Note that we must find the general solution of the non-homogeneous equation (including the particular solution) before we determine the constants.

Worked Examples

Example 23.1 Find the general solution of the recurrence equation

$$y_t - 6y_{t-1} + 5y_{t-2} = 0.$$

Solution

The auxiliary equation is $z^2 - 6z + 5 = 0$, that is $(z-5)(z-1) = 0$, with solutions 1 and 5. The general solution is therefore

$$y_t = A(1^t) + B(5^t) = A + B5^t,$$

for arbitrary constants A and B.

Example 23.2 Find the general solution of the recurrence equation

$$y_t - 2y_{t-1} + 4y_{t-2} = 0,$$

and determine the solution which satisfies the initial conditions $y_0 = 1, y_1 = 1 - \sqrt{3}$.

Solution

The auxiliary equation, $z^2 - 2z + 4 = 0$, has no solutions. Using the notation defined in Figure 23.2 we have $p = -a_1/2 = 1$ and $r = \sqrt{a_2} = 2$. It follows that $\cos\theta = p/r = 1/2$, so $\theta = \pi/3$. The general solution is therefore

$$y_t = 2^t\left[E\cos\left(\frac{\pi t}{3}\right) + F\sin\left(\frac{\pi t}{3}\right)\right].$$

Putting $t = 0$, and using the given initial condition $y_0 = 1$, we have $E = 1$. Similarly, $y_1 = 1 - \sqrt{3}$ implies that

$$2\left[E\cos\left(\frac{\pi}{3}\right) + F\sin\left(\frac{\pi}{3}\right)\right] = 2\left[\frac{1}{2} + F\frac{\sqrt{3}}{2}\right] = 1 - \sqrt{3},$$

so that $1 + \sqrt{3}F = 1 - \sqrt{3}$. Therefore, $F = -1$ and the required solution is

$$y_t = 2^t\left[\cos\left(\frac{\pi t}{3}\right) - \sin\left(\frac{\pi t}{3}\right)\right].$$

Example 23.3 Find the general solution of the recurrence equation

$$y_t - 5y_{t-1} - 14y_{t-2} = 18,$$

and determine the solution which satisfies the initial conditions $y_0 = -1, y_1 = 8$.

Solution

The auxiliary equation is

$$z^2 - 5z - 14 = (z+2)(z-7) = 0,$$

with solutions -2 and 7. The homogeneous equation

$$y_t - 5y_{t-1} - 14y_{t-2} = 0$$

therefore has general solution $y_t = A(-2)^t + B7^t$.

A particular solution of the non-homogeneous equation is the constant solution $y^* = 18/(1-5-14) = -1$, so this equation has general solution

$$y_t = -1 + A(-2)^t + B7^t.$$

To find the values of A and B we use the given values of y_0 and y_1. Since $y_0 = -1$, we must have $-1 + A + B = -1$ and since $y_1 = 8$, $-1 - 2A + 7B = 8$. Solving these, we obtain $A = -1$ and $B = 1$, and therefore

$$y_t = -1 - (-2)^t + 7^t.$$

Example 23.4 In the thirteenth century the Italian mathematician known as Fibonacci initiated the study of population growth. He postulated that rabbits form stable relationships on reaching maturity at the age of one year, and that in each subsequent year each pair produces precisely one male and one female offspring. He also assumed that rabbits never die!

If just one pair is miraculously 'created' when $t = 0$, show that the number y_t of pairs of rabbits after t years satisfies the recurrence equation

$$y_t = y_{t-1} + y_{t-2},$$

with the initial conditions $y_0 = 1$ and $y_1 = 1$. Hence find a general formula for y_t.

Solution

At the end of t years, the number of pairs in the population is the number at the end of the previous year, y_{t-1}, plus the number of newly born pairs. Since each pair over one year old produces one new pair, the number of newly born pairs is y_{t-2}. Thus we have the recurrence $y_t = y_{t-1} + y_{t-2}$. We are given that $y_0 = 1$, and clearly, $y_1 = 1$ since the initial pair does not reproduce during the first year.

Writing the recurrence as $y_t - y_{t-1} - y_{t-2} = 0$, we obtain the auxiliary equation $z^2 - z - 1 = 0$. This has two distinct solutions

$$\alpha = \frac{1-\sqrt{5}}{2} \quad \text{and} \quad \beta = \frac{1+\sqrt{5}}{2}.$$

The solution is $y_t = A\alpha^t + B\beta^t$, where A and B are suitable constants. Using the initial conditions we must have

$$A + B = 1 \quad \text{and} \quad \alpha A + \beta B = 1.$$

Therefore $(\beta - \alpha)B = 1 - \alpha$, or $\sqrt{5}B = (1+\sqrt{5})/2$, so

$$B = \frac{1+\sqrt{5}}{2\sqrt{5}} = \frac{\beta}{\sqrt{5}}.$$

Similarly

$$A = 1 - B = \frac{\sqrt{5}-1}{2\sqrt{5}} = -\frac{\alpha}{\sqrt{5}}.$$

Hence

$$y_t = \frac{1}{\sqrt{5}}\left(\beta^{t+1} - \alpha^{t+1}\right).$$

The formula looks rather complicated, especially when we observe that the sequence of values of y_t can be written down very easily using the recurrence. The resulting numbers, $1, 1, 2, 3, 5, 8, 13, 21, 34, \ldots$, are obtained by taking each term to be the sum of the preceding two terms. These are the famous *Fibonacci numbers*.

However, the formula is useful in one way. Noting that $|\alpha|$ is less than one, we see that the term in α^{t+1} tends to zero as t increases, and thus soon becomes negligible. This means that the tth Fibonacci number is approximately equal to

$$\frac{1}{\sqrt{5}}\left(\frac{1+\sqrt{5}}{2}\right)^{t+1}.$$

The formula tells us that the rabbit population grows like $(1.618)^t$.

Example 23.5 Use the identity $\cos(x + y) = \cos x \cos y - \sin x \sin y$ to show that, when r and θ are as defined in Figure 23.2:

$$r^2 \cos(s + 1)\theta + a_1 r \cos s\theta + a_2 \cos(s - 1)\theta = 0$$

for all positive integers s. Deduce that $y_t = r^t \cos(\theta t)$ is a solution of the recurrence $y_t + a_1 y_{t-1} + a_2 y_{t-2} = 0$.

Solution

We have

$$\begin{aligned}
&r^2 \cos(s + 1)\theta + a_1 r \cos s\theta + a_2 \cos(s - 1)\theta \\
&= r^2 \cos s\theta \cos\theta - r^2 \sin s\theta \sin\theta + a_1 r \cos s\theta + a_2 \cos s\theta \cos\theta + a_2 \sin s\theta \sin\theta \\
&= \left(r^2 \cos\theta + a_1 r + a_2 \cos\theta\right) \cos s\theta + \left(-r^2 \sin\theta + a_2 \sin\theta\right) \sin s\theta.
\end{aligned}$$

By definition, $r^2 = a_2$, so $-r^2 \sin\theta + a_2 \sin\theta = 0$. Furthermore, since $\cos\theta = p/r$ where $p = -a_1/2$, we have

$$r^2 \cos\theta + a_1 r + a_2 \cos\theta = r^2(p/r) + (-2p)r + r^2(p/r) = 0.$$

Thus the original expression is zero for all positive integers s. Putting $s = t - 1$ we obtain

$$r^2 \cos t\theta + a_1 r \cos(t - 1)\theta + a_2 \cos(t - 2)\theta = 0,$$

and multiplying by r^{t-2} we see that $y_t = r^t \cos(\theta t)$ is a solution of the recurrence.

Example 23.6 The sequences y_t and x_t are linked by the following equations, which hold for all $t \geq 1$:

$$y_t - y_{t-1} = 6x_{t-1},$$
$$x_t = y_{t-1} + 2.$$

Obtain a second-order recurrence equation for y_t. Find explicit expressions for y_t and x_t given that $y_0 = 1$ and $x_0 = 1/6$.

Solution
From the equations, for $t \geq 2$:

$$y_t - y_{t-1} = 6x_{t-1} = 6\left(y_{t-2} + 2\right),$$

so

$$y_t - y_{t-1} - 6y_{t-2} = 12.$$

The auxiliary equation is $z^2 - z - 6 = (z-3)(z+2) = 0$ and a particular solution is $12/(1-1-6) = -2$, so for some constants A and B

$$y_t = -2 + A3^t + B(-2)^t.$$

Given that $y_0 = 1$, $-2 + A + B = 1$. Since $x_0 = 1/6$, the first equation in the question gives $y_1 - y_0 = 6(1/6)$, so $y_1 = 2$. This means $-2 + 3A - 2B = 2$. Solving the equations for A and B, we obtain $A = 2, B = 1$. So

$$y_t = -2 + 2(3^t) + (-2)^t.$$

We can use either one of the two original equations to find an expression for x_t. Using the second:

$$x_t = y_{t-1} + 2 = -2 + 2(3^{t-1}) + (-2)^{t-1} + 2 = 2(3^{t-1}) + (-2)^{t-1}.$$

Main Topics

- basic macroeconomic concepts
- the multiplier-accelerator model
- general solution of a homogeneous second-order linear recurrence
- non-homogeneous equations
- using initial conditions

Key Terms, Notations and Formulae

- investment, I; production, Q; income, Y; consumption, C
- in equilibrium, $Q = Y$, $Y = C + I$
- sequences I_t, Q_t, Y_t, C_t

- multiplier-accelerator model, $C_t = c + bY_{t-1}, I_t = i + v(Q_{t-1} - Q_{t-2})$
- homogeneous linear second-order recurrence, $y_t + a_1 y_{t-1} + a_2 y_{t-2} = 0$
- auxiliary equation, $z^2 + a_1 z + a_2 = 0$
- solution of homogeneous equation depends on solutions of auxiliary:
  - two distinct solutions, α, β: $y_t = A\alpha^t + B\beta^t$
  - one solution, α: $y_t = (Ct + D)\alpha^t$
  - none: if $p = -\dfrac{a_1}{2}$, $r = \sqrt{a_2}$ and $\cos\theta = \dfrac{p}{r}$, $y_t = r^t(E\cos\theta t + F\sin\theta t)$
- initial values, initial conditions
- for a non-homogeneous equation, the general solution is

  particular solution + general solution of homogeneous equation
- particular solution of $y_t + a_1 y_{t-1} + a_2 y_{t-2} = k$ is $y^* = \dfrac{k}{1 + a_1 + a_2}$

Exercises

23.1 Find the general solution of the recurrence equation

$$y_t - y_{t-1} - 6y_{t-2} = 0.$$

23.2 Find the general solution of the recurrence equation

$$y_t - 5y_{t-1} + 4y_{t-2} = 0,$$

and determine the solution which satisfies $y_0 = 1$ and $y_1 = 10$.

23.3 Given that $y_0 = 2$ and $y_1 = 5$, find the solution of the recurrence equation

$$y_t - y_{t-1} + \frac{1}{4}y_{t-2} = 0.$$

23.4 A market for a commodity is modelled by taking the demand and supply functions to be

$$q^D(p) = 1 - p \quad \text{and} \quad q^S(p) = p.$$

The price change in year t (i.e., $p_t - p_{t-1}$) depends on the excess demand in the previous two years and is given by

$$p_t - p_{t-1} = \frac{7}{12}\left[q^D(p_{t-1}) - q^S(p_{t-1})\right] - \frac{1}{12}\left[q^D(p_{t-2}) - q^S(p_{t-2})\right].$$

Given that $p_0 = 5/2$ and $p_1 = 1/3$, find a formula for p_t.

Problems

23.1 Find the general solution of the recurrence equations

(a) $y_t - 7y_{t-1} - 18y_{t-2} = 0$,

(b) $y_t + 8y_{t-1} + 16y_{t-2} = 0$,

(c) $y_t - 3y_{t-1} + 9y_{t-2} = 0$.

23.2 Find the general solution of the recurrence equation

$$y_t + y_{t-1} - 6y_{t-2} = 0,$$

and determine the solution which satisfies $y_0 = 0$ and $y_1 = 5$.

23.3 Find the general solution of the recurrence equation

$$y_t - 8y_{t-1} + 16y_{t-2} = 0,$$

and determine the solution satisfying the initial conditions $y_0 = 0$, $y_1 = 8$.

23.4 Find the general solution of the recurrence equation

$$y_t = 9y_{t-1} - 27y_{t-2},$$

and determine the solution satisfying the initial conditions $y_0 = 0$, $y_1 = \sqrt{27}/2$.

23.5 Find the general solution of the recurrence equation

$$y_t + 12y_{t-1} + 11y_{t-2} = 24.$$

23.6 Find the general solution of the recurrence equation

$$y_t - 7y_{t-1} + 12y_{t-2} = 3.$$

Find the solution satisfying $y_0 = 7/2$, $y_1 = 21/2$.

23.7 Find the general solution of the recurrence equation

$$y_t - 10y_{t-1} + 25y_{t-2} = 32.$$

Find the solution satisfying $y_0 = 3$, $y_1 = 12$.

23.8 Find the general solution of the recurrence equation

$$y_t - 3y_{t-1} + 9y_{t-2} = 21.$$

Find the solution satisfying $y_0 = 4$, $y_1 = 9/2$.

23.9 Suppose that $y_t^{(1)}$ and $y_t^{(2)}$ are known to satisfy the homogeneous recurrence equation

$$y_t + a_1 y_{t-1} + a_2 y_{t-2} = 0.$$

Verify that $cy_t^{(1)}$ and $y_t^{(1)} + y_t^{(2)}$ also satisfy the equation.

23.10 The sequences y_t and x_t are linked by the following equations, which hold for all $t \geq 1$:

$$y_t - y_{t-1} = 4x_{t-1},$$
$$x_t = 3y_{t-1} + 3.$$

Obtain a second-order recurrence equation for y_t. Find explicit expressions for y_t and x_t given that $y_0 = 0$ and $x_0 = 5/2$.

24 Macroeconomic Applications

24.1 Recurrence Equations in Practice

In this chapter we shall consider the application to macroeconomic models of the techniques for solving second-order recurrences. We hope to obtain qualitative conclusions, such as whether the national income will increase steadily from year to year, or rise and fall periodically. In this context the general behaviour of a solution is more important than the explicit formula, although we often need the formula before we can determine the behaviour.

As we have seen, there are several different types of behaviour which may be exhibited by solutions of second-order recurrences.

Example Consider the solution $y_t = 3^t - 2^t$ to the first example in Section 23.3. As $t \to \infty$, we know (Section 3.3) that both 3^t and 2^t tend to infinity, but what happens to their difference? It appears that 3^t grows much faster than 2^t and thus we might expect that $y_t \to \infty$.

A good way to check this is to rewrite the formula as follows:

$$3^t - 2^t = 3^t \left(1 - \left(\frac{2}{3}\right)^t\right).$$

In this form it is easy to see that the solution tends to infinity. Since $(2/3)^t$ tends to zero, it follows that $1 - (2/3)^t$ tends to 1. The other factor 3^t tends to infinity, so the product tends to infinity, as claimed. ■

Example The solution to the second example in Section 23.3 is

$$y_t = \left(-\frac{2}{3}t + 1\right) 3^t = -\frac{2}{3}t3^t + 3^t.$$

Here the dominant term is $-(2/3)t3^t$, since we can write

$$y_t = -\frac{2}{3}t3^t \left(1 - \frac{3}{2t}\right),$$

and $1 - \dfrac{3}{2t} \to 1$ as $t \to \infty$. It follows that $y_t \to -\infty$ as $t \to \infty$. ■

24.2 Oscillatory Solutions

In many cases, the most useful qualitative observation to be made about the solution to a recurrence is that it is oscillatory, that is, increases and decreases repeatedly and continues to do so indefinitely. This kind of behaviour occurs, in particular, when the auxiliary equation of a second-order recurrence has no solutions. In that case the solution exhibits oscillatory behaviour because the sine and cosine functions are periodic. For example, the solution to the third example in Section 23.3 is $y_t = (\sqrt{2})^t \sin(\pi t/4)$. Working out the first few values of y_t, we obtain the sequence

$$0, 1, 2, 2, 0, \ -4, \ -8, \ -8, 0, 16, 32, 32, 0, \ldots .$$

So here it is clear that the solution oscillates with increasing magnitude.

Generally, the condition for the auxiliary equation $z^2 + a_1 z + a_2 = 0$ to have no solutions is that $a_1^2 < 4a_2$. Then the general solution has the form

$$r^t[E\cos(\theta t) + F\sin(\theta t)],$$

where $r = \sqrt{a_2}$. Since the values of both $\sin(\theta t)$ and $\cos(\theta t)$ always lie between -1 and 1, the oscillations increase in magnitude if r is greater than 1, that is if $a_2 > 1$, and decrease in magnitude if $a_2 < 1$.

Oscillatory behaviour can also occur in some cases where the auxiliary equation does have solutions, as the following example shows.

Example Consider the equation (Problem 23.2)

$$y_t + y_{t-1} - 6y_{t-2} = 0.$$

The solution satisfying $y_0 = 0$ and $y_1 = 5$ is $y_t = 2^t - (-3)^t$. The term 2^t increases, tending to infinity as t tends to infinity, but $(-3)^t$ oscillates; it is alternately positive and negative, of increasing magnitude. To see how y_t behaves, we rewrite the solution in the form

$$y_t = (-3)^t \left(\left(-\frac{2}{3} \right)^t - 1 \right).$$

The quantity in large brackets is negative for all values of t, and it approaches -1 as $t \to \infty$. The other factor, $(-3)^t$, is alternately positive and negative, so we deduce that y_t oscillates with increasing magnitude. ■

24.3 Business Cycles

We now return to the dynamics of the simplified national economy, as described by the multiplier-accelerator equations in Section 23.2. As we shall see, the solution to this model can exhibit oscillatory behaviour. When this happens, economists speak of 'business cycles'.

Recall that we obtained the following second-order recurrence equation for the national income Y_t in year t:

$$Y_t - (b+v)Y_{t-1} + vY_{t-2} = c + i.$$

This is a non-homogeneous equation, and it is easy to check that it has a particular solution $Y^* = (c+i)/(1-b)$, which is time-independent.

The behaviour of the general solution depends on the values of the parameters b and v. The auxiliary equation is $z^2 - (b+v)z + v = 0$, and there are three cases to consider, depending on the relative values of $(b+v)^2$ and $4v$.

Case 1. Suppose first that $(b+v)^2 > 4v$. Then the auxiliary equation has two distinct solutions

$$\alpha = \frac{(b+v) - \sqrt{(b+v)^2 - 4v}}{2} \quad \text{and} \quad \beta = \frac{(b+v) + \sqrt{(b+v)^2 - 4v}}{2}.$$

Since the constants b and v are positive, we have $0 < \alpha < \beta$. The solution of the homogeneous recurrence equation is $A\alpha^t + B\beta^t$, so

$$Y_t = Y^* + A\alpha^t + B\beta^t,$$

where Y^* is the time-independent solution.

The behaviour of this solution depends on the values of α and β and the constants A and B. Clearly, if $A = B = 0$, then $Y_t = Y^*$ is constant. Suppose then that $B = 0$ and $A \neq 0$, in which case the solution is $Y_t = Y^* + A\alpha^t$. If $\alpha > 1$ then $\alpha^t \to \infty$ as $t \to \infty$, so in this case $Y_t \to \infty$ if $A > 0$ and $Y_t \to -\infty$ if $A < 0$. On the other hand, if $\alpha < 1$ then $\alpha^t \to 0$ as $t \to \infty$ and so, regardless of the value of A, $Y_t \to Y^*$. When $\alpha = 1$, the solution is constant at $Y_t = Y^* + A$. If $A \neq 0$ and $B \neq 0$ we may use the method illustrated in the examples in Section 24.1; that is we write Y_t in the form

$$Y_t = Y^* + \beta^t \left(B + A \left(\frac{\alpha}{\beta} \right)^t \right).$$

Since the quantity in large brackets tends to B as t tends to infinity, Y_t behaves like $Y^* + B\beta^t$. The behaviour of Y_t depends on the values of β and the constant B. If $\beta > 1$ and $B > 0$, then $Y_t \to \infty$ as $t \to \infty$, whereas if $\beta > 1$ and $B < 0$, then $Y_t \to -\infty$ as $t \to \infty$. On the other hand, if $\beta < 1$, then $\beta^t \to 0$ and $Y_t \to Y^*$. The only remaining case is $\beta = 1$; here $Y_t \to Y^* + B$ as $t \to \infty$. (We should be aware of the possibility that Y_t may oscillate even though it tends to infinity, as in the sequence $1, 4, 3, 6, 5, 8, 7, 10, \ldots$ for example. But it is quite easy to show that this cannot happen here, at least in the long run. See Example 24.2.)

Case 2. Suppose $(b+v)^2 = 4v$. In this case, the only solution of the auxiliary equation is $\alpha = (b+v)/2$, and the solution to the recurrence equation is

$$Y_t = Y^* + (Ct + D)\alpha^t = Y^* + (Ct + D)\left(\frac{b+v}{2} \right)^t.$$

If $C = 0$ then $Y_t = Y^* + D\alpha^t$. If $\alpha > 1$ then $Y_t \to \infty$ if $D > 0$ and $Y_t \to -\infty$ if $D < 0$; if $\alpha < 1$ then $Y_t \to Y^*$; if $\alpha = 1$, the solution is constant, $Y_t = Y^* + D$. Suppose now that $C \neq 0$. Then it is helpful to write

$$Y_t = Y^* + t\alpha^t\left(C + \frac{D}{t}\right).$$

The quantity in parentheses tends to C as t tends to infinity. If $\alpha > 1$ then $Y_t \to \infty$ if $C > 0$, while if $C < 0$, $Y_t \to -\infty$. If, on the other hand, $\alpha < 1$, then, since $t\alpha^t \to 0$, $Y_t \to Y^*$ as $t \to \infty$. If $\alpha = 1$ and $C > 0$, then Y_t tends to ∞, while if $\alpha = 1$ and $C < 0$, $Y_t \to -\infty$.

Case 3. When $(b + v)^2 < 4v$, the auxiliary equation has no solutions and

$$Y_t = Y^* + r^t\left[E\cos(\theta t) + F\sin(\theta t)\right],$$

where $r = \sqrt{v}$ and $\cos\theta = (b + v)/2\sqrt{v}$.

This solution is oscillatory. If $v < 1$, the oscillations decrease in magnitude and $Y_t \to Y^*$ as $t \to \infty$, while if $v > 1$, the oscillations increase in magnitude and there is no limit as $t \to \infty$.

Thus we see that there are a number of types of behaviour which can result from this model of the national economy. The most striking qualitative feature is that, under certain conditions, the behaviour is oscillatory. Specifically, the national income Y_t will oscillate when the constants b and v satisfy $(b + v)^2 < 4v$. Furthermore, if $v > 1$, Y_t oscillates with increasing magnitude, while if $v < 1$, Y_t oscillates with decreasing magnitude and approaches the finite limit $(c+i)/(1-b)$. These two cases are illustrated in Figure 24.1.

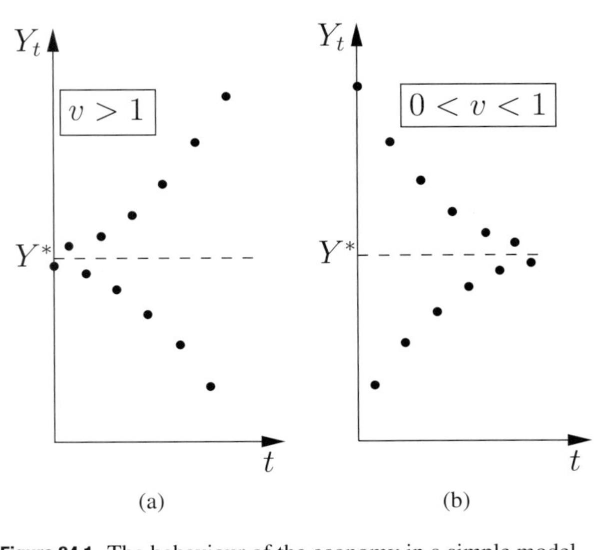

Figure 24.1 The behaviour of the economy in a simple model.

24.4 Improved Models of the Economy

You will recall that the multiplier-accelerator model was obtained by eliminating all variables except one from a number of equations. Some of these equations were equilibrium conditions, while others expressed assumptions about the behaviour of firms and households. It is the latter which 'drive' the economy and, depending on the assumptions made, we can obtain differing kinds of qualitative behaviour.

Of course, it is important to make the assumptions as realistic as possible, and one way of doing this is to take account of more factors. This is a topic which goes far beyond what can be discussed here, but we give one example.

Example In a modification of the multiplier-accelerator model where the interest rate ρ is taken into account, the constant i introduced in Section 23.2 is replaced by iR^t, where $R = 1 + \rho$. Thus it is assumed that the growth of the economy is driven by the equations

$$C_t = c + bY_{t-1} \quad \text{and} \quad I_t = v(Q_{t-1} - Q_{t-2}) + iR^t,$$

where c, b, v and i are positive constants.

Assuming the usual equilibrium conditions $Y_t = Q_t$ and $Y_t = C_t + I_t$, we have

$$\begin{aligned} Y_t &= C_t + I_t \\ &= c + bY_{t-1} + v\,(Q_{t-1} - Q_{t-2}) + iR^t \\ &= c + bY_{t-1} + v\,(Y_{t-1} - Y_{t-2}) + iR^t \\ &= (b+v)Y_{t-1} - vY_{t-2} + c + iR^t. \end{aligned}$$

Thus we obtain

$$Y_t - (b+v)Y_{t-1} + vY_{t-2} = c + iR^t.$$

Note that this is a non-homogeneous recurrence, and the right-hand side depends on t. In Chapter 23 we did not discuss how to find a particular solution in such circumstances, but here it can be done quite easily. Assume, for simplicity, that $c = 0$, so the equation is

$$Y_t - (b+v)Y_{t-1} + vY_{t-2} = iR^t.$$

To find a particular solution, let X be the constant

$$X = \frac{iR^2}{R^2 - (b+v)R + v},$$

and let $Z_t = XR^t$. Then

$$\begin{aligned} Z_t - (b+v)Z_{t-1} + vZ_{t-2} &= XR^t - (b+v)XR^{t-1} + vXR^{t-2} \\ &= XR^{t-2}\left(R^2 - (b+v)R + v\right) \\ &= R^{t-2}iR^2 = iR^t. \end{aligned}$$

This calculation shows that

$$Z_t = XR^t = \frac{iR^{t+2}}{R^2 - (b+v)R + v}$$

is a particular solution of the non-homogeneous equation.

The corresponding homogeneous equation is the same as the one we solved in the previous section. Since Z_t increases steadily with t, we conclude that oscillatory behaviour may occur only when it is the case that the general solution of the homogeneous equation exhibits such behaviour; that is if $(b+v)^2 < 4v$.

In this case the general solution of the non-homogeneous equation takes the form

$$Y_t = Z_t + r^t\left[E\cos(\theta t) + F\sin(\theta t)\right],$$

where r and θ are as before, in particular $r = \sqrt{v}$.

Since $Z_t = XR^t$, and the other term is a multiple of $(\sqrt{v})^t$, the behaviour of the solution depends on the relative values of R and $\sqrt{v}$. (Note that we may assume that $R > 1$, since $R = 1$ is just the case where the interest rate is zero, which has already been covered.) If $R > \sqrt{v}$ then the dominant term is Z_t and the solution is not, in the long run, oscillatory. On the other hand, if $R < \sqrt{v}$, the second term dominates and gives rise to oscillations with increasing magnitude. ■

Worked Examples

Example 24.1 Suppose that the auxiliary equation corresponding to a second-order recurrence has two distinct solutions, α and β, which are both negative. Show that, in general, the solution of the recurrence is oscillatory.

Solution

Suppose that $\alpha < \beta < 0$. Then the solution of the recurrence equation is $y_t = A\alpha^t + B\beta^t$. Suppose that $A \neq 0$. Then we may write

$$y_t = A\alpha^t\left(1 + \frac{B}{A}\left(\frac{\beta}{\alpha}\right)^t\right).$$

Since $\beta/\alpha < 1$, $(\beta/\alpha)^t \to 0$ as $t \to \infty$. It follows that the quantity in parentheses tends to 1. The solution y_t is therefore close to $A\alpha^t$, which is alternately positive and negative. If $A = 0$ but $B \neq 0$, then the solution to the recurrence is simply $y_t = B\beta^t$, which, since $\beta < 0$, is alternately positive and negative, and therefore oscillatory.

(The words ‘in general’ are put in the question to remind us that there is a trivial exception: when A and B are both zero, y_t is identically zero!)

Example 24.2 Suppose that y_t satisfies the recurrence

$$y_t + a_1 y_{t-1} + a_2 y_{t-2} = 0$$

and that the auxiliary equation has two distinct solutions which are both positive. Show that, in general, y_t is eventually either strictly increasing or strictly decreasing.

Solution

Suppose the general solution is $y_t = A\alpha^t + B\beta^t$, where $0 < \alpha < \beta$. Clearly we must consider the difference $y_{t+1} - y_t$. We have

$$y_{t+1} - y_t = A\alpha^{t+1} + B\beta^{t+1} - A\alpha^t - B\beta^t = A(\alpha - 1)\alpha^t + B(\beta - 1)\beta^t.$$

Now, provided $\beta \neq 1$ and $B \neq 0$, we may write this in the form

$$y_{t+1} - y_t = B(\beta - 1)\beta^t \left(1 + \frac{A(\alpha - 1)}{B(\beta - 1)} \left(\frac{\alpha}{\beta}\right)^t\right).$$

Since $\alpha < \beta$ we know that $(\alpha/\beta)^t \to 0$ as $t \to \infty$, and the expression in brackets is eventually close to 1. So $y_{t+1} - y_t$ behaves like $B(\beta - 1)\beta^t$, which is either always positive or always negative, depending on the sign of $B(\beta - 1)$. Hence y_t is eventually either increasing or decreasing. In particular, for large enough t there is no oscillatory behaviour.

If $\beta = 1$ or $B = 0$, $y_{t+1} - y_t = A(\alpha - 1)\alpha^t$, which is positive if $A(\alpha - 1) > 0$ and negative if $A(\alpha - 1) < 0$. Thus the same conclusion holds.

Example 24.3 Suppose that consumption this year is the average of this year's income and last year's consumption:

$$C_t = \frac{1}{2}(Y_t + C_{t-1}). \qquad (*)$$

Suppose also that the relationship between next year's income and current investment is $Y_{t+1} = kI_t$, for some positive constant k. Assuming that the usual equilibrium conditions hold, derive a second-order recurrence for Y_t.

Solution

We note first that substituting $I_t = Y_{t+1}/k$ in the equilibrium condition $Y_t = C_t + I_t$ gives

$$Y_t = C_t + \frac{Y_{t+1}}{k}.$$

We need to eliminate both C_t and C_{t-1} from the behavioural condition $(*)$ displayed above. We can do this by rearranging the preceding formula, and then replacing t by $t - 1$:

$$C_t = Y_t - \frac{Y_{t+1}}{k}, \quad \text{so that} \quad C_{t-1} = Y_{t-1} - \frac{Y_t}{k}.$$

Substituting in $(*)$ we get

$$Y_t - \frac{Y_{t+1}}{k} = \frac{1}{2}\left(Y_t + \left[Y_{t-1} - \frac{Y_t}{k}\right]\right).$$

Rearranging and replacing t by $t-1$, we obtain the second-order recurrence equation

$$Y_t - \left(\frac{k+1}{2}\right) Y_{t-1} + \frac{k}{2} Y_{t-2} = 0.$$

Example 24.4 In the model setup in the previous example, suppose that $k = 3$ and that the initial value Y_0 is positive. Show that Y_t oscillates with increasing magnitude.

Solution

The recurrence is

$$Y_t - 2Y_{t-1} + \frac{3}{2} Y_{t-2} = 0.$$

The auxiliary equation $z^2 - 2z + (3/2) = 0$ has no solutions, so the solution for Y_t is

$$Y_t = r^t \left[E \cos(\theta t) + F \sin(\theta t)\right],$$

where $r = \sqrt{3/2}$ and $\cos\theta = \sqrt{2/3}$. Since $Y_0 = E$ and Y_0 is positive, we have $E > 0$. Also $r > 1$, so Y_t oscillates with increasing magnitude.

Example 24.5 Find the values of k for which the model setup in Example 24.3 leads to 'business cycles', and determine whether or not the cycles increase in magnitude. (Remember, we are given that $k > 0$.)

Solution

The auxiliary equation for the recurrence with general k is

$$z^2 - \left(\frac{k+1}{2}\right) z + \frac{k}{2} = 0.$$

This has no solutions if

$$\left(\frac{k+1}{2}\right)^2 < 4\left(\frac{k}{2}\right), \quad \text{that is} \quad (k+1)^2 < 8k.$$

In this case the general solution is of the form

$$Y_t = \left(\sqrt{\frac{k}{2}}\right)^t (E \cos\theta t + F \sin\theta t).$$

This solution is oscillatory (e.g., as in the case $k = 3$ discussed above).

Suppose that $(k+1)^2 > 8k$. Then the solution is of the form

$$Y_t = A\alpha^t + B\beta^t,$$

where α and β are both positive (check!). It follows that in this case there can be no oscillatory behaviour. The same holds true when $(k+1)^2 = 8k$.

We have shown that business cycles occur when $(k+1)^2 < 8k$, in other words when k lies strictly between the roots of the equation $(k+1)^2 = 8k$. Rewriting this as the quadratic equation $k^2 - 6k + 1 = 0$, we find that the roots are

$$3 - 2\sqrt{2} \quad \text{and} \quad 3 + 2\sqrt{2}.$$

So the model predicts that, when k is between these two numbers, the national income Y_t will show business cycles.

Whether the oscillations increase or decrease in magnitude depends on k. Since the solution involves the factor $(\sqrt{k/2})^t$, the oscillations decrease if $\sqrt{k/2} < 1$, that is if $k < 2$, and increase if $k > 2$. See Figure 24.2.

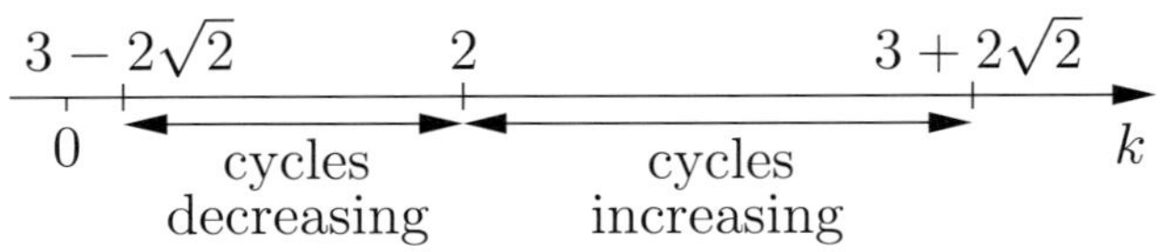

Figure 24.2 How the behaviour depends on k.

Main Topics

- the qualitative behaviour of solutions to recurrence equations
- the solution to the multiplier-accelerator equations
- business cycles
- more complex macroeconomic models

Key Terms, Notations and Formulae

- oscillatory solution
- if $a_1^2 < 4a_2$, the solution to $y_t + a_1 y_{t-1} + a_2 y_{t-2} = k$ is oscillatory with oscillations increasing if $a_2 > 1$, decreasing if $a_2 < 1$
- business cycles
- if $(b + v)^2 < 4v$, business cycles result from the multiplier-accelerator model

Exercises

24.1 Comment on the behaviour of the sequence

$$y_t = 3 + \left(-\frac{1}{2}\right)^t - \left(-\frac{1}{3}\right)^t.$$

24.2 Suppose that

$$C_t = 10 + \frac{1}{4}Y_{t-1} \quad \text{and} \quad I_t = 20 + \frac{1}{2}(Q_{t-1} - Q_{t-2}).$$

Assuming that the usual equilibrium conditions hold, find a second-order recurrence equation satisfied by Y_t.

24.3 Discuss the behaviour of the solutions to the recurrence found in Exercise 24.2.

Problems

24.1 Comment on the behaviour of each of the following sequences:

(a) $y_t = 2^t - \left(\frac{1}{2}\right)^t$;

(b) $y_t = (3t + 5)2^t$;

(c) $y_t = 2(-3)^t - 6(2^t)$.

24.2 The following equations refer to the simple economy described in Section 23.1:

$$C_t = \frac{3}{8}Y_{t-1} \quad \text{and} \quad I_t = 40 + \frac{1}{8}(Q_{t-1} - Q_{t-2}).$$

Assuming that the usual equilibrium conditions hold, show that Y_t satisfies the equation

$$Y_t - \frac{1}{2}Y_{t-1} + \frac{1}{8}Y_{t-2} = 40.$$

Given that $Y_0 = 65$ and $Y_1 = 64.5$, find Y_t and comment on the behaviour of this solution.

24.3 Examples 24.1 and 24.2 concern the behaviour of the general solution to a second-order recurrence for which the auxiliary equation has two distinct solutions, which are either both negative or both positive. Discuss the behaviour of the solution in the cases where one of the solutions is negative and the other positive.

24.4 Suppose that the auxiliary equation corresponding to a second-order recurrence has precisely one solution. Discuss the behaviour of the solution to the recurrence.

25 Areas and Integrals

25.1 The Consumer Surplus

A typical downward-sloping demand set D is illustrated in Figure 25.1. As the number of units of the good increases, the price consumers are prepared to pay for each unit decreases.

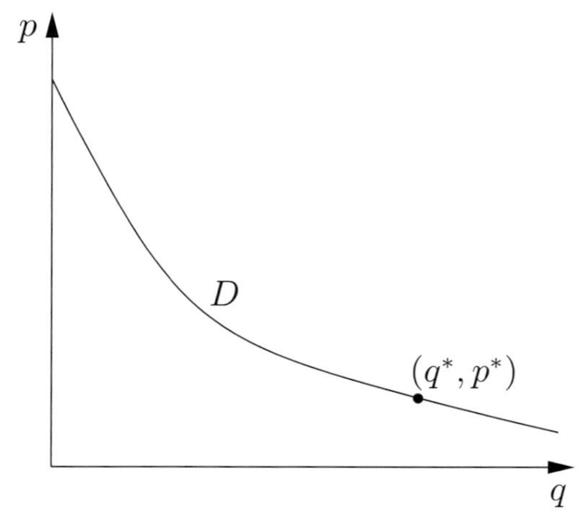

Figure 25.1 A typical demand set.

At equilibrium, when the quantity on the market is q^*, the selling price is $p^* = p^D(q^*)$. Thus the result of the operation of the market is that the consumers obtain q^* units, at a total cost of p^*q^*. But it can be argued that this is *less* than the total value the consumers place on q^* units of the good.

Suppose that, rather than the q^* units being made available all at once, the units are put on the market one after the other. When only one unit is available, the consumers are willing to pay a price $p^D(1)$ for that unit. So they obtain that unit at a cost of $p^D(1)$. If a second unit is made available the consumers would be prepared to pay only $p^D(2)$ for the second unit. And so on.

This argument suggests that the value consumers attach to the first unit is $p^D(1)$, the value they attach to the second is $p^D(2)$, and generally, the value they attach to the ith unit is $p^D(i)$. The total value to the consumers of q^* units is therefore

$$p^D(1) + p^D(2) + \cdots + p^D(q^*).$$

Figure 25.2 represents this situation in diagrammatic form. It shows the demand curve together with a series of q^* rectangles, each of width 1. The first rectangle has height $p^D(1)$, the next has height $p^D(2)$, and so on. The ith rectangle has height $p^D(i)$, and its width is 1, so its area is also $p^D(i)$. The total area of all the q^* rectangles is

$$p^D(1) + p^D(2) + \cdots + p^D(q^*),$$

which is the value to the consumers, as calculated above.

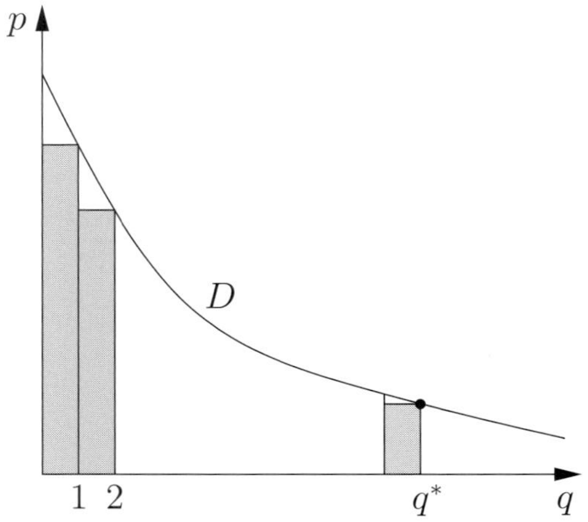

Figure 25.2 Diagrammatic representation of the value to the consumer.

If the units are small, this area is approximately equal to the area A of the shaded region in Figure 25.3a; that is the area bounded by the demand curve, the q-axis and the vertical lines $q = 0$ and $q = q^*$. So A represents the value to the consumers. On the other hand, the amount the consumers actually pay is p^*q^*, which is the area R of the shaded rectangular region in Figure 25.3b. The difference between these two areas (illustrated in Figure 25.3c) is a measure of the benefit consumers derive from the operation of the market.

It is known as the *consumer surplus*, $CS = A - R$.

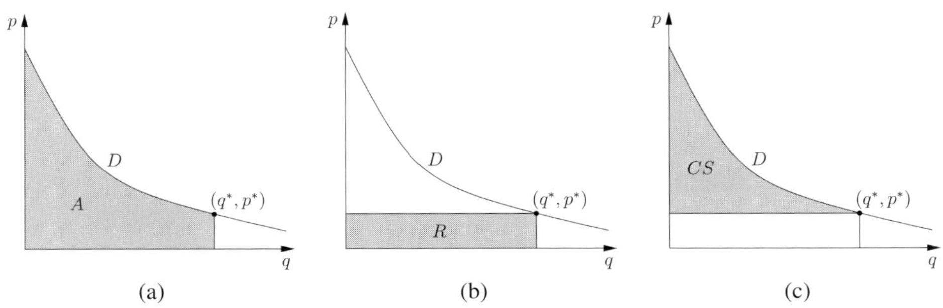

Figure 25.3 The consumer surplus, $CS = A - R$.

Example Suppose that the demand set for a given good is

$$D = \{(q, p) \mid q + 200p = 2600\},$$

and that the supply set is

$$S = \{(q, p) \mid q - 100p = -1000\}.$$

Then the equilibrium price is $p^* = 12$ and the equilibrium quantity is $q^* = 200$. The inverse demand function is

$$p^D(q) = \frac{2600 - q}{200} = 13 - \frac{q}{200}.$$

The consumer surplus is the area of the shaded region in Figure 25.4. This region is a (right-angled) triangle with base length 200 and height 1. (We note that the demand curve crosses the p-axis at $p = 13$.) Using the fact that the area of a triangle is half its base times its height, the consumer surplus is $(1/2)(200)(1) = 100$. ■

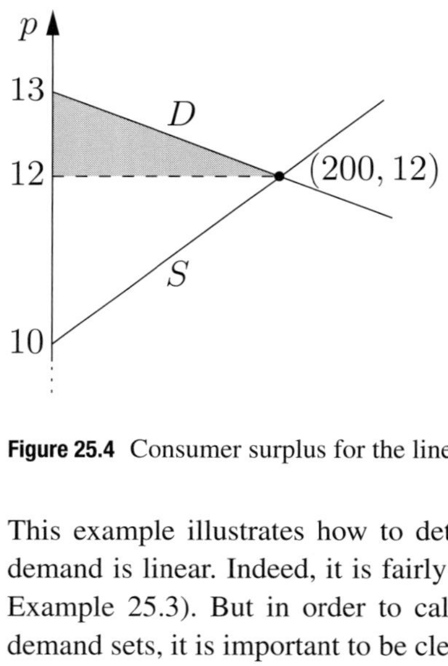

Figure 25.4 Consumer surplus for the linear demand set of the example.

This example illustrates how to determine the consumer surplus exactly when the demand is linear. Indeed, it is fairly easy to give a general analysis of this case (see Example 25.3). But in order to calculate the consumer surplus for more complex demand sets, it is important to be clear what is meant by the 'area under a curve', and to know how to calculate it.

25.2 The Concept of Area

Over a long period of time, mathematicians have worked out a satisfactory theory of curved areas. It is based on the same idea as the one we used in our discussion of the consumer surplus: begin with an approximation by rectangles, and then take the limit as the width of the rectangles tends to zero.

We shall not pursue this matter here: we simply take for granted the fact that there is a good definition of the area under a curve. Our problem is to find an effective way of calculating it.

Suppose that the function f is such that $f(x) \geq 0$ for all x in some interval $[0, X]$. Most functions occurring in economics have this property. For $t \leq X$, denote by $A(t)$ the area bounded by the curve $y = f(x)$, the x-axis and the vertical lines $x = 0$ and $x = t$ (Figure 25.5a). We refer to this as the area *under the graph of f from 0 to t*.

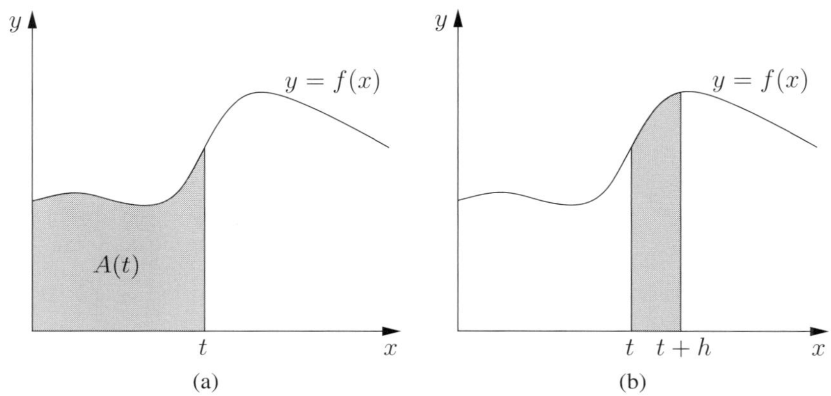

Figure 25.5 The area $A(t)$ and the difference $A(t+h) - A(t)$.

Consider the difference $A(t+h) - A(t)$ representing the area enclosed by the curve $y = f(x)$, the x-axis and the lines $x = t$ and $x = t+h$ (Figure 25.5b). For small values of h this is approximately equal to the area of a rectangle of width h and height $f(t)$, that is $A(t+h) - A(t) \simeq hf(t)$. Dividing by h, we get

$$\frac{A(t+h) - A(t)}{h} \simeq f(t).$$

This approximation becomes increasingly accurate for small h, and so

$$\lim_{h \to 0} \frac{A(t+h) - A(t)}{h} = f(t).$$

But the limit is just the derivative of A, as defined in Section 6.1. Consequently, we have shown that

$$A'(t) = f(t).$$

The conclusion is that the derivative of the function A, which measures the area under the graph of f, is just the given function f.

25.3 Anti-derivatives and Integrals

The result suggested by the preceding argument is that 'finding the area under a curve' may be thought of as the inverse of 'finding the derivative'. Precisely, if we are given the function f, the area A may be obtained by finding a function which, when differentiated, gives f.

We formalise this process as follows. Suppose the function f is given, and the function F is such that $F'(t) = f(t)$. Then we say that F is an *anti-derivative* of f. For example, t^3 is an anti-derivative of $3t^2$, and so is $t^3 + 1$.

If the functions F and G are both anti-derivatives of f, then the function defined by $H(t) = F(t) - G(t)$ satisfies

$$H'(t) = F'(t) - G'(t) = f(t) - f(t) = 0.$$

The only functions which have derivative equal to 0 are the constant functions, so it follows that $H(t) = c$, a constant, for all values of t. We have shown that any two anti-derivatives of a given function f differ only by a constant.

Example Consider $f(t) = 3t^2 + 2t + 1$. We can check (by differentiating) that the function $F(t) = t^3 + t^2 + t$ is an anti-derivative of f, as is $G(t) = t^3 + t^2 + t + 5$. *Any* anti-derivative of f is of the form $t^3 + t^2 + t + c$, for some constant c. ■

The general form of the anti-derivative of f is called the *indefinite integral* of $f(t)$, denoted by

$$\int f(t)\, dt.$$

Often we call it simply the *integral* of f. It is of the form $F(t) + c$, where F is *any particular* anti-derivative of f and c is an arbitrary constant, known as a *constant of integration*. Thus, for example, we write

$$\int (3t^2 + 2t + 1)\, dt = t^3 + t^2 + t + c.$$

Example The derivative of t^r is rt^{r-1} for $r \neq 0$; so an anti-derivative of rt^{r-1} is t^r. Putting $n = r - 1$ and rearranging we have the result that, for $n \neq -1$, an anti-derivative of t^n is $t^{n+1}/(n + 1)$. Thus, for $n \neq -1$, the integral of t^n is given by

$$\int t^n\, dt = \frac{1}{n+1} t^{n+1} + c.$$

■

The process of finding the indefinite integral of f is usually known as *integrating* f, and f is known as the *integrand*.

Two important properties of integrals are easily verified:

$$\int [f(t) + g(t)]\, dt = \int f(t)\, dt + \int g(t)\, dt,$$

for any functions f and g, and

$$\int kf(t)\, dt = k \int f(t)\, dt,$$

for any constant k.

Example These two rules justify the following calculation:

$$\begin{aligned}\int \left(5t^2 + 7t + 3\right) dt &= \int 5t^2\, dt + \int 7t\, dt + \int 3\, dt \\ &= 5\int t^2\, dt + 7\int t\, dt + 3\int 1\, dt \\ &= \frac{5}{3}t^3 + \frac{7}{2}t^2 + 3t + c.\end{aligned}$$

■

25.4 Definite Integrals

Let f be a function with an anti-derivative F. The *definite integral* of the function f over the interval $[a,b]$ is

$$\int_a^b f(t)\, dt = F(b) - F(a).$$

Note that any anti-derivative $G(t)$ of f is of the form $G(t) = F(t) + c$, for some constant c, so that $G(b) - G(a) = F(b) - F(a)$. Thus, whichever anti-derivative of f is chosen, the quantity on the right-hand side of the definition is the same. In calculations, the notation $[F(t)]_a^b$ is often used as shorthand for $F(b) - F(a)$.

Example The definite integral of t^2 over $[1,2]$ is

$$\int_1^2 t^2\, dt = \left[\frac{t^3}{3}\right]_1^2 = \frac{8}{3} - \frac{1}{3} = \frac{7}{3}.$$

■

There is an obvious relationship between the definite integral and the area under a curve. Recall that we denoted the area under the graph of f from 0 to t by $A(t)$. We showed that A is an anti-derivative of f, and clearly $A(0) = 0$, so the definition of the definite integral says that

$$\int_0^x f(t)\, dt = A(x) - A(0) = A(x).$$

In other words, the definite integral over $[0,x]$ is the area from 0 to x.

We can easily extend this to an arbitrary interval $[a,b]$, because it is clear that the area between the lines $t = a$ and $t = b$ is the difference of the areas $A(b)$ and $A(a)$. It follows that

- *the area enclosed by the curve $y = f(t)$, the t-axis and the vertical lines $t = a$ and $t = b$ is equal to $\int_a^b f(t)\, dt$.*

This is a very useful result: in order to calculate the area under the graph of f between a and b, all that is needed is to find an anti-derivative F of f and compute $F(b) - F(a)$.

Example What is the area under the curve $y = t^3$ between $t = 2$ and $t = 4$?

Since t^3 is the derivative of $t^4/4$, an anti-derivative of t^3 is $t^4/4$. So the area is

$$\int_2^4 t^3\,dt = \left[\frac{t^4}{4}\right]_2^4 = \frac{4^4}{4} - \frac{2^4}{4} = 64 - 4 = 60.$$

■

25.5 Standard Integrals

In the previous section we showed that in order to calculate the area under a curve $y = f(x)$ it is only necessary to integrate f, that is to find an anti-derivative of f. In theory, this is straightforward enough. In practice, however, integration is not always that simple.

In the next chapter we shall develop a toolbox of techniques for integration. As a first step, we can write down a number of 'standard integrals' just by inverting standard results about derivatives. We encountered one important standard integral in Section 25.3, when we observed that

$$\int t^n\,dt = \frac{1}{n+1}t^{n+1} + c \quad (n \neq -1).$$

Other useful standard integrals which may be obtained similarly are

$$\int e^t\,dt = e^t + c,$$

$$\int \sin t\,dt = -\cos t + c,$$

$$\int \cos t\,dt = \sin t + c,$$

$$\int \frac{1}{t}\,dt = \ln t + c.$$

There is one point to watch, especially when we come to evaluate areas by means of these formulae. All our results have been obtained on the assumption that the functions concerned have positive values. However, functions like t^3, $\sin t$ and $1/t$ are negative for some values of t. What happens in that case is largely a matter of common sense, but if the point is overlooked, nonsensical results may be obtained!

There is a subtle point concerning the integral of $1/t$. Of course, $\ln t$ is only defined for positive t. But if $t < 0$, then the function $\ln(-t)$ is defined and its derivative is $1/t$. So, for that reason, it would be more precise to say that the integral of $1/t$ is $\ln |t|$ (where $|t| = t$ if $t \geq 0$ and $|t| = -t$ if $t < 0$). We shall, however, proceed as indicated above.

Example In the example of Section 25.1, we computed the consumer surplus in the case where the inverse demand function is

$$p^D(q) = \frac{2600 - q}{200} = 13 - \frac{q}{200}.$$

The area $A(200)$ enclosed by the curve $p = p^D(q)$, the q-axis and the lines $q = 0$ and $q = 200$ is easily calculated using elementary geometry. It is the area of the triangular region (i.e., the consumer surplus) plus the area of the rectangle of width 200 and height 12 lying underneath it. Thus

$$A = 100 + (200)(12) = 2500.$$

The theory of Section 25.4 asserts that $A = \int_0^{200} p^D(q)\,dq$. We now verify that this 'works'. It is easy to check (using the standard integrals, or by differentiating it) that an anti-derivative of p^D is the function

$$F(q) = 13q - \frac{q^2}{400}.$$

So, $\int_0^{200} p^D(q)\,dq$ is

$$\int_0^{200} \left(13 - \frac{q}{200}\right) dq = F(200) - F(0) = \left(13(200) - \frac{(200)^2}{400}\right) - 0 = 2500,$$

in agreement with the elementary calculation. ■

Recall that in general the consumer surplus is defined to be $CS = A - R$, where A is the area under the demand curve from 0 to q^* and $R = p^*q^*$. Using the integral formula for A, we have the expression

$$CS = A - R = \int_0^{q^*} p^D(q)\,dq - p^*q^*.$$

Example Suppose the demand set for tins of caviar is, as in Section 6.2:

$$D = \{(q,p) \mid p^3 q = 8000\},$$

and suppose the supply set is

$$S = \{(q,p) \mid q = 500p\}.$$

We shall calculate the consumer surplus. First, we find the equilibrium quantity and price. The equilibrium price p^* satisfies the equation

$$\frac{8000}{(p^*)^3} = 500p^*,$$

so $p^* = 2$. The equilibrium quantity is therefore $q^* = 500p^* = 1000$. The inverse demand function is $p^D(q) = 20q^{-1/3}$, so the consumer surplus is

$$\begin{aligned} CS &= \int_0^{q^*} p^D(q)\, dq - p^* q^* \\ &= \int_0^{1000} 20q^{-1/3}\, dq - 2(1000) \\ &= \left[30q^{2/3}\right]_0^{1000} - 2000. \end{aligned}$$

Since $1000^{2/3} = 100$, this is $3000 - 2000 = 1000$. ■

Worked Examples

Example 25.1 Find the area enclosed by the lines $t = 1$, $t = 2$, the t-axis and the graph of the function $f(t) = e^t$.

Solution

The required area is equal to the definite integral $\int_1^2 e^t\, dt$. Now, $\int e^t\, dt$ is one of the list of standard integrals: $\int e^t\, dt = e^t + c$. In other words, e^t is an anti-derivative of e^t. Hence

$$\text{Area} = \int_1^2 e^t\, dt = \left[e^t\right]_1^2 = e^2 - e = 4.67077\ldots .$$

Example 25.2 Show that when $t > -1$, the derivative of $\ln(t+1)$ is $1/(t+1)$. Show also that

$$\frac{t^2}{t+1} = t - 1 + \frac{1}{t+1},$$

and hence find the indefinite integral

$$\int \frac{t^2}{t+1}\, dt.$$

Evaluate the definite integral

$$\int_2^3 \frac{t^2}{t+1}\, dt.$$

Solution

The function $\ln(t+1)$ is the composition of $\ln x$ and $x(t) = t + 1$. The derivative of $\ln x$ is $1/x$ and the derivative of $x(t)$ is 1, so the 'function of a function rule' tells us that the derivative is $1/x(t) \times 1$, that is $1/(t+1)$.

To check the required identity, we start with the right-hand side and simplify it by taking a common denominator:

$$t - 1 + \frac{1}{t+1} = \frac{t(t+1) - 1(t+1) + 1}{t+1} = \frac{t^2}{t+1}.$$

It follows that

$$\begin{aligned}\int \frac{t^2}{t+1}\,dt &= \int \left(t - 1 + \frac{1}{t+1}\right) dt \\ &= \int t\,dt - \int 1\,dt + \int \frac{1}{t+1}\,dt.\end{aligned}$$

From the list of standard integrals, $\int t\,dt = t^2/2 + c$ and $\int 1\,dt = t + c$. Also, we have established that the derivative of $\ln(t+1)$ is $1/(t+1)$, so

$$\int \frac{1}{t+1}\,dt = \ln(t+1) + c.$$

Hence

$$\int \frac{t^2}{t+1}\,dt = \frac{t^2}{2} - t + \ln(t+1) + c.$$

To determine the definite integral, we have

$$\begin{aligned}\int_2^3 \frac{t^2}{t+1}\,dt &= \left[\frac{t^2}{2} - t + \ln(t+1)\right]_2^3 \\ &= \left(\frac{9}{2} - 3 + \ln 4\right) - \left(\frac{4}{2} - 2 + \ln 3\right),\end{aligned}$$

which simplifies to $3/2 + \ln(4/3)$.

Example 25.3 Suppose that the demand set and supply set for a good are

$$D = \{(q,p) \mid q = c - dp\} \quad \text{and} \quad S = \{(q,p) \mid q = bp - a\},$$

where a, b, c, d are positive constants. Find an expression for the consumer surplus.

Solution

We first calculate the consumer surplus by elementary methods, using only the fact that the area of a triangle is half its base times its height. The supply and demand sets are illustrated in Figure 25.6. As shown in previous chapters, the equilibrium price is

$$p^* = \frac{c+a}{b+d},$$

and the equilibrium quantity is

$$q^* = c - dp^* = \frac{bc - ad}{b+d}.$$

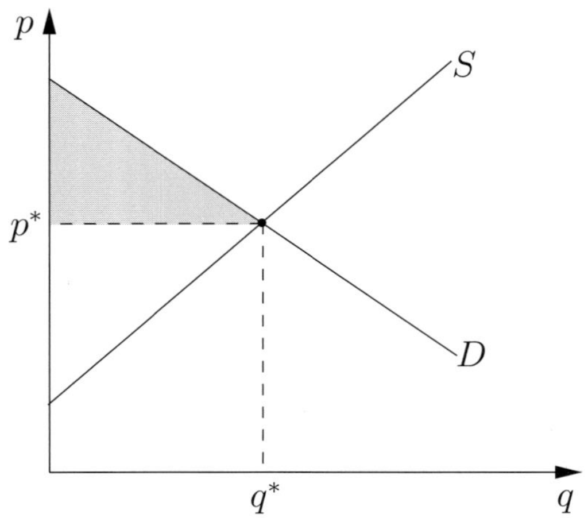

Figure 25.6 Consumer surplus for general linear supply and demand sets.

The inverse demand function is

$$p^D(q) = \frac{c-q}{d}.$$

The consumer surplus is the area of the triangular region bounded by the lines $p = p^*$ and $q = 0$, and by the demand curve. Since the demand curve crosses the p-axis at $(0, p^D(0))$, this area is

$$CS = \frac{1}{2}(p^D(0) - p^*)q^* = \frac{1}{2}\left(\frac{c}{d} - \frac{c+a}{b+d}\right)\left(\frac{bc-ad}{b+d}\right).$$

That is

$$CS = \frac{1}{2}\left(\frac{cb+cd-cd-ad}{d(b+d)}\right)\left(\frac{bc-ad}{b+d}\right) = \frac{(bc-ad)^2}{2d(b+d)^2}.$$

We can also calculate the consumer surplus by definite integration. (This turns out to be more difficult in this particular case, but, as noted earlier, definite integration really is needed for demand functions more complex than the simple linear one discussed here.) We have

$$CS = \int_0^{q^*} p^D(q)\,dq - p^*q^* = \int_0^{q^*} \frac{c-q}{d}\,dq - p^*q^*.$$

Now

$$\int \frac{c-q}{d}\,dq = \int \left(\frac{c}{d} - \frac{1}{d}q\right)\,dq = \frac{c}{d}q - \frac{1}{2d}q^2 + c,$$

so

$$\begin{aligned}
CS &= \left(\frac{c}{d}q^* - \frac{1}{2d}(q^*)^2\right) - (0) - p^*q^* \\
&= \frac{q^*}{2d}\left(2c - q^* - 2p^*d\right) \\
&= \frac{q^*}{2d}\left(2c - \frac{bc-ad}{b+d} - 2\frac{c+a}{b+d}d\right) \\
&= \frac{q^*}{2d(b+d)}\,(2cb + 2cd - bc + ad - 2cd - 2ad) \\
&= \frac{bc-ad}{2d(b+d)^2}(bc-ad) \\
&= \frac{(bc-ad)^2}{2d(b+d)^2}
\end{aligned}$$

as above.

Example 25.4 Find the area enclosed by the curves $y = 1/t^2$ and $y = t^3$, the t-axis and the lines $t = 1/2$ and $t = 2$.

Solution

The region described is pictured in Figure 25.7; the curves $y = 1/t^2$ and $y = t^3$ intersect when $t^5 = 1$ which, in the positive quadrant, means $t = 1$.

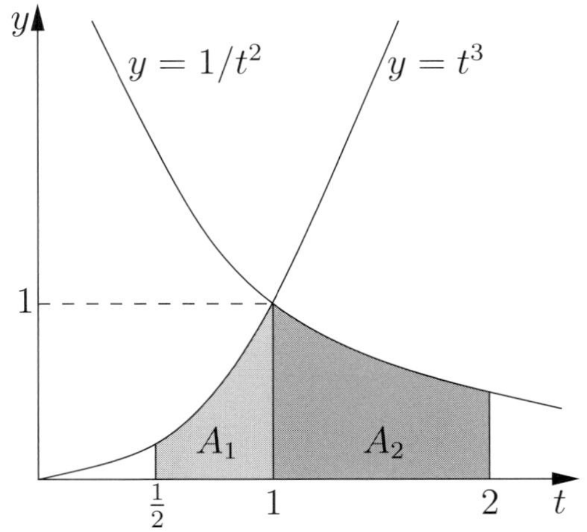

Figure 25.7 The required area.

The easiest way to compute the area, A, of the region is to calculate each of the areas A_1 and A_2 separately; then $A = A_1 + A_2$. By the theory of Section 25.4:

$$A_1 = \int_{1/2}^{1} t^3\,dt = \left[\frac{t^4}{4}\right]_{1/2}^{1} = \frac{1}{4}(1)^4 - \frac{1}{4}\left(\frac{1}{2}\right)^4 = \frac{15}{64}$$

and

$$A_2 = \int_1^2 \frac{1}{t^2}\, dt = \left[-\frac{1}{t}\right]_1^2 = -\frac{1}{2} - \left(-\frac{1}{1}\right) = \frac{1}{2}.$$

It follows that

$$A = A_1 + A_2 = \frac{15}{64} + \frac{1}{2} = \frac{47}{64}.$$

Example 25.5 The demand set for a commodity is

$$D = \{(q,p) \mid p(q+1) = 231\},$$

and the supply set is

$$S = \{(q,p) \mid p - q = 11\}.$$

Determine the consumer surplus.

Solution

The inverse demand function is $p^D(q) = 231/(q+1)$. The equilibrium quantity q^* is the solution to the equation

$$\frac{231}{q+1} = q + 11$$

obtained by equating $p^D(q)$ and $p^S(q)$. So, q^* satisfies the equation

$$(q+11)(q+1) = 231, \quad q^2 + 12q - 220 = 0, \quad (q+22)(q-10) = 0.$$

Since q^* cannot be negative, $q^* = 10$. The equilibrium price is $p^* = q^* + 11 = 21$. The consumer surplus is then

$$CS = \int_0^{q^*} p^D(q)\, dq - p^* q^* = \int_0^{10} \frac{231}{q+1}\, dq - (21)(10).$$

An anti-derivative of $231/(q+1)$ is $231 \ln(q+1)$. (This follows from the result mentioned in Example 25.2.) It follows that

$$CS = 231 \ln(11) - 231 \ln(1) - 210 = 231 \ln(11) - 210,$$

which is approximately 343.9.

Main Topics

- the consumer surplus
- the area under a curve
- anti-derivatives and indefinite integrals
- definite integrals
- area as a definite integral
- standard integrals
- calculating consumer surplus

Key Terms, Notations and Formulae

- consumer surplus, CS
- $A(t)$, area under the graph of $f(x)$ from 0 to t: $A'(t) = f(t)$
- anti-derivative of f: any function F such that $F'(t) = f(t)$
- (indefinite) integral of f: $\int f(t)\,dt = F(t) + c$
- $\int (f(t) + g(t))\,dt = \int f(t)\,dt + \int g(t)\,dt, \quad \int kf(t)\,dt = k\int f(t)\,dt$
- definite integral, $\int_a^b f(t)\,dt = [F(t)]_a^b = F(b) - F(a)$
- area enclosed by $y = f(t)$, t-axis, $t = a$, $t = b$ equals $\int_a^b f(t)\,dt$
- standard integrals:

$$\int t^n\,dt = \frac{1}{n+1}t^{n+1} + c \ (n \neq -1)$$

$$\int \frac{1}{t}\,dt = \ln t + c$$

$$\int e^t\,dt = e^t + c$$

$$\int \sin t\,dt = -\cos t + c$$

$$\int \cos t = \sin t + c$$

- consumer surplus, $CS = A - R = \int_0^{q^*} p^D(q)\,dq - p^*q^*$

Exercises

25.1 If a function f has derivative $t^2 + 2\sin t$, and $f(0) = 1$, what is the function?

25.2 Evaluate the definite integrals

$$\int_0^2 t^2\,dt \quad \text{and} \quad \int_{-1}^1 e^t\,dt.$$

25.3 The inverse demand function for a product is

$$p^D(q) = \frac{100}{q+5}$$

and the supply function is

$$q^S(p) = 5p - 45.$$

Determine the equilibrium price and quantity and the consumer surplus.

Problems

25.1 Show that when $t > -3$ the derivative of $\ln(t+3)$ is $1/(t+3)$. Show also that

$$\frac{t^2 + 4t + 4}{t+3} = t + 1 + \frac{1}{t+3}$$

and hence find the indefinite integral

$$\int \frac{t^2 + 4t + 4}{t + 3} \, dt.$$

Evaluate the definite integral

$$\int_1^2 \frac{t^2 + 4t + 4}{t + 3} \, dt.$$

25.2 The demand and supply sets for a good are

$$D = \{(q,p) \mid q = 170 - p\} \quad \text{and} \quad S = \{(q,p) \mid 2q + 140 = p\}.$$

Find the equilibrium price and quantity and the consumer surplus.

25.3 Find the area enclosed by the lines $t = 0$, $t = \pi$, the t-axis and the graph of the function $f(t) = 2 + \sin t$.

25.4 In Example 7.2 it was shown that the derivative of the function $f(x) = \ln\left(x + \sqrt{x^2 + 1}\right)$ is $f'(x) = 1/\sqrt{x^2 + 1}$. Use this fact to determine the area enclosed by the graph of $f(t) = 1/\sqrt{t^2 + 1}$, the t-axis and the lines $t = 1$ and $t = 3$.

25.5 Find the area enclosed by the curves $y = 8/t$, $y = \sqrt{t}$, the t-axis and the lines $t = 1$ and $t = 8$.

25.6 Show that the derivative of the function $\ln(t^2 + 1)$ is $2t/(t^2 + 1)$.
Suppose that the inverse demand function for a good is

$$p^D(q) = \frac{2q}{q^2 + 1},$$

and that the equilibrium quantity is 10. Find the consumer surplus.

26 Techniques of Integration

26.1 Integration by Substitution

In Chapters 6 and 7 we discussed several methods for finding the derivative of a function. We obtained some specific rules for simple functions, together with general methods, such as the composite function rule and the product rule. In a similar vein, we shall now discuss two general methods for integrating certain combinations of functions. The first method is the rule for *integration by substitution*, which we shall introduce by means of an example.

Example Suppose that we are asked to find the indefinite integral

$$\int (7x+2)^{11}\,dx.$$

We note that if we substitute $7x+2=t$ the integrand becomes t^{11}, which we know how to integrate. Unfortunately, it is not quite as simple as that, because we have changed the variable of integration. Originally we integrated with respect to x, signified by dx in the integral, now we must integrate with respect to t.

So how is dx related to dt? This is easy if we think about how x is related to t. The substitution $7x+2=t$ is the same as saying that

$$x=\frac{1}{7}t-\frac{2}{7},$$

so $dx/dt=1/7$. This suggests that when we replace $7x+2$ by t we should replace dx by $(1/7)dt$, giving

$$\int (7x+2)^{11}\,dx=\int t^{11}\left(\frac{1}{7}\right)dt=\frac{1}{7}\left(\frac{t^{12}}{12}\right)+c.$$

We need the answer in terms of x, the original variable. Since $t = 7x+2$ the integral is

$$\frac{1}{84}(7x+2)^{12}+c.$$

■

Of course, the justification given in the example was a bit unsatisfactory. The rule suggested there is that when we change the variable by putting $x=x(t)$ in the integral of $f(x)$ with respect to x, we must replace dx by $(dx/dt)dt$. In other words, in order to obtain

$$\int f(x)\,dx$$

we can work out

$$\int f(x(t))\,x'(t)\,dt,$$

and then put back x for $x(t)$.

Fortunately, we can show that this rule is indeed correct. We need only remember that an indefinite integral is just an anti-derivative, and use the rule for finding the derivative of a composite function.

Suppose that $\int f(x)\,dx = F(x)$. This simply means that the derivative of F is f, that is $F' = f$. So

$$\int f(x(t))\,x'(t)\,dt = \int F'(x(t))\,x'(t)\,dt.$$

Now $F'(x(t))\,x'(t)$ is the derivative of the composite function $F(x(t))$, and so $F(x(t))$ is the integral of $F'(x(t))\,x'(t)$:

$$\int F'(x(t))\,x'(t)\,dt = F(x(t)) + c.$$

If we put back x for $x(t)$, we get $F(x)$, which was defined to be $\int f(x)\,dx$.

It is often better to think of this technique as a 'change of variable', rather than a 'substitution'. In practice we tend to overlook the distinction between x as a function of t and the inverse function, relying on the fact (Section 6.5) that dt/dx is equal to $1/(dx/dt)$. This allows us to write 'shorthand' statements like

$$7x + 2 = t \quad \text{therefore} \quad 7\,dx = dt,$$

which determines both dt/dx and dx/dt.

Of course there is an art as well as a science in finding a good substitution. The aim is to make the integral simpler, but many substitutions will make it worse! However, a well-chosen substitution can be extremely effective.

Example Suppose we wish to find the integral

$$\int x\sqrt{3x^2+5}\,dx.$$

The difficulty here is the messy expression $\sqrt{3x^2+5}$. We can try to overcome this by making the substitution $3x^2 + 5 = t$. Then, using the conventional shorthand, we have

$$3x^2 + 5 = t \quad \text{therefore} \quad 6x\,dx = dt.$$

Now we can argue as follows:

$$\int x\sqrt{3x^2+5}\,dx = \int \sqrt{t}\left(\frac{1}{6}\right)dt$$

$$= \frac{1}{6}\int t^{1/2}\,dt$$
$$= \frac{1}{6}\left(\frac{2}{3}t^{3/2}\right) + c$$
$$= \frac{1}{9}(3x^2+5)^{3/2} + c.$$

Remember that the final step is needed to revert to a function of x. ■

26.2 Definite Integrals by Substitution

In the case of a definite integral there is no need to revert to the original variable before evaluating the anti-derivative: we simply use the appropriate values of the new variable. If we change from the variable x to the variable t, and the interval of integration for x was $[a,b]$, the interval for t will be $[\alpha,\beta]$, where α and β are the values of t which correspond to $x = a$ and $x = b$, respectively. Formally

$$\int_{x=a}^{x=b} f(x)\,dx = \int_{t=\alpha}^{t=\beta} f(x(t))\,x'(t)\,dt,$$

where $x(\alpha) = a$ and $x(\beta) = b$. This result holds provided that t increases steadily from α to β as x goes from a to b.

Example In order to evaluate the definite integral

$$\int_2^3 x\sqrt{3x^2+5}\,dx,$$

we can use the substitution $3x^2+5 = t$, as before. As x goes from 2 to 3, $t = 3x^2+5$ increases from 17 to 32. Hence

$$\int_{x=2}^{x=3} x\sqrt{3x^2+5}\,dx = \frac{1}{6}\int_{t=17}^{t=32} t^{1/2}\,dt = \frac{1}{6}\left[\frac{2}{3}t^{3/2}\right]_{17}^{32} = \frac{1}{9}\left(32^{3/2} - 17^{3/2}\right).$$

■

Example Evaluate the definite integral

$$\int_0^5 \frac{x+1}{x^2+2x+5}\,dx.$$

The difficulty here arises from the complicated denominator. So, a possible substitution is $x^2+2x+5 = t$, which implies $(2x+2)\,dx = dt$. For the *indefinite* integral, we have

$$\int \frac{1}{x^2+5x+2}(x+1)\,dx = \frac{1}{2}\int \frac{1}{t}\,dt = \frac{1}{2}\ln t + c.$$

As x goes from 0 to 5, $t = x^2+2x+5$ increases from 5 to 40. Therefore

$$\int_0^5 \frac{x+1}{x^2+2x+5}\,dx = \frac{1}{2}\int_5^{40} \frac{1}{t}\,dt = \frac{1}{2}\Big[\ln t\Big]_5^{40} = \frac{1}{2}(\ln 40 - \ln 5).$$

This is $\frac{1}{2}\ln(40/5)$, that is $\frac{1}{2}\ln 8$. ■

26.3 Integration by Parts

The product rule for differentiation says that the derivative of $u(t)v(t)$ is $u'(t)v(t) + u(t)v'(t)$. Hence the anti-derivative of $u'(t)v(t) + u(t)v'(t)$ is $u(t)v(t)$, or equivalently

$$\int u'(t)v(t)\,dt + \int u(t)v'(t)\,dt = u(t)v(t).$$

Rearranging, we get the rule for *integration by parts*:

$$\int u'(t)v(t)\,dt = u(t)v(t) - \int u(t)v'(t)\,dt.$$

The significance of this rule is that we can express an integral of the form $\int u'(t)v(t)\,dt$ as a known function minus another integral. The second integral may be easier than the first.

Example Consider the integral

$$\int t\ln t\,dt.$$

Taking $u'(t) = t$ and $v(t) = \ln t$ in the integration by parts rule, we have

$$\begin{aligned}\int u'(t)v(t)\,dt &= u(t)v(t) - \int u(t)v'(t)\,dt \\ &= \left(\frac{1}{2}t^2\right)\ln t - \int \left(\frac{1}{2}t^2\right)\frac{1}{t}\,dt \\ &= \frac{1}{2}t^2\ln t - \frac{1}{2}\int t\,dt \\ &= \frac{1}{2}t^2\ln t - \frac{1}{4}t^2 + c.\end{aligned}$$

■

Note that the procedure is successful when we get a simpler integral as a result of replacing one 'part' $u'(t)$ by its integral $u(t)$ and the other 'part' $v(t)$ by its derivative $v'(t)$. Of course, we must be careful to choose the 'parts' correctly in order to make this work.

Example Consider the integral

$$\int te^t\,dt.$$

If we replace t by its derivative (namely, the constant 1) and e^t by its integral (namely, the function e^t), we have $1 \times e^t = e^t$, which is a function we can integrate immediately. Therefore we use integration by parts, taking $u'(t) = e^t$ and $v(t) = t$. We have

$$\int te^t\,dt = te^t - \int 1e^t\,dt = te^t - \int e^t\,dt = te^t - e^t + c.$$

■

In some cases it may be necessary to use both substitution and integration by parts, as in Example 26.4. There can be no fixed rules about how to proceed, because integration is essentially a problem of 'working backwards': we have to find a function which, when differentiated, produces a given result.

26.4 Partial Fractions

This is not, strictly speaking, a technique for integration; rather, it is an algebraic method for rewriting particular expressions in a simpler form which, it turns out, makes them easier to integrate.

Example Consider the integral

$$\int \frac{t+1}{t^2+t-2}\,dt.$$

The integrand is of the form $p(t)/q(t)$, where $p(t) = t+1$ and $q(t) = t^2+t-2$. Further, $q(t)$ factorises as $(t-1)(t+2)$. We claim that we can find constants A_1 and A_2 such that

$$\frac{t+1}{t^2+t-2} = \frac{t+1}{(t-1)(t+2)} = \frac{A_1}{t-1} + \frac{A_2}{t+2}.$$

Multiplying through by $(t-1)(t+2)$, we obtain

$$A_1(t+2) + A_2(t-1) = t+1.$$

Taking $t = 1$ gives $3A_1 + 0A_2 = 2$, so that $A_1 = 2/3$. Taking $t = -2$ gives $-3A_2 = -1$ and $A_2 = 1/3$. The identity

$$\frac{t+1}{t^2+t-2} = \frac{2/3}{t-1} + \frac{1/3}{t+2}$$

is called an expansion in *partial fractions*. It can easily be checked by multiplying out.

Now we can determine the integral, as follows:

$$\begin{aligned}\int \frac{t+1}{t^2+t-2}\,dt &= \frac{1}{3}\int\left(\frac{2}{t-1} + \frac{1}{t+2}\right)dt\\ &= \frac{2}{3}\ln(t-1) + \frac{1}{3}\ln(t+2) + c.\end{aligned}$$

■

As suggested by the preceding example, we can often rewrite an expression of the form $p(t)/q(t)$, where p and q are polynomials, as a sum of simpler terms, called partial fractions. For our purposes it is sufficient to consider what happens when the degree of $p(t)$ (the largest power of t occurring) is less than the degree of $q(t)$, and $q(t)$ can be completely factorised into distinct linear factors.

Thus we shall consider the case when

$$q(t) = C(t - a_1)(t - a_2) \cdots (t - a_m),$$

where C is the coefficient of t^m and $a_1, a_2, \ldots, a_m$ are distinct numbers. In this case there is a simple expansion in partial fractions. That is, there are numbers $A_1, A_2, \ldots, A_m$ such that

$$\frac{p(t)}{q(t)} = \frac{p(t)}{C(t - a_1)(t - a_2) \cdots (t - a_m)} = \frac{A_1}{t - a_1} + \frac{A_2}{t - a_2} + \cdots + \frac{A_m}{t - a_m}.$$

One way of determining the constants $A_1, A_2, \ldots, A_m$ was illustrated in the previous example. We simply multiply through by the denominator $q(t) = C(t - a_1)(t - a_2) \ldots (t - a_m)$ and substitute the values $t = a_1$, $t = a_2, \ldots, t = a_m$ in turn. In fact it is quite possible to obtain a general formula for the constants A_i in this way, but in practice it is simpler to do the algebra explicitly, rather than try to remember the formula.

Given an expression $p(t)/q(t)$ which admits a partial fraction expansion of this kind, it is easy to work out the integral. We have

$$\int \frac{p(t)}{q(t)}\, dt = \int \left(\frac{A_1}{t - a_1} + \frac{A_2}{t - a_2} + \cdots + \frac{A_m}{t - a_m} \right) dt,$$

and the integral of $A_i/(t - a_i)$ is $A_i \ln(t - a_i) + c$.

Example Suppose we wish to find the area enclosed by the x-axis, the lines $x = 1$ and $x = 3$, and the graph of the function

$$f(x) = \frac{2x + 1}{x^2 + 5x + 6}.$$

Note first that f is positive throughout the interval $[1, 3]$. It follows that the required area is the definite integral $\int_1^3 f(x)\, dx$, which we can calculate using partial fractions because $f(x) = p(x)/q(x)$, where $p(x) = 2x + 1$ and $q(x) = x^2 + 5x + 6 = (x + 2)(x + 3)$.

We rewrite $f(x)$ in the form

$$\frac{2x + 1}{x^2 + 5x + 6} = \frac{A_1}{x + 2} + \frac{A_2}{x + 3},$$

for some constants A_1 and A_2. Multiplying by the denominator gives

$$2x + 1 = A_1(x + 3) + A_2(x + 2).$$

Taking $x = -2$, we obtain $-3 = A_1$, and taking $x = -3$ gives $-5 = -A_2$, or $A_2 = 5$. Thus we can calculate as follows:

$$\begin{aligned}
\int_1^3 f(x)\,dx &= \int_1^3 \left(\frac{-3}{x+2} + \frac{5}{x+3} \right) dx \\
&= \Big[-3\ln(x+2) + 5\ln(x+3) \Big]_1^3 \\
&= (-3\ln 5 + 5\ln 6) - (-3\ln 3 + 5\ln 4) \\
&= 5\ln(3/2) - 3\ln(5/3).
\end{aligned}$$

■

Worked Examples

Example 26.1 Find the integrals

$$\int \frac{2t+1}{t^2+t+1}\,dt \quad \text{and} \quad \int_0^2 \frac{2t+1}{t^2+t+1}\,dt.$$

Solution

The denominator does not factorise, so we cannot use partial fractions. However, we can try substitution. Let $t^2 + t + 1 = u$, so that $(2t+1)\,dt = du$. The indefinite integral is therefore equal to

$$\int \frac{1}{u}\,du = \ln u + c = \ln(t^2 + t + 1) + c.$$

Note that $t^2 + t + 1$ is positive for all values of t, so the logarithm is valid.

For the definite integral, note that as t goes from 0 to 2, $u = t^2 + t + 1$ goes from 1 to 7. Hence

$$\int_0^2 \frac{2t+1}{t^2+t+1}\,dt = \int_1^7 \frac{1}{u}\,du = \Big[\ln u \Big]_1^7 = \ln 7.$$

This example is a special case of a rule given in Example 26.6.

Example 26.2 Find an anti-derivative of $1/(t^2 - 1)$ and hence calculate the definite integral

$$\int_2^7 \frac{1}{t^2 - 1}\,dt.$$

Solution

Note that $t^2 - 1 = (t-1)(t+1)$, so the method of partial fractions is applicable. We write

$$\frac{1}{t^2 - 1} = \frac{A_1}{t-1} + \frac{A_2}{t+1},$$

and multiply through by $(t-1)(t+1)$ to obtain

$$1 = A_1(t+1) + A_2(t-1).$$

Taking $t = 1$ we get $2A_1 = 1$, so $A_1 = 1/2$. Taking $t = -1$ we get $-2A_2 = 1$, so $A_2 = -1/2$. Thus

$$\frac{1}{t^2-1} = \frac{1/2}{t-1} - \frac{1/2}{t+1}$$

and

$$\int \frac{1}{t^2-1}\,dt = \frac{1}{2}\int \left(\frac{1}{t-1} - \frac{1}{t+1}\right) dt = \frac{1}{2}\Big(\ln(t-1) - \ln(t+1)\Big) + c,$$

provided t is in a range where the logarithms are defined; that is $t > 1$.

Since $\frac{1}{2}\left(\ln(t-1) - \ln(t+1)\right)$ is an anti-derivative of $1/(t^2-1)$:

$$\begin{aligned}\int_2^7 \frac{1}{t^2-1}\,dt &= \left[\frac{1}{2}\Big(\ln(t-1) - \ln(t+1)\Big)\right]_2^7 \\ &= \frac{1}{2}\left(\ln 6 - \ln 8\right) - \frac{1}{2}\left(\ln 1 - \ln 3\right).\end{aligned}$$

This can be simplified as follows (remembering that $\ln 1 = 0$):

$$\frac{1}{2}\left(\ln 6 - \ln 8 + \ln 3\right) = \frac{1}{2}\ln\left(\frac{6\times 3}{8}\right) = \frac{1}{2}\ln\left(\frac{9}{4}\right) = \ln\left(\frac{3}{2}\right).$$

Example 26.3 Evaluate the definite integral

$$\int_0^1 t^2 e^t\,dt.$$

Solution

Here we use integration by parts. The idea is to reduce the degree of the t^2 part, by taking $u'(t) = e^t$ and $v(t) = t^2$. This gives

$$\int_0^1 t^2 e^t\,dt = \left[t^2 e^t\right]_0^1 - \int_0^1 (2t)e^t\,dt = e - 2\int_0^1 te^t\,dt.$$

The integral on the right is 'less complicated' than the one we started with, and it too can be integrated by parts; in fact, we obtained the indefinite integral in Section 26.3. Using that result

$$\int_0^1 te^t\,dt = \left[te^t - e^t\right]_0^1 = (e-e) - (0-1) = 1,$$

and we have

$$\int_0^1 t^2 e^t\,dt = e - 2(1) = e - 2.$$

Example 26.4 Find the area under the curve $f(t) = t^3 e^{t^2}$ between the lines $t = 1$ and $t = 3$.

Solution

We note first that $f(t)$ is positive for $1 \leq t \leq 3$, so the area is given by the definite integral of $f(t)$ over the interval $[1, 3]$.

The most complicated part of $f(t)$ is the t^2 in the exponential. We therefore try the substitution $u = t^2$, which implies that $du = 2t\,dt$. We have

$$\int t^3 e^{t^2}\,dt = \int t^2 e^{t^2}\,t\,dt = \int ue^u \left(\frac{1}{2}\right) du = \frac{1}{2}\int ue^u\,du.$$

Once again, recall that in Section 26.3 we used integration by parts to show that

$$\int ue^u\,du = ue^u - e^u + c.$$

It follows that the definite integral is

$$\int_1^3 t^3 e^{t^2}\,dt = \frac{1}{2}\int_1^9 ue^u\,du = \frac{1}{2}\Big[ue^u - e^u\Big]_1^9 = \frac{1}{2}\Big((9e^9 - e^9) - (e - e)\Big) = 4e^9.$$

Example 26.5 Evaluate the definite integral

$$\int_1^2 \frac{\sin(\ln x)}{x}\,dx.$$

Solution

The most awkward part of the integrand is $\sin(\ln x)$, complicated by the $\ln x$. Motivated by this observation, we try the substitution $\ln x = t$. Then $(1/x)\,dx = dt$ and we have, for the indefinite integral:

$$\int \frac{\sin(\ln x)}{x}\,dx = \int \sin(\ln x)\frac{1}{x}\,dx = \int \sin t\,dt = -\cos t + c = c - \cos(\ln x).$$

So it follows that the definite integral is equal to

$$\Big[-\cos(\ln x)\Big]_1^2 = -\cos(\ln 2) - \Big(-\cos(\ln 1)\Big) = -\cos(\ln 2) + \cos 0 = 1 - \cos(\ln 2),$$

which is approximately 0.231.

Example 26.6 Show that if $f(x)$ and $f'(x)$ are both positive in the interval $[a, b]$, then

$$\int_a^b \frac{f'(x)}{f(x)}\,dx = \ln\left(\frac{f(b)}{f(a)}\right).$$

Solution

This is a generalisation of a method we have used before (see Example 26.1). We make the substitution

$$f(x) = t, \quad \text{so that} \quad f'(x)\,dx = dt.$$

The range of integration for t is from $f(a)$ to $f(b)$, and, since $f'(x)$ is positive, t increases steadily as x goes from a to b. It follows that

$$\int_a^b \frac{f'(x)}{f(x)}\,dx = \int_{f(a)}^{f(b)} \frac{1}{t}\,dt = \Big[\ln t\Big]_{f(a)}^{f(b)} = \ln f(b) - \ln f(a) = \ln\left(\frac{f(b)}{f(a)}\right).$$

Main Topics

- indefinite and definite integration using substitution
- integration by parts
- partial fractions

Key Terms, Notations and Formulae

- substitution: $\int F'(x(t))\,x'(t)\,dt = F(x(t)) + c$
- substitution and definite integrals: set $x = x(t)$, then

$$\int_a^b f(x)\,dx = \int_\alpha^\beta f(x(t))\,x'(t)\,dt \quad \text{where} \quad x(\alpha) = a \text{ and } x(\beta) = b$$

- integration by parts: $\displaystyle\int u'(t)v(t)\,dt = u(t)v(t) - \int u(t)v'(t)\,dt$
- partial fractions:

$$\frac{p(t)}{q(t)} = \frac{p(t)}{C(t-a_1)(t-a_2)\cdots(t-a_m)} = \frac{A_1}{t-a_1} + \frac{A_2}{t-a_2} + \cdots + \frac{A_m}{t-a_m}$$

Exercises

26.1 Find the indefinite integral

$$\int x(2x^2+2)^{1/2}\,dx.$$

26.2 Find the indefinite integral

$$\int x\sqrt{x-1}\,dx$$

using the substitution (a) $u = x - 1$ and (b) $u = \sqrt{x-1}$.

(You should, of course, get the same answer. The point we're emphasising here is that there can be more than one appropriate substitution.)

26.3 Find the indefinite integral

$$\int \frac{\sqrt{x}}{1+\sqrt{x}}\,dx.$$

26.4 Evaluate the definite integral

$$\int_0^1 \frac{x+2}{x^2+4x+5}\,dx.$$

26.5 Find the indefinite integral

$$\int \frac{dx}{x^2+4x+3}.$$

26.6 Find the indefinite integral

$$\int \frac{e^x}{e^{2x}-1}\,dx.$$

[*Hint:* start by substituting $u = e^x$.]

26.7 Find the indefinite integral

$$\int \frac{\ln(1+x^2)}{x^3}\,dx.$$

[*Hint:* it might be useful to start with a substitution.]

Problems

26.1 Find the indefinite integrals

$$\int \frac{t^2+1}{3t+t^3}\,dt, \quad \int \frac{1}{t^2-4t+3}\,dt \quad \text{and} \quad \int te^{t^2}\,dt.$$

26.2 Evaluate the definite integrals

$$\int_1^2 \frac{t^2+1}{3t+t^3}\,dt, \quad \int_9^{10} \frac{1}{t^2-4t+3}\,dt \quad \text{and} \quad \int_0^1 te^{t^2}\,dt.$$

26.3 Show that

$$\frac{x^3+2}{x^2-1} = x + \frac{x+2}{x^2-1}.$$

Using this result and the method of partial fractions, determine

$$\int \frac{x^3+2}{x^2-1}\,dx.$$

26.4 Find the indefinite integrals

$$\int (\sin\theta)^2\cos\theta\,d\theta, \quad \int \frac{\ln x}{\sqrt{x}}\,dx \quad \text{and} \quad \int t^5 e^{t^3}\,dt.$$

26.5 Find the indefinite integral

$$\int \frac{\sqrt{t-1}}{t-2}\,dt$$

by using the substitution $u = \sqrt{t-1}$.

26.6 Using substitution and the method of partial fractions, find

$$\int \frac{1}{\sqrt{x}(1-\sqrt{x})(2-\sqrt{x})}\,dx.$$

In a similar way, find

$$\int \frac{1}{x^{1/2}-x^{3/2}}\,dx.$$

[*Hint:* for the second integral, it is useful to note that $x^{1/2} - x^{3/2} = \sqrt{x}(1-x)$ and that $1 - x = (1-\sqrt{x})(1+\sqrt{x})$.]

26.7 Find the area bounded by the lines $t = 1$ and $t = 2$, the t-axis and the graph of the function $f(t) = te^{2t}$.

26.8 Find the area bounded by the lines $t = 0$ and $t = 2$, the t-axis and the graph of the function $f(t) = te^{t^2}$.

26.9 The inverse demand function for a good is

$$p^D(q) = \frac{192}{q^2 + 4q + 3}$$

and the equilibrium price is $p^* = 4$. Find the equilibrium quantity and the consumer surplus.

26.10 The inverse demand function for a good is

$$p^D(q) = \frac{1000}{(q+10)\ln(q+10)}$$

and the equilibrium quantity is $q^* = 10$. Find the equilibrium price and the consumer surplus.

27 First-Order Differential Equations

27.1 Continuous-Time Models

In Chapter 23 we looked at the dynamics of the economy, using what are known as *discrete-time* models. This means that the time periods involved were taken to be successive calendar years (for instance), and the various quantities were measured over those periods. For example, I_t denoted the total investment during the tth calendar year, and we considered the behaviour of the sequence of values $I_0, I_1, I_2, \ldots$.

In this chapter we take a different approach, in that we shall use what are known as *continuous-time* models. Again, we have in mind some given time period, let us say 1 year. But instead of considering a sequence of values taken over successive calendar years (from 1 January to 31 December), we now look at the values taken over the period of 1 year immediately preceding time t, where t is a continuously varying parameter (e.g., one value of t might represent 0737 hrs on 15 March). Thus we denote by $I(t)$ the total investment in the year preceding time t, and we have to consider the behaviour of the function $I(t)$.

Recall that in Section 23.1 we discussed the relationships between investment, production, income and consumption. In the continuous model these are represented by $I(t)$, $Q(t)$, $Y(t)$ and $C(t)$, respectively. As in the discrete case, we have the *equilibrium conditions*

$$Q(t) = Y(t) \quad \text{and} \quad Y(t) = C(t) + I(t).$$

In addition, we have certain *behavioural conditions* which describe reasonable hypotheses about the relationships between the various quantities.

Example Suppose that the consumption $C(t)$ and income $Y(t)$ are linked by the equation

$$C(t) = c + bY(t) \quad (c, b \text{ constants}).$$

This implies that the rate of change of C with respect to time is b times the rate of change of Y with respect to time:

$$\frac{dC}{dt} = b\frac{dY}{dt}.$$

If we think of C as a function of Y, this means that

$$\frac{dC}{dY} = b,$$

which explains why economists refer to the constant b as the *marginal propensity to consume*.

Using the equilibrium condition $Y = C + I$, it follows that

$$I = Y - C = Y - c - bY = -c + sY,$$

where $s = 1 - b$ is equal to dI/dY and is known as the *marginal propensity to invest.*

In order to determine how the economy grows, we need one more behavioural condition. A very simple assumption is that production increases at a rate proportional to investment. That is

$$\frac{dQ}{dt} = \rho I,$$

where ρ is a constant. ■

How do we use these equations to determine the behaviour of the economy? The first step is to get an equation which involves only one of the basic functions, in this case I. Using the results written down above, we can argue as follows:

$$\frac{dI}{dt} = \frac{dI}{dY}\frac{dY}{dt} = s\frac{dY}{dt} = s\frac{dQ}{dt} = s\rho I.$$

We have obtained the equation

$$\frac{dI}{dt} = s\rho I,$$

which involves I only. If we can find a function $I(t)$ which satisfies this equation, we shall be able to provide an explicit description of the growth of the economy.

The equation for I is an example of a *differential equation.* In this case, we can solve it 'by inspection', because it says that the derivative of I is a constant multiple of I itself. Now, it follows from our standard rules for derivatives that

$$\frac{d}{dt}\exp(mt) = m\exp(mt),$$

if m is constant. Here the constant is $s\rho$, and so a function which satisfies the differential equation is $I(t) = \exp(s\rho t)$. In fact, the result remains true if we multiply by any constant A, so we can say that the general solution of the differential equation is

$$I(t) = A\exp(s\rho t).$$

Of course, it is possible to make more complicated behavioural assumptions about how the economy grows. Such assumptions lead to more complicated differential equations, as in Example 27.6, for instance. In order to solve these equations we shall need some mathematical techniques.

27.2 Some Types of Differential Equations

The equation

$$\frac{dy}{dt} + 2y = t^2$$

is an example of a *first-order* differential equation. It involves only the first derivative dy/dt of an unknown function y, together with y itself and the variable t.

Generally, a differential equation expresses a relationship between a function, its derivatives and the independent variable. For example, here is another differential equation for a function y of t:

$$\frac{d^2y}{dt^2} + 6\frac{dy}{dt} + 9y = t^2e^t.$$

This is known as a *second-order* equation since it involves both the first and second derivatives of y.

Both the examples given are known as *linear* differential equations, because they contain only linear functions of the derivatives – there are no terms in the square of $y'(t)$, and so on. We shall discuss the simplest types of linear differential equations, both first-order and second-order.

The very simplest type of first-order equation is one which can be written in the form

$$\frac{dy}{dt} = f(t),$$

where f is some known function of t. The solution is a function y whose derivative is f: in other words, an anti-derivative of f. So this differential equation is just another way of stating the problem of indefinite integration, since we have to find

$$y(t) = \int f(t)\,dt.$$

As we shall see, the solution of more complicated differential equations is often a matter of reducing the problem to one involving indefinite integration.

27.3 Separable Differential Equations

A first-order linear differential equation is *separable* if it can be written in the form

$$\frac{dy}{dt} = f(y)g(t),$$

where f and g are given functions. The significant point is that the right-hand side is the product of two functions, one depending *only* on y and the other *only* on t. For example, the equation

$$\frac{dy}{dt} = y^2e^t$$

is separable, whereas

$$\frac{dy}{dt} = t^2 y + ty^2$$

is not, because it is not possible to write $t^2y + ty^2$ in the form $f(y)g(t)$.

The solution of a separable differential equation requires us to think of the equation for dy/dt as a relationship between dy and dt, in much the same way as we do when substituting in an integral. In fact, the argument given in Section 26.1 justifies this procedure. So we resort to the conventional shorthand and rewrite the equation

$$\frac{dy}{dt} = f(y)g(t) \quad \text{as} \quad \frac{1}{f(y)}dy = g(t)\,dt.$$

This is called 'separating the variables'. Integrating formally, we have

$$\int \frac{1}{f(y)}\,dy = \int g(t)\,dt.$$

Of course, the integrals still have to be found.

Example The equation

$$\frac{dy}{dt} = \frac{t^2}{y}$$

is separable, the right-hand side being the product of $1/y$ and t^2. Separating the variables, we obtain

$$y\,dy = t^2\,dt, \quad \text{so} \quad \int y\,dy = \int t^2\,dt.$$

In this case we can work out both integrals, giving

$$\frac{y^2}{2} + c_1 = \frac{t^3}{3} + c_2 \quad (c_1, c_2 \text{ constants}),$$

and we can amalgamate the constants, so that $y^2 = \frac{2}{3}t^3 + c$. ■

Often it is convenient to leave an answer in the form given above, although it does not strictly define y as a function of t. There is an undetermined constant c, and there are two possible square roots of the right-hand side. We shall refer to an answer of this kind as the *general solution* of the relevant equation.

Generally, in order to determine any arbitrary constants occurring in a general solution, we have to make use of *initial conditions*. In this way we may determine a *particular solution*, by which we mean one in which there are no arbitrary features. Although the term 'initial condition' suggests that we are given the value of the unknown function y when $t = 0$, the term is also used when the value of y at some other value of t is given.

Example Consider the differential equation

$$\frac{dy}{dt} = \frac{(y^2+1)e^t}{2y},$$

and suppose we are given that $y(0) = 1$.

The equation is separable. Separating and integrating, we obtain

$$\int \frac{2y}{y^2+1}\,dy = \int e^t\,dt,$$

that is

$$\ln(y^2+1) = e^t + c.$$

The logarithm is valid, since $y^2 + 1$ is positive for all y.

We now use the fact that $y(0) = 1$; this means that the last equation holds if we substitute $y = 1$ and $t = 0$. Thus

$$\ln(1^2+1) = e^0 + c, \qquad \ln 2 = 1 + c, \qquad c = \ln 2 - 1,$$

and inserting this value of c we get

$$\ln(y^2+1) = e^t + \ln 2 - 1.$$

Taking the exponential of both sides:

$$y^2 + 1 = \exp(e^t + \ln 2 - 1),$$

and solving for y there are two possibilities:

$$y = \sqrt{\exp(e^t + \ln 2 - 1) - 1} \quad \text{or} \quad y = -\sqrt{\exp(e^t + \ln 2 - 1) - 1}.$$

Since $y(0) = 1$ is positive, we must take the first of these options. Simplifying (check!), the solution is

$$y(t) = \sqrt{2\exp(e^t - 1) - 1}.$$

■

27.4 A Continuous-Time Model of Price Adjustment

In Chapter 5 we discussed the cobweb model of market stability. That is a discrete-time model, leading to a recurrence equation which describes the sequence (p_t) of yearly prices when the initial price is not the equilibrium value.

It is also possible to use continuous-time models in this situation, and such models lead to differential equations, rather than recurrence equations. To do this we must consider the price to be given by $p(t)$, a function of the continuously varying time parameter t.

Suppose that, initially, the price p is less than the equilibrium value. How does the price adjust to market forces? We might expect that since the consumers wish to buy more units, $q^D(p)$, than the suppliers will provide, $q^S(p)$, the price will increase; and a reasonable assumption is that the *rate* of increase depends on the *excess* of demand over supply, $x(p) = q^D(p) - q^S(p)$ (see Figure 27.1).

This assumption means that the derivative p' is equal to some given function of p; in other words, we have a differential equation for $p(t)$.

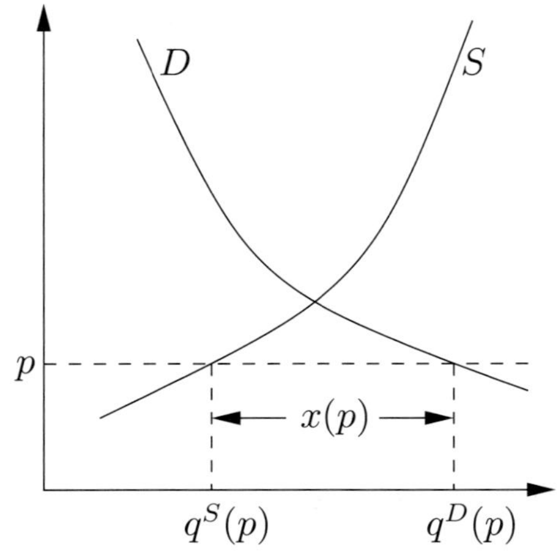

Figure 27.1 The excess of demand over supply, $x(p)$.

Example Suppose the market is governed by the supply and demand sets

$$D = \{(q,p) \mid 3q + 2p = 10\} \quad \text{and} \quad S = \{(q,p) \mid q - 2p = -6\}.$$

The usual calculation shows that the equilibrium price is $7/2$.

Suppose also that when $t = 0$ we have $p = 3$, and the law of price adjustment is $p' = [3x(p)]^3$. Explicitly, we have

$$\begin{aligned}\frac{dp}{dt} &= 27\left[q^D(p) - q^S(p)\right]^3\\ &= 27\left[\frac{1}{3}(10-2p) - (2p-6)\right]^3\\ &= (28-8p)^3\,.\end{aligned}$$

This is a (rather trivially) separable equation. Separating and integrating, we have

$$\int \frac{1}{(28-8p)^3}\,dp = \int 1\,dt,$$

from which we obtain

$$\frac{1}{16}\frac{1}{(28-8p)^2} = t + c.$$

Rearranging:

$$(28-8p)^2 = \frac{1}{16t+16c},$$

so that

$$\text{either} \quad 28 - 8p = \sqrt{\frac{1}{16t+16c}} \quad \text{or} \quad 28 - 8p = -\sqrt{\frac{1}{16t+16c}}.$$

We are given that $p = 3$ when $t = 0$. Substituting these values, it is clear that the first of the alternatives holds, because $28 - 8 \times 3$ is positive. Furthermore, we have

$$28 - 8 \times 3 = 4 = \sqrt{\frac{1}{16 \times 0 + 16c}},$$

so $c = 1/256$ and

$$28 - 8p = \sqrt{\frac{1}{16t + 1/16}}.$$

Rewriting this equation, we get the price function

$$p(t) = \frac{7}{2} - \frac{1}{2}\sqrt{\frac{1}{256t + 1}}.$$

Observe that, as $t \to \infty$, the square-root term tends to zero, and so $p(t)$ approaches the equilibrium price $7/2$. ■

27.5 Cash Flows under Continuous Compounding

One way of describing how the balance $A(t)$ of an account varies with time under continuous compounding is to say that the rate of increase is proportional to the amount. This translates into the differential equation

$$\frac{dA}{dt} = rA,$$

where r is the interest rate. If the initial balance (i.e., the principal P) is positive, what is the amount at time t?

The equation

$$\frac{dA}{dt} = rA$$

is separable. Separating and integrating:

$$\int \frac{1}{A}\,dA = \int r\,dt, \quad \text{that is} \quad \ln A = rt + c.$$

(Since the principal is positive, $A(t)$ is positive for all t.) Taking the exponential of both sides:

$$A = e^{rt}e^{c} = Ke^{rt},$$

where $K = e^{c}$. Since $A(0) = P$, we have $P = K$ and so $A(t) = Pe^{rt}$, in agreement with the formula obtained in Chapter 7.

The situation just described is a fairly straightforward one, in which some money is invested and allowed to grow. More generally, we can consider a situation in which money is continuously added to the account at rate $f(t)$. (Negative values of $f(t)$ would correspond to withdrawals.) Then the amount $A(t)$ of money in the account at time t satisfies the differential equation

$$\frac{dA}{dt} = f(t) + rA(t).$$

As a specific example, let us suppose $f(t) = 1$ for all t. So

$$\frac{dA}{dt} = 1 + rA(t).$$

This is *not* a separable equation, so even this slight modification has complicated matters. We shall show how to solve this, and describe the general underlying method in the next section.

This equation can be written as

$$\frac{dA}{dt} - rA(t) = 1.$$

Suppose we multiply both sides of the equation by e^{-rt}. Then

$$\frac{dA}{dt}e^{-rt} - rA(t)e^{-rt} = e^{-rt}.$$

Now, the left-hand side is the derivative of the function $A(t)e^{-rt}$, just by the product rule and the function of a function rule. So we may write this as

$$\frac{d}{dt}\left(A(t)e^{-rt}\right) = e^{-rt}.$$

We can now integrate each side, as follows:

$$A(t)e^{-rt} = \int e^{-rt}\,dt = -\frac{1}{r}e^{-rt} + c.$$

It follows that

$$A(t) = -\frac{1}{r} + ce^{rt}$$

where, because $A(0) = P$, we have $P = -(1/r) + c$ and hence $c = P + (1/r)$. Finally

$$A(t) = Pe^{rt} + \frac{e^{rt} - 1}{r}.$$

27.6 Integrating Factors for Linear Equations

The process just undertaken to solve the differential equation of the previous section is mysterious. How did we know multiplying by e^{-rt} would help? As you might have anticipated, this is an example of a general technique for solving first-order linear equations. The method applies to equations written very specifically in the form

$$\frac{dy}{dt} + yP(t) = Q(t),$$

where $P(t)$ and $Q(t)$ are functions only of the variable t. An equation might not have this form but could be rewritten in this way, and then the method could be deployed.

There is a general technique for solving such equations, based on the notion of an *integrating factor*. This is a function $\mu(t)$ of t with the property that if we multiply our equation through by $\mu(t)$, then the left hand-side takes the form

$$\frac{d}{dt}\left(\mu(t)y\right).$$

For example, in the previous section, for the equation

$$\frac{dA}{dt} - rA(t) = 1,$$

in which y is A, $\mu(t) = e^{-rt}$ is an integrating factor because

$$\frac{d}{dt}\left(e^{-rt}A\right) = \frac{dA}{dt}e^{-rt} - rA(t)e^{-rt} = e^{-rt}\left(\frac{dA}{dt} - rA(t)\right) = \mu(t)\left(\frac{dA}{dt} - rA(t)\right).$$

The natural question to ask is: how do we find an integrating factor? Fortunately, there is a simple recipe. If we have the linear equation

$$\frac{dy}{dt} + yP(t) = Q(t),$$

then an integrating factor is

$$\mu(t) = e^{\int P(t)\,dt}.$$

That is, we integrate P and take the exponential of the result. There is no need to include a constant of integration when integrating P, since we just need one integrating factor. When we multiply through by $\mu(t)$, the equation becomes

$$\mu(t)\frac{dy}{dt} + \mu(t)P(t)y = \mu(t)Q(t),$$

and the left-hand side of the equation turns out to be

$$\frac{d}{dt}\left(\mu(t)y\right) = \mu(t)Q(t).$$

Let us see why. We have

$$\frac{d}{dt}\left(\mu(t)y\right) = \frac{d\mu}{dt}y + \mu\frac{dy}{dt}.$$

Now, $\mu(t) = \exp(\int P(t)\,dt)$ so, by the chain rule:

$$\frac{d\mu}{dt} = e^{\int P(t)\,dt}\frac{d}{dt}\left(\int P(t)\,dt\right) = e^{\int P(t)\,dt}P(t) = \mu(t)P(t)$$

since, clearly, the derivative of the integral of $P(t)$ is $P(t)$. So

$$\frac{d}{dt}\left(\mu(t)y\right) = \frac{d\mu}{dt}y + \mu\frac{dy}{dt} = \mu(t)P(t)y + \mu(t)\frac{dy}{dt},$$

which is exactly the left-hand side.

Therefore, since

$$\frac{d}{dt}\left(\mu(t)y\right) = \mu(t)Q(t),$$

we have

$$\mu(t)y = \int \mu(t)Q(t)\,dt.$$

To find y, we integrate $\mu(t)Q(t)$, remembering the constant of integration, and then divide by $\mu(t)$.

Example Let us find the function $y(t)$ which satisfies $y(1) = 3$ and

$$t\frac{dy}{dt} + 2y = t^2.$$

This is *not* in the standard linear form to which our method applies, as there is a function (t) in front of the 'dy/dt'. But we can rewrite the equation as

$$\frac{dy}{dt} + \frac{2}{t}y = t,$$

and we now have an equation in the standard format, with $P(t) = 2/t$ and $Q(t) = t$. To calculate an integrating factor, we first note that

$$\int P(t)\,dt = \int \frac{2}{t}\,dt = 2\ln t = \ln t^2.$$

So an integrating factor is

$$\mu(t) = e^{\ln t^2} = t^2,$$

and the solution y satisfies

$$\mu(t)y = t^2y = \int \mu(t)Q(t)\,dt = \int t^3\,dt = \frac{1}{4}t^4 + c,$$

for some constant c of integration. Hence

$$y = \frac{1}{4}t^2 + \frac{c}{t^2}.$$

Given that $y(1) = 3$, we have $3 = 1/4 + c$ and so $c = 11/4$ and

$$y = \frac{1}{4}t^2 + \frac{11}{4t^2}.$$

■

Worked Examples

Example 27.1 Find the solution $y(t)$ of the differential equation

$$\frac{dy}{dt} = \frac{ty^2}{\sqrt{1+t^2}},$$

given that $y(0) = -2$.

Solution

The equation is separable. Separating and integrating, we obtain

$$\int \frac{1}{y^2}\,dy = \int \frac{t}{(1+t^2)^{1/2}}\,dt.$$

The integral on the left is $-1/y + c_1$, where c_1 is a constant of integration. To evaluate the integral on the right, let $r = 1 + t^2$. Then $dr = 2t\,dt$ and

$$\int \frac{t}{(1+t^2)^{1/2}}\,dt = \int \frac{1}{r^{1/2}}\frac{1}{2}\,dr = r^{1/2} + c_2 = \sqrt{1+t^2} + c_2,$$

where c_2 is a constant of integration. We therefore have

$$-\frac{1}{y} + c_1 = \sqrt{1+t^2} + c_2,$$

so

$$y(t) = -\frac{1}{\sqrt{1+t^2}+c},$$

where $c = c_2 - c_1$. Since $y(0) = -2$, we must have $-2 = -1/(1+c)$, so $c = -1/2$ and

$$y(t) = -\frac{1}{\sqrt{1+t^2} - 1/2} = \frac{-2}{2\sqrt{1+t^2} - 1}.$$

Example 27.2 Recall that the elasticity of demand is defined to be

$$\varepsilon(p) = -\frac{q'p}{q},$$

where $q = q^D(p)$ is the demand function. Suppose that there is a constant r such that $\varepsilon(p) = r$ for all p. Show that the demand function must take the form

$$q^D(p) = \frac{K}{p^r}, \quad K \text{ a positive constant.}$$

Solution

The equation $\varepsilon(p) = r$ may be written as

$$-\frac{p}{q}\frac{dq}{dp} = r,$$

or

$$\frac{dq}{dp} = -r\frac{q}{p}.$$

Separating and integrating:

$$\int \frac{1}{q}\,dq = \int -\frac{r}{p}\,dp,$$

so that

$$\ln q = -r \ln p + c,$$

where c is some constant. Since p and q are positive, the logarithms are valid and the equation can be written $\ln q = \ln\left(p^{-r}\right) + c$. Taking the exponential of each side:

$$q = p^{-r}e^{c} = Kp^{-r} = \frac{K}{p^{r}},$$

where we have denoted the positive constant e^{c} by K.

In Chapter 9, we showed that if the demand function has this form then the elasticity is constant. We now have the *converse* result: if the elasticity is constant, then the demand function has this special form.

Example 27.3 The elasticity of demand for a good is

$$\varepsilon(p) = \frac{2p^2}{p^2+1}.$$

Given that $q = 4$ when $p = 1$, find the demand function $q^D(p)$.

Solution

We have

$$-\frac{p}{q}\frac{dq}{dp} = \frac{2p^2}{p^2+1},$$

where $q = q^D(p)$ is the demand function. This may be written as a separable differential equation

$$\frac{dq}{dp} = -\frac{2p}{p^2+1}q.$$

Separating and integrating:

$$\int \frac{1}{q}\,dq = \int -\frac{2p}{p^2+1}\,dp.$$

We can work out the integrals (using the substitution $r = p^2 + 1$ in the second one) and obtain

$$\ln q = -\ln(p^2+1) + c,$$

which is valid since q and p^2+1 are positive. Taking the exponential of both sides:

$$q = e^c e^{-\ln(p^2+1)} = \frac{e^c}{p^2+1} = \frac{K}{p^2+1},$$

where $K = e^c$ is some positive constant. Given that $q(1) = 4$, we have $4 = K/2$, so $K = 8$. Therefore the demand function is

$$q^D(p) = \frac{8}{p^2+1}.$$

Example 27.4 The supply and demand functions for a good are

$$q^S(p) = \frac{2}{3}p - 4 \quad \text{and} \quad q^D(p) = 20 - 2p.$$

Assuming that the initial price is $p(0) = 6$, and the price adjusts over time according to the equation $p' = [x(p)]^3$, where $x(p)$ is the excess of demand over supply, find a formula for $p(t)$.

Solution

We have

$$x(p) = q^D(p) - q^S(p) = 20 - 2p - \left(\frac{2}{3}p - 4\right) = 24 - \frac{8}{3}p.$$

Hence the differential equation for $p(t)$ is

$$\frac{dp}{dt} = \left(24 - \frac{8}{3}p\right)^3,$$

which is separable. Separating and integrating:

$$\int \frac{1}{\left(24 - \frac{8}{3}p\right)^3}\, dp = \int 1\, dt,$$

so

$$\frac{3}{16}\frac{1}{\left(24 - \frac{8}{3}p\right)^2} = t + c.$$

Therefore

$$\text{either} \quad 24 - \frac{8}{3}p = \sqrt{\frac{3}{16t + 16c}} \quad \text{or} \quad 24 - \frac{8}{3}p = -\sqrt{\frac{3}{16t + 16c}}.$$

Now, $p(0) = 6$ and so when $t = 0$, $24 - (8/3)p = 8$. This shows that we must take the positive square root, and $8 = \sqrt{3/16c}$. Therefore, $16c = 3/64$ and

$$24 - \frac{8}{3}p = \sqrt{\frac{3}{16t + 3/64}}.$$

Rearranging, we obtain the following formula for $p(t)$:

$$p(t) = 9 - \frac{3}{8}\sqrt{\frac{192}{3 + 1024t}}.$$

Note that, as $t \to \infty$, the price approaches 9, which is the equilibrium price.

Example 27.5 In one kind of continuous-time model of economic growth, the level of national production Q is taken to be a function $Q(K, L)$ of the amounts of capital K and labour L available, by analogy with the situation for a single firm. By definition, investment I is the rate of increase of capital with respect to time, that is $I(t) = K'(t)$.

Using the same notation as in Section 27.1, suppose we assume that the equilibrium conditions $Q = Y$ and $Y = C + I$ hold, as well as the following behavioural conditions:

(i) $C(t) = bY(t)$ (in other words, $c = 0$);
(ii) $Q(K, L) = K^{1/2}L$;
(iii) $L(t) = \alpha t + \beta$, that is the amount of labour available grows linearly.

Find a differential equation for $K(t)$, and solve it.

Solution

Using the equilibrium conditions and the behavioural conditions, we have

$$I = Y - C = Y - bY = sY = sQ = sK^{1/2}L.$$

Here, $s = 1 - b$ is the marginal propensity to invest. Also $I = dK/dt$, so

$$\frac{dK}{dt} = sK^{1/2}L = sK^{1/2}(\alpha t + \beta).$$

This is a separable differential equation for K. Separating and integrating, we have

$$\int \frac{1}{K^{1/2}}\, dK = \int s(\alpha t + \beta)\, dt.$$

That is

$$2K^{1/2} = \frac{s\alpha}{2}t^2 + s\beta t + k,$$

for some constant k. Equivalently

$$K(t) = \left(\frac{s\alpha}{4}t^2 + \frac{s\beta}{2}t + \frac{k}{2}\right)^2.$$

Example 27.6 Use an integrating factor to solve the equation

$$\frac{dy}{dt} + \frac{t}{(1+t^2)}y = \sqrt{1+t^2}.$$

Solution

Here, $P(t) = t/(1 + t^2)$ and

$$\int P(t)\,dt = \int \frac{t}{1+t^2}\,dt = \frac{1}{2}\ln(1+t^2) + c,$$

so an integrating factor is

$$\mu(t) = \exp\left(\frac{1}{2}\ln(1+t^2)\right) = \exp\left(\ln(\sqrt{1+t^2})\right) = \sqrt{1+t^2}.$$

Therefore, since

$$\frac{d}{dt}(y\mu(t)) = \mu(t)Q(t),$$

we have

$$\frac{d}{dt}\left(y\sqrt{1+t^2}\right) = \sqrt{1+t^2}\sqrt{1+t^2} = (1+t^2),$$

and

$$y\sqrt{1+t^2} = \int (1+t^2)\,dt = t + \frac{t^3}{3} + c.$$

Finally

$$y = \frac{t}{\sqrt{1+t^2}} + \frac{t^3}{3\sqrt{1+t^2}} + \frac{c}{\sqrt{1+t^2}}$$

for some constant c.

Example 27.7 In a market, the price of a share t seconds after the market opens is $p(t)$ dollars. An analyst constructs a model which predicts that the price of this share should vary continuously with time according to the differential equation

$$\frac{dp}{dt} = p(t) - e^{-t}.$$

If the initial price is \$1, solve this differential equation to find $p(t)$.

Show that, according to the model, $p(t)$ is increasing for all $t \geq 0$.

Solution

We are asked to solve the differential equation

$$\frac{dp}{dt} - p(t) = -e^{-t}$$

and, as this is linear, we solve it by comparing it to the standard form

$$\frac{dp}{dt} + P(t)p(t) = Q(t),$$

to get $P(t) = -1$ and $Q(t) = -e^{-t}$. So, following the standard method, the integrating factor is

$$e^{\int P(t)\,dt} = e^{\int -1\,dt} = e^{-t},$$

and we have

$$\frac{d}{dt}(\mu(t)p(t)) = \frac{d}{dt}(e^{-t}p(t)) = \mu(t)Q(t) = -e^{-2t}.$$

So

$$e^{-t}p(t) = \int \left(-e^{-2t}\right) dt = \frac{1}{2}e^{-2t} + c.$$

Then, multiplying both sides of this expression by e^t, this gives us

$$p(t) = \frac{1}{2}e^{-t} + c\,e^t$$

as the general solution. However, we are told that $p(0) = 1$ and so we have

$$1 = \frac{1}{2} + c,$$

which means $c = 1/2$. So

$$p(t) = \frac{e^{-t} + e^t}{2}$$

is the particular solution we seek.

To see that $p(t)$ is increasing for all $t \geq 0$ in this model, we note that

$$p'(t) = \frac{e^t - e^{-t}}{2} > 0$$

as $e^t > e^{-t}$ for all $t \geq 0$.

Main Topics

- continuous-time macroeconomic models and differential equations
- separable first-order differential equations with initial conditions
- continuous-time price adjustment
- solving linear first-order differential equations with integrating factors

Key Terms, Notations and Formulae

- continuous-time macroeconomic model:
  $I(t)$, the investment in the year up to time t; similarly, $Q(t), Y(t), C(t)$
- equilibrium conditions, $Q(t) = Y(t)$, $Y(t) = C(t) + I(t)$
- marginal propensity to consume, $b = \dfrac{dC}{dY}$
- marginal propensity to invest, $s = \dfrac{dI}{dY} = 1 - b$
- separable differential equation, $\dfrac{dy}{dt} = f(y)g(t)$:
  solve by separating and integrating, $\displaystyle\int \frac{1}{f(y)}\,dy = \int g(t)\,dt$
- general solution; particular solution using initial conditions
- price adjustment: $\dfrac{dp}{dt}$ depends on excess, $x(p) = q^D(p) - q^S(p)$

- integrating factor for equation $\frac{dy}{dt} + yP(t) = Q(t)$ is $\mu(t) = e^{-\int P(t)\,dt}$ and equation becomes $\frac{d}{dt}(\mu(t)y) = \mu(t)Q(t)$.

Exercises

27.1 Find the general solution of the differential equation

$$\frac{dy}{dt} = -\frac{t^3y^2}{\sqrt{1+t^4}}.$$

27.2 The elasticity of demand for a good is

$$\varepsilon(p) = \frac{p^2}{p^2+5p+6}.$$

Given that $q = 9$ when $p = 1$, find the demand function $q^D(p)$.

27.3 The balance of an account varies with time according to the differential equation

$$\frac{dA}{dt} = t + rA(t),$$

and the initial balance is P. Find an expression for $A(t)$.

27.4 The value, $v(t)$, of an asset varies continuously with time according to the differential equation

$$\frac{dv}{dt} = v(t) + e^{-2t}$$

and the initial value is $v(0) = 2/3$. By solving this differential equation, find $v(t)$.

Problems

27.1 Find the general solution of the differential equation

$$\frac{dy}{dt} = \frac{\sqrt{1-y^2}}{y(1-t^2)}.$$

27.2 Find the general solution of the differential equation

$$\frac{dy}{dt} = \frac{t^2(y^3-5)}{y^2(t^3+1)}.$$

27.3 Find a function $y(t)$ such that $y(0) = 1$, $y(t) > 0$ for all t, and $y(t)$ satisfies the differential equation

$$\frac{dy}{dt} = \frac{ty}{t^2+1}.$$

27.4 Solve the differential equation

$$\frac{dy}{dt} = \frac{t^3y^2}{\sqrt{1-t^4}},$$

and find the particular solution satisfying $y(0) = 1$.

27.5 Solve the differential equation

$$2t\frac{dy}{dt} + \frac{1}{y} = y$$

by first writing it explicitly as a separable equation, and find the solution for which $y(1) = 4$.

27.6 The supply and demand functions for a good are

$$q^S(p) = 3p - 2 \quad \text{and} \quad q^D(p) = 2 - p.$$

Show that the equilibrium price is 1. Assume that the price adjusts according to the equation $p' = [x(p)]^5$, where $x(p)$ is the excess of demand over supply, and that the initial price is $p(0) = 1/2$. Find a formula for $p(t)$.

27.7 The elasticity of demand for a good is

$$\varepsilon(p) = 1 - \frac{1}{p+1}.$$

Given that $q = 8$ when $p = 1$, find the demand function $q^D(p)$.

27.8 The elasticity of demand for a good is

$$\varepsilon(p) = \frac{p^2}{\sqrt{p^2+2}}.$$

Given that $q = 5$ when $p = 2$, find the demand function $q^D(p)$.

27.9 Using the model of economic growth in Example 27.6, suppose we make the following specific assumptions:

(i) $C(t) = bY(t)$;
(ii) $Q(K, L) = K^{1/3}L$;
(iii) $L(t) = \alpha t^2 + \beta t + \gamma$, that is labour grows quadratically.

Find a differential equation for $K(t)$, and solve it.

27.10 Find the function, $y(t)$, that satisfies $y(0) = 1$ and the differential equation

$$\frac{dy}{dt} + \frac{2ty}{t^2+1} = t^3.$$

27.11 Find the function $f(x)$ that satisfies the differential equation

$$\frac{df}{dx} - \frac{f}{x+1} = \frac{1}{x-2}$$

and the initial condition $f(0) = 0$.

27.12 The value of an asset, $V(t)$, varies continuously with time according to the differential equation

$$\frac{dV}{dt} = \frac{V}{5} + 4,$$

and the initial value is A. Find $V(t)$.

28 Second-Order Differential Equations

28.1 Market Trends and Consumer Demand

In markets such as the housing market, consumers often try to anticipate trends. They base their demand not simply on the current selling price, but on how fast this price has been rising (or falling) and whether its rate of change is slowing down or speeding up. For example, if people who are thinking about buying a house see that the price is falling fast, they may decide to wait before entering the market. As a result, the demand for property is reduced.

To analyse this sort of problem, we assume that the price is a function of time, $p(t)$, so that the techniques of calculus and differential equations may be applied. The first derivative $p'(t)$ is the rate at which the price is increasing, and the second derivative $p''(t)$ measures the rate of change of p'. For example, if $p'(t) > 0$ and $p''(t) < 0$, then consumers will see that the price is increasing, but that its rate of increase is diminishing, and they may well anticipate that the price will level off in due course. Similarly, it is quite likely that suppliers (people with houses to sell) will also try to anticipate market trends in this way.

Example Suppose that the consumers' demand is given by

$$q = 9 - 2p + 6\frac{dp}{dt} - 2\frac{d^2p}{dt^2}.$$

This corresponds to a linear demand function, modified by taking account of the trend. Suppose also that the quantity supplied is determined in a similar way, and is given by

$$q = -3 + 4p - \frac{dp}{dt} - \frac{d^2p}{dt^2}.$$

If the market is in equilibrium at any given time, then we may equate the supply and demand to obtain an equation for the market price p^*:

$$9 - 2p^* + 6\frac{dp^*}{dt} - 2\frac{d^2p^*}{dt^2} = -3 + 4p^* - \frac{dp^*}{dt} - \frac{d^2p^*}{dt^2},$$

or $$\frac{d^2p^*}{dt^2} - 7\frac{dp^*}{dt} + 6p^* = 12.$$

■

As we have already explained, this is called a *second-order* differential equation, because it contains the second derivative of the unknown function. In this chapter we shall describe some techniques for solving such equations.

28.2 Linear Equations with Constant Coefficients

The second-order differential equations we consider are those of the form

$$\frac{d^2y}{dt^2}+a_1\frac{dy}{dt}+a_2y=f(t),$$

where a_1 and a_2 are constants. These are known as *linear* equations with *constant coefficients*.

The method for solving these equations is very similar to that used in Chapter 23 to solve second-order recurrence equations. We consider first the *homogeneous* equation, obtained by taking $f(t)$ to be identically zero:

$$\frac{d^2y}{dt^2}+a_1\frac{dy}{dt}+a_2y=0.$$

We observe that if $y_1(t)$ and $y_2(t)$ are solutions of the homogeneous equation, then so is $C_1y_1(t)+C_2y_2(t)$, for any constants C_1 and C_2. If $y_1(t)$ and $y_2(t)$ are essentially different (i.e., if $y_2(t)$ is not a constant multiple of $y_1(t)$), then we say that $C_1y_1(t)+C_2y_2(t)$ is the *general solution* of the homogeneous equation.

Suppose that $y^*(t)$ is a particular solution of the non-homogeneous equation and that $g(t)$ is the general solution of the homogeneous equation. Let

$$y(t)=y^*(t)+g(t).$$

Then $y(t)$ is also a solution of the non-homogeneous equation, since

$$\begin{aligned}\frac{d^2y}{dt^2}+a_1\frac{dy}{dt}+a_2y&=\left(\frac{d^2y^*}{dt^2}+\frac{d^2g}{dt^2}\right)+a_1\left(\frac{dy^*}{dt}+\frac{dg}{dt}\right)+a_2\left(y^*+g\right)\\&=\left(\frac{d^2y^*}{dt^2}+a_1\frac{dy^*}{dt}+a_2y^*\right)+\left(\frac{d^2g}{dt^2}+a_1\frac{dg}{dt}+a_2g\right)\\&=f(t)+0\\&=f(t).\end{aligned}$$

We have established the same principle as for second-order recurrence equations (Section 23.4), that is:

- *General solution of the non-homogeneous equation =*
  particular solution + general solution of the homogeneous equation.

28.3 Solution of Homogeneous Equations

The key to solving the homogeneous equation

$$\frac{d^2y}{dt^2} + a_1\frac{dy}{dt} + a_2y = 0$$

is to look for solutions of the form $y(t) = e^{zt}$, where z is a constant. For this function we have

$$\frac{d^2y}{dt^2} + a_1\frac{dy}{dt} + a_2y = z^2e^{zt} + a_1ze^{zt} + a_2e^{zt} = (z^2 + a_1z + a_2)e^{zt}.$$

So we have a solution whenever z satisfies

$$z^2 + a_1z + a_2 = 0.$$

This is called the *auxiliary equation*. The form of the general solution depends on whether the auxiliary equation has two distinct solutions, just one solution, or no solutions at all. We consider each possibility in turn.

Case 1: The Auxiliary Equation Has Two Distinct Solutions
Suppose the auxiliary equation has two distinct solutions α and β. Then $e^{\alpha t}$ and $e^{\beta t}$ are solutions of the differential equation, and the general solution is

$$Ae^{\alpha t} + Be^{\beta t} \quad (A, B \text{ constants}).$$

The constants A and B may be determined if some additional information is given, such as the initial values of y and y'.

Example To find the general solution of the differential equation

$$\frac{d^2y}{dt^2} - 4\frac{dy}{dt} + 3y = 0,$$

we consider the auxiliary equation

$$z^2 - 4z + 3 = 0.$$

Since $z^2 - 4z + 3 = (z-1)(z-3)$ there are two solutions, 1 and 3. The general solution is therefore

$$y(t) = Ae^t + Be^{3t},$$

where A and B are arbitrary constants.

If we are given that $y(0) = 1$ and $y'(0) = 2$, the constants can be determined as follows. Substituting $t = 0$, we obtain

$$1 = y(0) = Ae^0 + Be^0 = A + B.$$

Further, the derivative of y is $y'(t) = Ae^t + 3Be^{3t}$, so the condition $y'(0) = 2$ means that

$$2 = y'(0) = Ae^0 + 3Be^0 = A + 3B.$$

Solving the equations $A + B = 1$ and $A + 3B = 2$, we have

$$A = \frac{1}{2} \quad \text{and} \quad B = \frac{1}{2}.$$

The required solution is therefore

$$y(t) = \frac{1}{2}e^t + \frac{1}{2}e^{3t}.$$

■

Case 2: The Auxiliary Equation Has Exactly One Solution
The auxiliary equation has only one solution when $a_1^2 = 4a_2$, and it is $\alpha = -a_1/2$. In this case, in addition to the solution $y_1(t) = e^{\alpha t}$, we claim that $y_2(t) = te^{\alpha t}$ is also a solution. This claim can easily be checked. Since

$$\frac{dy_2}{dt} = (\alpha t + 1)e^{\alpha t} \quad \text{and} \quad \frac{d^2y_2}{dt^2} = (\alpha^2 t + 2\alpha)e^{\alpha t},$$

we have

$$\begin{aligned} \frac{d^2y_2}{dt^2} + a_1\frac{dy_2}{dt} + a_2y_2 &= (\alpha^2 t + 2\alpha)e^{\alpha t} + a_1(\alpha t + 1)e^{\alpha t} + a_2te^{\alpha t} \\ &= (\alpha^2 + a_1\alpha + a_2)te^{\alpha t} + (2\alpha + a_1)e^{\alpha t}, \end{aligned}$$

which is zero since α satisfies the auxiliary equation $\alpha^2 + a_1\alpha + a_2 = 0$ and $2\alpha + a_1 = 0$.

So in this case the general solution is

$$(Ct + D)e^{\alpha t} \quad (C, D \text{ constants}).$$

Example Find the general solution of the differential equation

$$\frac{d^2y}{dt^2} + 6\frac{dy}{dt} + 9y = 0,$$

and the solution which satisfies $y(0) = 1$ and $y'(0) = 3$.

The auxiliary equation is $z^2 + 6z + 9 = 0$, that is $(z + 3)^2 = 0$. This has only one solution, $\alpha = -3$. The general solution is therefore

$$y(t) = (Ct + D)e^{-3t},$$

where C and D are arbitrary constants. To find the solution which satisfies the given initial conditions, we note first that $y(0) = 1$ means $D = 1$. Since

$$y'(t) = (C - 3D - 3Ct)\,e^{-3t},$$

$y'(0) = 3$ implies $C - 3D = 3$, so $C = 6$ and the required solution is

$$y(t) = (6t + 1)e^{-3t}.$$

■

Case 3: The Auxiliary Equation Has No Solutions

When $a_1^2 < 4a_2$, the auxiliary equation has no solutions. In this case, as for recurrence equations, the general solution involves trigonometrical functions. We shall simply give the formula here (see also Problem 28.6).

If the equation $z^2 + a_1 z + a_2 = 0$ has no solutions, we define

$$\gamma = -\frac{a_1}{2} \quad \text{and} \quad \delta = \frac{\sqrt{4a_2 - a_1^2}}{2}.$$

Then the general solution is

$$y(t) = e^{\gamma t}\left[E\cos(\delta t) + F\sin(\delta t)\right],$$

where E and F are arbitrary constants.

Example Find the general solution of the equation

$$\frac{d^2y}{dt^2} - 4\frac{dy}{dt} + 13y = 0,$$

and the solution which satisfies $y(0) = 1$ and $y'(0) = 0$.

The auxiliary equation is $z^2 - 4z + 13 = 0$, which has no solutions. We have

$$\gamma = \frac{4}{2} = 2 \quad \text{and} \quad \delta = \frac{\sqrt{52 - 16}}{2} = 3.$$

Therefore the general solution is

$$y = e^{2t}[E\cos(3t) + F\sin(3t)],$$

where E and F are arbitrary constants. Given the initial conditions, we can determine the constants E and F. First, since $y(0) = 1$, we have

$$1 = e^0(E\cos 0 + F\sin 0),$$

so $E = 1$. The derivative of y is

$$y'(t) = e^{2t}\left[2E\cos(3t) + 2F\sin(3t) - 3E\sin(3t) + 3F\cos(3t)\right].$$

Hence the condition $y'(0) = 0$ implies $2E + 3F = 0$, and so $F = -2/3$. The required solution is therefore

$$y(t) = e^{2t}\left(\cos(3t) - \frac{2}{3}\sin(3t)\right).$$

■

28.4 Non-homogeneous Equations

We now return to the non-homogeneous equation

$$\frac{d^2y}{dt^2} + a_1\frac{dy}{dt} + a_2 y = f(t).$$

Recall that, if we have found the general solution of the corresponding homogeneous equation, in order to solve the non-homogeneous equation it remains only to find a *particular solution* $y^*(t)$. We shall explain how this can be done in many cases.

Finding a particular solution is very easy if the right-hand side $f(t)$ is constant, say $f(t) = k$. In this case the particular solution is also a constant: $y^* = k/a_2$. Since the derivatives of a constant are all zero, it is easy to see that this satisfies the equation.

Example We can now solve the differential equation obtained in Section 28.1:

$$\frac{d^2y}{dt^2} - 7\frac{dy}{dt} + 6y = 12.$$

Recall that this equation determines the *equilibrium* price $p^*(t)$; we have changed the notation to $y(t)$ for convenience.

This is a non-homogeneous equation, and the right-hand side is a constant. Clearly, a particular solution is $y^*(t) = 2$. The relevant auxiliary equation is $z^2 - 7z + 6 = 0$, or $(z - 1)(z - 6) = 0$, which has solutions 1 and 6. It follows that the general solution is

$$y(t) = 2 + Ae^t + Be^{6t},$$

for some constants A and B. ■

As a rule, to find a particular solution $y^*(t)$ of a non-homogeneous equation we look for a function of the same 'general form' as $f(t)$. For instance, if $f(t) = kt$, we try to find a particular solution of the form $y^*(t) = at + b$, where a and b are constants to be determined. Note that we do not simply try for a particular solution of the form at: even though there is no constant term in $f(t)$, we might need one in the particular solution. Similarly, if $f(t)$ is of the form kt^2 for some constant k, we should try $y^*(t) = ct^2 + dt + e$, and if $f(t) = e^{kt}$, we try $y^*(t) = ce^{kt}$. The following example illustrates how the constants can be determined. (Further illustrations are given in Examples 28.2 to 28.4.)

Example Consider the differential equation

$$\frac{d^2y}{dt^2} - \frac{dy}{dt} - 2y = 4t^2.$$

The right-hand side is a quadratic, so we try for a particular solution of the 'general' quadratic form $y^*(t) = ct^2 + dt + e$. We have

$$\frac{d^2y^*}{dt^2} - \frac{dy^*}{dt} - 2y^* = 2c - (2ct + d) - 2(ct^2 + dt + e),$$

so $y^*(t)$ is a solution if

$$(-2c)t^2 + (-2c - 2d)t + (2c - d - 2e) = 4t^2.$$

Since this must hold for all values of t, we must have

$$-2c = 4, \quad -2c - 2d = 0 \quad \text{and} \quad 2c - d - 2e = 0,$$

which gives $c = -2, d = 2, e = -3$. A particular solution is therefore

$$y^* = -2t^2 + 2t - 3.$$

To find the general solution to the corresponding homogeneous equation, we examine the auxiliary equation, $z^2 - z - 2 = 0$, that is $(z-2)(z+1) = 0$. This has solutions 2 and -1, so the general solution to the homogeneous equation is $Ae^{2t} + Be^{-t}$. Finally, the general solution to the original equation is

$$y(t) = -2t^2 + 2t - 3 + Ae^{2t} + Be^{-t}.$$

■

28.5 Behaviour of Solutions

In practice, we are often less interested in the exact solution of a differential equation than in how the solution evolves as t tends to infinity. Suppose, for example, that the auxiliary equation has two solutions α and β. Then the general solution takes the form

$$y(t) = y^*(t) + Ae^{\alpha t} + Be^{\beta t}$$

and its behaviour depends on the particular solution y^* and on the functions $e^{\alpha t}$ and $e^{\beta t}$. For instance, if α and β are both negative then, as $t \to \infty$, $e^{\alpha t}$ and $e^{\beta t}$ both tend to 0 as $t \to \infty$, and the general solution of the differential equation behaves in the same way as $y^*(t)$, in the long run.

If the auxiliary equation has no solutions, so that the general solution is of the form

$$y(t) = y^*(t) + e^{\gamma t}\left[E\cos(\delta t) + F\sin(\delta t)\right],$$

then (provided at least one of E and F is nonzero) the solution will exhibit some form of oscillatory behaviour. For example, suppose the solution is

$$y(t) = e^{2t} + e^{-t}\left[5\cos(3t) + 4\sin(3t)\right].$$

Then, since $e^{-t} \to 0$ as $t \to \infty$, the oscillatory part of the solution decreases in significance as t increases; in other words, the solution is e^{2t} with some oscillations which fade away in the long run. On the other hand, if the solution were

$$y(t) = e^{2t} + e^{4t}\left[5\cos(3t) + 4\sin(3t)\right],$$

then the oscillations would increase in magnitude.

Worked Examples

Example 28.1 Find the general solutions of the differential equations

$$\frac{d^2y}{dt^2} - 3\frac{dy}{dt} + 2y = 0 \quad \text{and} \quad \frac{d^2y}{dt^2} + 2\frac{dy}{dt} + 5y = 0.$$

Solution

The auxiliary equation for the first of these is $z^2 - 3z + 2 = 0$, which factorises as $(z - 1)(z - 2) = 0$ and so has solutions 1 and 2. The general solution is therefore

$$y(t) = Ae^t + Be^{2t}$$

for arbitrary constants A and B.

The auxiliary equation for the second equation, $z^2 + 2z + 5 = 0$, has no roots. In the notation used earlier,

$$\gamma = -\frac{2}{2} = -1 \quad \text{and} \quad \delta = \frac{\sqrt{20-4}}{2} = 2.$$

The general solution is therefore

$$y(t) = e^{-t}\left[E\cos(2t) + F\sin(2t)\right],$$

for arbitrary constants E and F.

Example 28.2 Find the general solution of

$$\frac{d^2y}{dt^2} + \frac{dy}{dt} - 2y = e^{3t}.$$

Solution

This is a non-homogeneous equation. The auxiliary equation is $z^2 + z - 2 = 0$, which factorises as $(z - 1)(z + 2) = 0$ and so has solutions 1 and -2. The corresponding homogeneous equation therefore has general solution

$$Ae^t + Be^{-2t}.$$

To find a particular solution $y^*(t)$, we assume that $y^*(t)$ is of the form ce^{3t}, and determine the value of c by substitution. The given y^* is a solution of the equation if and only if

$$9ce^{3t} + 3ce^{3t} - 2ce^{3t} = e^{3t}.$$

This gives $10ce^{3t} = e^{3t}$, so $c = 1/10$. The general solution is therefore

$$\frac{1}{10}e^{3t} + Ae^t + Be^{-2t}.$$

Example 28.3 Find the solution of the equation

$$\frac{d^2y}{dt^2} + 4\frac{dy}{dt} + 4y = 8t^2 + 8$$

which satisfies $y(0) = 1$ and $y'(0) = 3$. How does this solution behave as t tends to infinity?

Solution

First, we find the solution of the corresponding homogeneous equation. The auxiliary equation is $z^2 + 4z + 4 = 0$, which is $(z+2)^2 = 0$. This has only one solution, -2, so the general solution to the homogeneous equation is

$$(Ct + D)e^{-2t},$$

where C and D are constants. In order to find a particular solution, noting that the right-hand side of the equation is a quadratic, we try $y^*(t) = ct^2 + dt + e$. Then, y^* is a solution if and only if

$$(2c) + 4(2ct + d) + 4(ct^2 + dt + e) = 8t^2 + 8,$$

or, equivalently:

$$(4c)t^2 + (8c + 4d)t + (2c + 4d + 4e) = 8t^2 + 8,$$

from which it follows that

$$4c = 8, \quad 8c + 4d = 0 \quad \text{and} \quad 2c + 4d + 4e = 8.$$

Solving, we find $c = 2$, $d = -4$ and $e = 5$, so a particular solution is

$$y^*(t) = 2t^2 - 4t + 5.$$

The general solution to the differential equation is therefore

$$y(t) = 2t^2 - 4t + 5 + (Ct + D)e^{-2t}.$$

To determine the solution satisfying the given initial conditions, we note that

$$y'(t) = 4t - 4 + (C - 2D - 2Ct)e^{-2t}.$$

So the initial conditions tell us that

$$5 + D = 1 \quad \text{and} \quad -4 + C - 2D = 3,$$

from which we obtain $C = -1$ and $D = -4$. The required solution is therefore

$$y(t) = 2t^2 - 4t + 5 - (t + 4)e^{-2t}.$$

As $t \to \infty$, the term $(t+4)e^{-2t}$ tends to 0. Thus, for large t, $y(t)$ is very close to the quadratic $2t^2 - 4t + 5$. It follows that $y(t) \to \infty$ as $t \to \infty$; technically we say that $y(t)$ is *asymptotically equal* to $2t^2 - 4t + 5$.

Example 28.4 Determine constants a and b such that $(at + b)e^{3t}$ is a solution of the differential equation

$$\frac{d^2y}{dt^2} - 3\frac{dy}{dt} + 2y = te^{3t},$$

and hence find the general solution. Also find the solution satisfying $y(0) = 0$ and $y'(0) = 1$.

Solution

Let $y^*(t) = (at + b)e^{3t}$. Then

$$\frac{dy^*}{dt} = ae^{3t} + 3(at + b)e^{3t} = (3at + 3b + a)e^{3t}$$

and

$$\frac{d^2y^*}{dt^2} = 3ae^{3t} + 3(3at + 3b + a)e^{3t} = (9at + 6a + 9b)e^{3t}.$$

If y^* is a solution of the equation then

$$(9at + 6a + 9b)e^{3t} - 3(3at + 3b + a)e^{3t} + 2(at + b)e^{3t} = te^{3t},$$

so

$$2at + (3a + 2b) = t.$$

We must have $2a = 1$ and $3a + 2b = 0$, so $a = 1/2$ and $b = -3/4$. A particular solution is therefore

$$y^*(t) = \left(\frac{1}{2}t - \frac{3}{4}\right)e^{3t}.$$

To find the general solution, we note that the corresponding homogeneous equation

$$\frac{d^2y}{dt^2} - 3\frac{dy}{dt} + 2y = 0$$

has solution $Ae^t + Be^{2t}$ (Example 28.1). The general solution is therefore

$$y(t) = \left(\frac{1}{2}t - \frac{3}{4}\right)e^{3t} + Ae^t + Be^{2t}.$$

The initial condition $y(0) = 0$ means that $-3/4 + A + B = 0$. The derivative of y is

$$y'(t) = \frac{1}{2}e^{3t} + 3\left(\frac{1}{2}t - \frac{3}{4}\right)e^{3t} + Ae^t + 2Be^{2t},$$

so, since $y'(0) = 1$, we must have $-7/4 + A + 2B = 1$. Solving these equations for A and B, we find $A = -5/4$ and $B = 2$. Therefore the required solution is

$$y(t) = \left(\frac{1}{2}t - \frac{3}{4}\right)e^{3t} - \frac{5}{4}e^t + 2e^{2t}.$$

Example 28.5 Suppose that consumer demand q depends on the price trend according to the equation

$$q = 11 - p - 4\frac{dp}{dt} + \frac{d^2p}{dt^2}.$$

If the supply function is $q^S(p) = -4 + 4p$, write down the condition for equilibrium and determine the equilibrium price $p(t)$ given that $p(0) = 5$ and $p'(0) = 4$. Describe in words the behaviour of $p(t)$.

Solution

The condition for equilibrium is that

$$11 - p - 4\frac{dp}{dt} + \frac{d^2p}{dt^2} = -4 + 4p,$$

or

$$\frac{d^2p}{dt^2} - 4\frac{dp}{dt} - 5p = -15.$$

A particular solution of this equation is the constant solution $p^*(t) = 3$. The auxiliary equation is

$$z^2 - 4z - 5 = (z-5)(z+1) = 0,$$

which has roots 5 and -1. Therefore the solution to the corresponding homogeneous equation is $Ae^{5t} + Be^{-t}$ and the general solution for $p(t)$ is

$$p(t) = 3 + Ae^{5t} + Be^{-t},$$

where the constants A and B are determined by the initial conditions.

Using the fact that $p(0) = 5$, we have $3 + A + B = 5$, or $A + B = 2$. The derivative of p is $5Ae^{5t} - Be^{-t}$, so the condition $p'(0) = 4$ means that $5A - B = 4$. Solving these equations for A and B gives $A = 1$ and $B = 1$, so

$$p(t) = 3 + e^{5t} + e^{-t}.$$

As t tends to infinity, e^{-t} tends to 0, but e^{5t} tends to infinity; therefore, $p(t)$ tends to infinity.

Example 28.6 A disease has two stages, A and B. A person in stage A of the disease may or may not develop the more serious stage B condition. Epidemiologists believe that a person in stage A or B of the disease can, by infection, pass on the disease to another person, who then develops stage A. In their model of the spread of the disease, they formalise this by assuming that if $y_A(t)$ is the number of people in stage A of the disease and $y_B(t)$ the number in stage B at time t, then

$$\frac{dy_A}{dt} = ay_A + by_B,$$

where a and b are positive constants. Furthermore, their model assumes that

$$\frac{dy_B}{dt} = cy_A,$$

for a positive constant c. Find expressions for $y_A(t)$ and $y_B(t)$, in terms of the constants a, b and c, given that there are initially no people in stage B of the disease and 300 in stage A. Comment on the behaviour of y_A and y_B.

Health officials believe that public awareness of the seriousness of the disease depends on the ratio of the number of people in stage B to those in stage A. How does this ratio behave in the long run?

Solution

We have two first-order differential equations. In order to solve them we differentiate the first one, obtaining

$$\frac{d}{dt}\left(\frac{dy_A}{dt}\right) = \frac{d}{dt}(ay_A + by_B) = a\frac{dy_A}{dt} + b\frac{dy_B}{dt},$$

that is

$$\frac{d^2y_A}{dt^2} = a\frac{dy_A}{dt} + b\frac{dy_B}{dt}.$$

Now, using the second equation for dy_B/dt, we obtain

$$\frac{d^2y_A}{dt^2} = a\frac{dy_A}{dt} + bcy_A,$$

or

$$\frac{d^2y_A}{dt^2} - a\frac{dy_A}{dt} - bcy_A = 0.$$

This is a second-order differential equation for y_A. The auxiliary equation is $z^2 - az - bc = 0$, and it has two distinct solutions

$$\alpha = \frac{a}{2} - \frac{\sqrt{a^2+4bc}}{2} \quad \text{and} \quad \beta = \frac{a}{2} + \frac{\sqrt{a^2+4bc}}{2},$$

since the constants a, b and c are all positive. For future reference, we note that $\beta > a$ and α is negative.

It follows that y_A has the form

$$y_A(t) = Xe^{\alpha t} + Ye^{\beta t}$$

for some constants X and Y. We can now use the first of the given equations to obtain a formula for y_B:

$$y_B(t) = \frac{1}{b}\left(\frac{dy_A}{dt} - ay_A\right) = \left(\frac{\alpha - a}{b}\right)Xe^{\alpha t} + \left(\frac{\beta - a}{b}\right)Ye^{\beta t}.$$

The initial conditions are $y_A(0) = 300$ and $y_B(0) = 0$. These lead to the equations

$$X + Y = 300 \quad \text{and} \quad \left(\frac{\alpha - a}{b}\right)X + \left(\frac{\beta - a}{b}\right)Y = 0,$$

and solving these we obtain

$$X = \frac{300(\beta - a)}{\beta - \alpha} \quad \text{and} \quad Y = \frac{300(a - \alpha)}{\beta - \alpha}.$$

Note that X and Y are both positive, since $\beta > a$ and α is negative.

Both $y_A(t)$ and $y_B(t)$ have terms in $e^{\beta t}$ with positive coefficients, and so it follows that they grow exponentially.

The behaviour of the ratio $y_B(t)/y_A(t)$ is also determined by the ratio of the respective terms in $e^{\beta t}$. Since

$$y_A(t) = Ye^{\beta t} + \cdots \quad \text{and} \quad y_B(t) = \left(\frac{\beta - a}{b}\right) Ye^{\beta t} + \cdots,$$

the limit of the ratio is

$$\frac{\beta - a}{b} = \frac{\sqrt{a^2 + 4bc} - a}{2b}.$$

Note that this number is positive, as we should expect since y_A and y_B represent numbers of people.

Main Topics

- anticipation of market trends in a continuous-time market
- second-order differential equations
- general solution of homogeneous equations
- non-homogeneous equations and finding particular solutions
- behaviour of solutions

Key Terms, Notations and Formulae

- anticipating market trends: demand, supply can depend on $p'(t), p''(t)$
- linear equation with constant coefficients:

$$\frac{d^2y}{dt^2} + a_1\frac{dy}{dt} + a_2y = f(t)$$

- for a non-homogeneous equation the general solution is
  particular solution + general solution of homogeneous equation
- solution of homogeneous equation depends on solutions of auxiliary:
  - two distinct solutions, α, β: $y(t) = Ae^{\alpha t} + Be^{\beta t}$
  - one solution, α: $y(t) = (Ct + D)e^{\alpha t}$
  - no solutions: if $\gamma = -\dfrac{a_1}{2}, \delta = \dfrac{\sqrt{4a_2 - a_1^2}}{2}$, $y(t) = e^{\gamma t}(E\cos\delta t + F\sin\delta t)$
- particular solution, $y^*(t)$, to non-homogeneous equation
- initial conditions

Exercises

28.1 Find a function $f(x)$ which satisfies the differential equation

$$\frac{d^2f}{dx^2} - 2\frac{df}{dx} - 3f = 1 + x,$$

and is such that $f(0) = 1$ and $f'(0) = 2$.

28.2 Find a particular solution, of the form $a \sin t + b \cos t$, of the differential equation

$$\frac{d^2y}{dt^2} - 5\frac{dy}{dt} + 6y = 18 \sin t,$$

and hence find the general solution. Also find the solution satisfying $y(0) = 0$ and $y'(0) = 1$.

Problems

28.1 Find the general solution of the differential equations

(a) $\dfrac{d^2y}{dt^2} - 5\dfrac{dy}{dt} + 6y = 0,$

(b) $\dfrac{d^2y}{dt^2} - 4\dfrac{dy}{dt} + 4y = 0,$

(c) $\dfrac{d^2y}{dt^2} + 2\dfrac{dy}{dt} + 2y = 0.$

28.2 Find a particular solution, of the form $(at + b)e^{3t}$, of the differential equation

$$\frac{d^2y}{dt^2} - 8\frac{dy}{dt} + 12y = te^{3t}.$$

What is the general solution of this differential equation?

28.3 Find a function $y(t)$ which satisfies the differential equation

$$\frac{d^2y}{dt^2} + y = e^t$$

and is such that $y(0) = 1$ and $y'(0) = 5$.

28.4 Suppose that consumer demand depends on the price trend according to the formula

$$q = 60 - 20p - 7\frac{dp}{dt} + \frac{d^2p}{dt^2}.$$

If the supply function is $q^S(p) = -12 + 10p$, write down the condition for equilibrium and determine the equilibrium price $p(t)$ if $p(0) = 27/5$ and $p'(0) = 17$. Describe in words the behaviour of $p(t)$.

28.5 Suppose that consumer demand depends on the price trend according to the formula

$$q = 40 - 10p + 10\frac{dp}{dt} + \frac{d^2p}{dt^2}.$$

Suppose also that the quantity available is determined in a similar way, by the formula

$$q = -4 + 14p - 4\frac{dp}{dt} + 2\frac{d^2p}{dt^2}.$$

Write down the condition for equilibrium and determine the equilibrium price $p(t)$ if $p(0) = 17/6$ and $p'(0) = 22$. Describe in words the behaviour of $p(t)$.

28.6 With the notation used in Section 28.3, suppose that the auxiliary equation $z^2 + a_1 z + a_2 = 0$ has no solutions. Let

$$\gamma = -\frac{a_1}{2} \quad \text{and} \quad \delta = \frac{\sqrt{4a_2 - a_1^2}}{2}.$$

Show that the functions

$$y_1(t) = e^{\gamma t} \cos \delta t \quad \text{and} \quad y_2(t) = e^{\gamma t} \sin \delta t$$

are solutions of the relevant homogeneous differential equation.

28.7 Suppose the auxiliary equation corresponding to a second-order homogeneous differential equation has one positive solution and one negative solution. Discuss the behaviour of the solution to the differential equation.

29 Coupled Systems and Diagonalisation

29.1 Population Modelling

Assume that there is a fixed population in a certain metropolitan area and, each year, 15% of the city dwellers move to the suburbs and 10% of suburban dwellers move to the city. Suppose, to start with, there are 1 million people living in the city and 1 million in its suburbs. Planners would certainly want to know about the long-term distribution of the population between the city and its suburbs.

Suppose we let x_t and y_t be the number of people living in the city and suburbs, respectively, in year t. Let us consider x_t, the city population in year t. Of those who live in the city in the previous year, year $t-1$, 85% will remain living in the city, amounting to $0.85x_{t-1}$ people. In addition, the city will receive some new inhabitants: 10% of those who live in the suburbs will move to the city, an amount of $0.1y_{t-1}$ people. So the city population in year t will be the sum of these two quantities: $x_t = 0.85x_{t-1}+0.1y_{t-1}$. A similar analysis of the suburban population leads to $y_t = 0.15x_{t-1} + 0.9y_{t-1}$. So the changes in the population distribution of the city can be modelled by the following *system* of recurrence equations:

$$x_t = 0.85x_{t-1} + 0.10y_{t-1},$$
$$y_t = 0.15x_{t-1} + 0.90y_{t-1}.$$

This is an example of a *coupled* system of first-order recurrence equations. (In fact, this specific example is a special type of system of recurrence equations, known as a Markov chain, an important concept in probability theory.)

The system is said to be coupled because the two equations linking the sequences each involve both sequences. What is challenging about solving such systems is that, at first glance, it appears almost impossible because we cannot determine x_t without knowing y_t (as the first equation indicates) and we cannot determine y_t without knowing x_t (as the second equation indicates).

29.2 Coupled Systems of Recurrence Equations

Generally, we want to explore how to solve (by which we mean find explicit expressions for x_t and y_t) systems of the following form:

$$x_t = a_{11}x_{t-1} + a_{12}y_{t-1},$$
$$y_t = a_{21}x_{t-1} + a_{22}y_{t-1}.$$

We can write the system in *matrix form* as $\mathbf{x}_t = A\mathbf{x}_{t-1}$, where

$$\mathbf{x}_t = \begin{pmatrix} x_t \\ y_t \end{pmatrix} \quad \text{and} \quad A = \begin{pmatrix} a_{11} & a_{12} \\ a_{21} & a_{22} \end{pmatrix}.$$

The population modelling problem above provides an example of such a system. Here is another.

Example The sequences x_t and y_t are related as follows: $x_0 = 5$, $y_0 = -2$ and

$$\begin{aligned} x_t &= 2x_{t-1} + 4y_{t-1}, \\ y_t &= 3x_{t-1} + 3y_{t-1}, \end{aligned}$$

or, in matrix form:

$$\mathbf{x}_t = \begin{pmatrix} x_t \\ y_t \end{pmatrix} = \begin{pmatrix} 2 & 4 \\ 3 & 3 \end{pmatrix} \begin{pmatrix} x_{t-1} \\ y_{t-1} \end{pmatrix} = A\mathbf{x}_{t-1}.$$

■

How might we solve such a system? We will look at some approaches.

29.3 Reduction to Second-Order Recurrence

One way in which a coupled system of recurrence equations can be solved is by reducing, or translating, it into a corresponding single second-order recurrence equation. We illustrate with our example.

Example Suppose, as above, that we have sequences x_t and y_t which satisfy $x_0 = 5$, $y_0 = -2$, and which are related as follows:

$$\begin{aligned} x_t &= 2x_{t-1} + 4y_{t-1}, \\ y_t &= 3x_{t-1} + 3y_{t-1}. \end{aligned}$$

We have, from the first equation, and using the second equation with $t - 1$ in place of t:

$$x_t = 2x_{t-1} + 4y_{t-1} = 2x_{t-1} + 4(3x_{t-2} + 3y_{t-2}) = 2x_{t-1} + 12x_{t-2} + 12y_{t-2}.$$

Now, to get an equation just involving x_t, x_{t-1} and x_{t-2}, we will express y_{t-2} in terms of the sequence x_t. From the first of the two original equations, we have $y_{t-1} = \frac{1}{4}(x_t - 2x_{t-1})$. It follows (on replacing t by $t - 1$) that

$$y_{t-2} = \frac{1}{4}(x_{t-1} - 2x_{t-2}).$$

Therefore

$$\begin{aligned} x_t &= 2x_{t-1} + 12x_{t-2} + 12y_{t-2} \\ &= 2x_{t-1} + 12x_{t-2} + 12\left(\frac{x_{t-1} - 2x_{t-2}}{4}\right) \\ &= 5x_{t-1} + 6x_{t-2} \end{aligned}$$

and we have the following second-order recurrence equation for x_t:

$$x_t - 5x_{t-1} - 6x_{t-2} = 0.$$

The auxiliary equation, $z^2 - 5z - 6 = 0$, has solutions 6 and -1, so, for some constants A and B:

$$x_t = A(6^t) + B(-1)^t.$$

Now, as already observed, $y_{t-1} = \frac{1}{4}(x_t - 2x_{t-1})$, so (replacing $t-1$ by t and, correspondingly, t by $t+1$):

$$\begin{aligned} y_t &= \frac{1}{4}(x_{t+1} - 2x_t) \\ &= \frac{1}{4}\left[A(6^{t+1}) + B(-1)^{t+1} - 2\left[A(6^t) + B(-1)^t\right]\right] \\ &= \frac{1}{4}\left[6A(6^t) - B(-1)^t - 2A(6^t) - 2B(-1)^t\right] \\ &= \frac{1}{4}\left[4A(6^t) - 3B(-1)^t\right] \\ &= A(6^t) - \frac{3}{4}B(-1)^t. \end{aligned}$$

Now we use the initial conditions $x_0 = 5$, $y_0 = -2$. The first gives $A + B = 5$ and the second $A - \frac{3}{4}B = -2$. Subtracting the second of these equations from the first, we obtain $\frac{7}{4}B = 7$, so $B = 4$ and $A = 1$. The solutions are therefore

$$x_t = 6^t + 4(-1)^t \quad \text{and} \quad y_t = 6^t - 3(-1)^t.$$

We have therefore solved the coupled system. ■

What we will do next is present an entirely different method, which has more applicability (in particular, to more general systems where the number of variables and equations is larger). This is based on the concepts of *eigenvalues* and *eigenvectors* and the important related method known as *diagonalisation*.

29.4 Eigenvalues and Eigenvectors

One of the most useful techniques in applications of matrices is *diagonalisation*. Before discussing this, we have to look at the topic of *eigenvalues and eigenvectors*. We shall be primarily interested in eigenvalues, eigenvectors and diagonalisation in the case of 2×2 matrices, but the analysis and methods can be extended to deal with larger matrices.

Suppose that A is a square matrix. The number λ is said to be an *eigenvalue* of A if for some *non-zero* vector $\mathbf{x}$, $A\mathbf{x} = \lambda\mathbf{x}$. Any *non-zero* vector $\mathbf{x}$ for which this equation holds is called an *eigenvector for eigenvalue* λ or an *eigenvector of A corresponding to eigenvalue* λ.

To determine whether λ is an eigenvalue of A, we need to determine whether there are any non-zero solutions to the matrix equation $A\mathbf{x} = \lambda\mathbf{x}$. Note that the matrix equation $A\mathbf{x} = \lambda\mathbf{x}$ is not of standard form, since the right-hand side is not a fixed vector $\mathbf{b}$, but depends explicitly on $\mathbf{x}$. However, we can rewrite it in standard form. Note that $\lambda\mathbf{x} = \lambda I\mathbf{x}$, where I is, as usual, the identity matrix. So, the equation is equivalent to $A\mathbf{x} = \lambda I\mathbf{x}$, or $A\mathbf{x} - \lambda I\mathbf{x} = \mathbf{0}$, which is equivalent to $(A - \lambda I)\mathbf{x} = \mathbf{0}$.

In Chapter 20, we noted that a system of linear equations $M\mathbf{x} = \mathbf{b}$ has a unique solution if and only if the determinant of M is non-zero. (We are using M to denote a matrix here rather than A, simply because we're already using A in this discussion.) To find eigenvalues, we are interested in the system $M\mathbf{x} = \mathbf{0}$, where $M = A - \lambda I$ and $\mathbf{0}$ is the zero vector. What we want to know is whether, for a given value of λ, this system has a non-zero solution $\mathbf{x}$. Clearly, the zero vector $\mathbf{0}$ is a solution, so we are trying to determine when this is *not* the unique solution. It follows that the system $M\mathbf{x} = \mathbf{0}$ will have a non-zero solution $\mathbf{x}$ (which means λ will be an eigenvalue and $\mathbf{x}$ a corresponding eigenvector), when the determinant $|M| = |A - \lambda I|$ is zero.

So, to find the eigenvalues, we solve the equation $|A - \lambda I| = 0$ for λ. This determinant, $p(\lambda) = |A - \lambda I|$, is known as the *characteristic polynomial* of A, since it turns out to be a polynomial in the variable λ. To find the eigenvalues, we solve the equation $|A - \lambda I| = 0$. Let us illustrate with an example.

Example Suppose that

$$A = \begin{pmatrix} 2 & 4 \\ 3 & 3 \end{pmatrix},$$

the matrix we had in the earlier example. We have

$$|A - \lambda I| = \begin{vmatrix} 2-\lambda & 4 \\ 3 & 3-\lambda \end{vmatrix} = (2-\lambda)(3-\lambda) - 12 = \lambda^2 - 5\lambda - 6.$$

To determine the eigenvalues, we therefore seek the solutions to $\lambda^2 - 5\lambda - 6 = 0$. You can use the formula for the solutions of a quadratic, but in this case it is quite easy to spot that the quadratic factorises as $(\lambda - 6)(\lambda + 1)$, so the eigenvalues are 6 and -1. ■

To find an eigenvector corresponding to eigenvalue λ, we want to find a solution to $(A - \lambda I)\mathbf{x} = \mathbf{0}$ *other than the zero vector*. (It's important to remember that eigenvectors cannot be the zero vector.) This is easy, since for a particular value of λ, all we need to do is solve a simple linear system. We illustrate by finding eigenvectors for the matrix of the example just given.

Example Let us continue with the matrix

$$A = \begin{pmatrix} 2 & 4 \\ 3 & 3 \end{pmatrix}.$$

To find an eigenvector for eigenvalue 6, we solve the system $(A - 6I)\mathbf{x} = \mathbf{0}$; that is

$$\begin{pmatrix} -4 & 4 \\ 3 & -3 \end{pmatrix} \begin{pmatrix} x_1 \\ x_2 \end{pmatrix} = \begin{pmatrix} 0 \\ 0 \end{pmatrix}.$$

This could be solved using row operations. (Note that it cannot be solved by using inverse matrices since A is not invertible. In fact, inverse matrix techniques or Cramer's rule will never be of use here since λ being an eigenvalue means that $A - \lambda I$ is not invertible.) However, we can solve this fairly directly just by looking at the equations. The system says

$$-4x_1 + 4x_2 = 0 \quad \text{and} \quad 3x_1 - 3x_2 = 0.$$

These are clearly equivalent to the single equation $x_1 = x_2$, so (setting $x_2 = 1$) a suitable eigenvector is

$$\begin{pmatrix} 1 \\ 1 \end{pmatrix}.$$

The choice $x_2 = 1$ was arbitrary; we could have chosen any non-zero number, so, for example, we could also have taken

$$\begin{pmatrix} 2 \\ 2 \end{pmatrix} \quad \text{or} \quad \begin{pmatrix} 5.2 \\ 5.2 \end{pmatrix}$$

as eigenvectors for 0. There are infinitely many suitable eigenvectors: for each $\alpha \neq 0$,

$$\begin{pmatrix} \alpha \\ \alpha \end{pmatrix}$$

is an eigenvector corresponding to eigenvalue 6. But be careful not to think that you can choose $\alpha = 0$; for then $\mathbf{x}$ becomes the zero vector, and this is never an eigenvector, simply by definition.

To find an eigenvector for eigenvalue -1 we solve the system $(A - (-1)I)\mathbf{x} = \mathbf{0}$; that is

$$\begin{pmatrix} 3 & 4 \\ 3 & 4 \end{pmatrix} \begin{pmatrix} x_1 \\ x_2 \end{pmatrix} = \begin{pmatrix} 0 \\ 0 \end{pmatrix}.$$

This is equivalent to the single equation $3x_1 + 4x_2 = 0$, so a suitable eigenvector is

$$\begin{pmatrix} 4 \\ -3 \end{pmatrix},$$

if we take $x_1 = 4$. (Again, any non-zero multiple of this will also do.) ■

It should not be assumed that a 2×2 matrix always has two distinct eigenvalues. It could have just one, or it could have none.

For example, consider the zero matrix

$$A = \begin{pmatrix} 0 & 0 \\ 0 & 0 \end{pmatrix}.$$

An eigenvalue would satisfy

$$|A - \lambda I| = \begin{vmatrix} -\lambda & 0 \\ 0 & -\lambda \end{vmatrix} = 0,$$

which is $\lambda^2 = 0$, so the only solution (and hence the only eigenvalue) is 0. Every non-zero vector $\mathbf{x}$ is a corresponding eigenvector, since $A\mathbf{x} = \mathbf{0} = 0\mathbf{x}$.

Consider now the matrix

$$A = \begin{pmatrix} 0 & -1 \\ 1 & 0 \end{pmatrix}.$$

To determine eigenvalues, we solve

$$|A - \lambda I| = \begin{vmatrix} -\lambda & -1 \\ 1 & -\lambda \end{vmatrix} = \lambda^2 + 1 = 0,$$

which has no (real) solutions for λ.

29.5 Diagonalisation of a Square Matrix

A matrix D is said to be *diagonal* if all entries not on the main diagonal are 0. (Explicitly, if d_{ij} is the entry in row i and column j, then $d_{ij} = 0$ whenever $i \neq j$.) The matrix A is said to be *diagonalisable* if there is a diagonal matrix D and an invertible matrix P such that $P^{-1}AP = D$.

Suppose that the 2×2 matrix A is diagonalisable, and that $P^{-1}AP = D$, where D is the diagonal matrix

$$D = \begin{pmatrix} \lambda_1 & 0 \\ 0 & \lambda_2 \end{pmatrix}$$

for some numbers λ_1 and λ_2. The equation $P^{-1}AP = D$ implies $AP = PD$. If the columns of P are the vectors $\mathbf{v}_1$ and $\mathbf{v}_2$, then, because of the way matrix multiplication works, we have

$$AP = A(\mathbf{v}_1 \ \mathbf{v}_2) = (A\mathbf{v}_1 \ A\mathbf{v}_2)$$

and

$$PD = (\mathbf{v}_1 \ \mathbf{v}_2) \begin{pmatrix} \lambda_1 & 0 \\ 0 & \lambda_2 \end{pmatrix} = (\lambda_1\mathbf{v}_1 \ \lambda_2\mathbf{v}_2).$$

So, $AP = PD$ means that

$$A\mathbf{v}_1 = \lambda_1\mathbf{v}_1, \quad A\mathbf{v}_2 = \lambda_2\mathbf{v}_2.$$

The fact that P^{-1} exists means that neither of the vectors $\mathbf{v}_1$ or $\mathbf{v}_2$ is the zero vector (for, otherwise, there could be no inverse of P). So, for $i = 1, 2$, this means that λ_i is an eigenvalue of A and $\mathbf{v}_i$ is a corresponding eigenvector.

So we have established that A is diagonalisable if we can find an invertible matrix P whose columns are eigenvectors of A.

This analysis extends to larger, $n \times n$, matrices. It is not always the case that such a matrix P can be found, but in the 2×2 case the issue is quite simple: if $\mathbf{v}_1$ and $\mathbf{v}_2$ are eigenvectors which are not multiples of each other, then the matrix P with columns $\mathbf{v}_1$ and $\mathbf{v}_2$ will be as required.

Example Consider again the matrix

$$A = \begin{pmatrix} 2 & 4 \\ 3 & 3 \end{pmatrix}.$$

We have seen that the eigenvalues of A are 6 and -1 and examples of corresponding eigenvectors are

$$\begin{pmatrix} 1 \\ 1 \end{pmatrix} \quad \text{and} \quad \begin{pmatrix} 4 \\ -3 \end{pmatrix},$$

respectively. If we let

$$P = \begin{pmatrix} 1 & 4 \\ 1 & -3 \end{pmatrix},$$

then P^{-1} exists and we should have $P^{-1}AP = D$ where

$$D = \begin{pmatrix} 6 & 0 \\ 0 & -1 \end{pmatrix}.$$

It is easy to check that this is so, by calculating P^{-1} and then evaluating $P^{-1}AP$, seeing that this product equals D. ■

Of course, this is only one of the many possible pairs of matrices that we could choose for P and D; others are possible depending on which eigenvectors we choose when we form the columns of P and the order in which we choose to place them in P. For instance, choosing the other order for the eigenvectors we found above, we can see that

$$P = \begin{pmatrix} 4 & 1 \\ -3 & 1 \end{pmatrix} \quad \text{and} \quad D = \begin{pmatrix} -1 & 0 \\ 0 & 6 \end{pmatrix}$$

is another possible answer here.

In the next chapter, we show how the method of diagonalisation can be used to solve coupled recurrence equations.

Worked Examples

Example 29.1 Find the sequences x_t and y_t that satisfy $x_0 = 1$, $y_0 = -1$ and

$$x_t = 5x_{t-1} - y_{t-1},$$
$$y_t = -3x_{t-1} + 7y_{t-1}.$$

Solution

We have, from the first equation and using the second equation with $t-1$ in place of t:

$$x_t = 5x_{t-1} - y_{t-1} = 5x_{t-1} - (-3x_{t-2} + 7y_{t-2}) = 5x_{t-1} + 3x_{t-2} - 7y_{t-2}.$$

Now, to get an equation just involving x_t, x_{t-1} and x_{t-2}, we express y_{t-2} in terms of the sequence x_t. From the first of the two original equations, we have $y_{t-1} = 5x_{t-1} - x_t$. It follows (on replacing t by $t-1$) that

$$y_{t-2} = 5x_{t-2} - x_{t-1}.$$

Therefore

$$\begin{aligned} x_t &= 5x_{t-1} + 3x_{t-2} - 7y_{t-2} \\ &= 5x_{t-1} + 3x_{t-2} - 7\,(5x_{t-2} - x_{t-1}) \\ &= 12x_{t-1} - 32x_{t-2} \end{aligned}$$

and we have the following second-order recurrence equation for x_t:

$$x_t - 12x_{t-1} + 32x_{t-2} = 0.$$

The auxiliary equation, $z^2 - 12z + 32 = 0$, has solutions 4 and 8, so, for some constants A and B:

$$x_t = A(4^t) + B(8^t).$$

Now, as already observed, $y_{t-1} = 5x_{t-1} - x_t$, so (replacing $t-1$ by t and, correspondingly, t by $t+1$):

$$\begin{aligned} y_t &= 5x_t - x_{t+1} \\ &= 5\left[A(4^t) + B(8^t)\right] - \left[A(4^{t+1}) + B(8^{t+1})\right] \\ &= 5A(4^t) + 5B(8^t) - \left[4A(4^t) + 8B(8^t)\right] \\ &= A(4^t) - 3B(8^t). \end{aligned}$$

Now we use the initial conditions $x_0 = 1$, $y_0 = -1$. The first gives $A + B = 1$ and the second $A - 3B = -1$. Solving these, we obtain $A = B = 1/2$. The solutions are therefore

$$x_t = \frac{4^t + 8^t}{2} \quad \text{and} \quad y_t = \frac{4^t - 3(8^t)}{2}.$$

Example 29.2 Find the eigenvalues of the matrix

$$\begin{pmatrix} 2 & 1 \\ 3 & 1 \end{pmatrix}.$$

Solution

Denoting the matrix by A, we have

$$|A - \lambda I| = \begin{vmatrix} 2-\lambda & 1 \\ 3 & 1-\lambda \end{vmatrix} = (2-\lambda)(1-\lambda) - 3 = \lambda^2 - 3\lambda - 1.$$

To determine the eigenvalues, we therefore seek the solutions to $\lambda^2 - 3\lambda - 1 = 0$. By the formula for the roots of a quadratic, we have solutions

$$\lambda = \frac{3 \pm \sqrt{13}}{2},$$

so the eigenvalues are $(3 + \sqrt{13})/2$ and $(3 - \sqrt{13})/2$.

Example 29.3 Find the eigenvalues of the matrix

$$A = \begin{pmatrix} 1 & 1 \\ 2 & 2 \end{pmatrix},$$

and determine an eigenvector corresponding to each eigenvalue. Find an invertible matrix P and a diagonal matrix D such that $P^{-1}AP = D$.

Solution

Denoting the matrix by A, we have

$$|A - \lambda I| = \begin{vmatrix} 1-\lambda & 1 \\ 2 & 2-\lambda \end{vmatrix} = (1-\lambda)(2-\lambda) - 2 = \lambda^2 - 3\lambda.$$

The eigenvalues are the solutions of $\lambda^2 - 3\lambda = \lambda(\lambda - 3) = 0$, which are 0 and 3.

We now find eigenvectors. To find an eigenvector for eigenvalue 0, we solve the system $(A - 0I)\mathbf{x} = \mathbf{0}$; that is $A\mathbf{x} = \mathbf{0}$ or

$$\begin{pmatrix} 1 & 1 \\ 2 & 2 \end{pmatrix} \begin{pmatrix} x_1 \\ x_2 \end{pmatrix} = \begin{pmatrix} 0 \\ 0 \end{pmatrix}.$$

We have to solve

$$x_1 + x_2 = 0 \qquad \text{and} \qquad 2x_1 + 2x_2 = 0.$$

Clearly both equations are equivalent. From either one, we obtain $x_1 = -x_2$. We can choose x_2 to be any non-zero number we like. Let's take $x_2 = 1$; then we need $x_1 = -x_2 = -1$. It follows that an eigenvector for 0 is

$$\mathbf{x} = \begin{pmatrix} -1 \\ 1 \end{pmatrix}.$$

(Again, any non-zero multiple of this vector is also an eigenvector for eigenvalue 0.)

To find an eigenvector for 3, we solve $(A - 3I)\mathbf{x} = \mathbf{0}$, which is

$$\begin{pmatrix} -2 & 1 \\ 2 & -1 \end{pmatrix} \begin{pmatrix} x_1 \\ x_2 \end{pmatrix} = \begin{pmatrix} 0 \\ 0 \end{pmatrix}.$$

This is equivalent to the equations

$$-2x_1 + x_2 = 0 \qquad \text{and} \qquad 2x_1 - x_2 = 0,$$

which are together equivalent to the single equation $x_2 = 2x_1$. If we choose $x_1 = 1$, we obtain the eigenvector

$$\mathbf{x} = \begin{pmatrix} 1 \\ 2 \end{pmatrix}.$$

(Again, any non-zero multiple of this vector is also an eigenvector for eigenvalue 3.)

We have seen that the eigenvalues of A are 0 and 3 and examples of corresponding eigenvectors are

$$\begin{pmatrix} -1 \\ 1 \end{pmatrix} \quad \text{and} \quad \begin{pmatrix} 1 \\ 2 \end{pmatrix},$$

respectively. If we let

$$P = \begin{pmatrix} -1 & 1 \\ 1 & 2 \end{pmatrix},$$

then P^{-1} exists and we should have $P^{-1}AP = D$ where

$$D = \begin{pmatrix} 0 & 0 \\ 0 & 3 \end{pmatrix}.$$

It is easy to check that this is so, by calculating P^{-1} and then evaluating $P^{-1}AP$, seeing that this product equals D.

Example 29.4 Find the eigenvalues and eigenvectors of the matrix

$$A = \begin{pmatrix} 1 & a^2 \\ 4 & 1 \end{pmatrix}$$

where a is a constant. Hence, for each value of a, find an invertible matrix P and a diagonal matrix D such that $P^{-1}AP = D$, or explain why this is not possible.

Solution

To find the eigenvalues of this matrix, we solve the equation $|A - \lambda I| = 0$, or

$$\begin{vmatrix} 1-\lambda & a^2 \\ 4 & 1-\lambda \end{vmatrix} = 0,$$

which is $(1-\lambda)^2 - 4a^2 = 0$, giving us $(1-\lambda)^2 = 4a^2$; so $1 - \lambda = \pm 2a$. The eigenvalues are $1 + 2a$ and $1 - 2a$. To find the corresponding eigenvectors, we seek a non-zero vector $\mathbf{x}$, which is a solution to the equation $(A - \lambda I)\mathbf{x} = \mathbf{0}$.
For $\lambda = 1 + 2a$, we solve

$$\begin{pmatrix} -2a & a^2 \\ 4 & -2a \end{pmatrix} \begin{pmatrix} x \\ y \end{pmatrix} = \mathbf{0},$$

which is equivalent to $4x - 2ay = 0$. So, with $y = 2$, an eigenvector is

$$\begin{pmatrix} x \\ y \end{pmatrix} = \begin{pmatrix} a \\ 2 \end{pmatrix}.$$

For $\lambda = 1 - 2a$, we solve

$$\begin{pmatrix} 2a & a^2 \\ 4 & 2a \end{pmatrix} \begin{pmatrix} x \\ y \end{pmatrix} = \mathbf{0},$$

which is equivalent to $4x + 2ay = 0$. So, with $y = -2$, an eigenvector is

$$\begin{pmatrix} x \\ y \end{pmatrix} = \begin{pmatrix} -a \\ 2 \end{pmatrix}.$$

Consequently, if $a \neq 0$, we see that

$$P = \begin{pmatrix} a & -a \\ 2 & 2 \end{pmatrix}$$

is an invertible matrix, as

$$|P| = \begin{vmatrix} a & -a \\ 2 & 2 \end{vmatrix} = 4a \neq 0,$$

and so, together with the diagonal matrix

$$D = \begin{pmatrix} 1 + 2a & 0 \\ 0 & 1 - 2a \end{pmatrix},$$

we have $P^{-1}AP = D$.

However, if $a = 0$, we see that $\lambda = 1$ is the only eigenvalue and all of the eigenvectors are non-zero multiples of the vector

$$\begin{pmatrix} 0 \\ 1 \end{pmatrix}.$$

As such, we find that any matrix P we construct will have the form

$$P = \begin{pmatrix} 0 & 0 \\ r & s \end{pmatrix},$$

and this is not an invertible matrix as

$$|P| = \begin{vmatrix} 0 & 0 \\ r & s \end{vmatrix} = 0.$$

That is, if $a = 0$, there is no invertible matrix P and diagonal matrix D such that $P^{-1}AP = D$.

Main Topics

- coupled systems of first-order recurrence equations
- reduction of coupled systems of first-order recurrence equations to second-order recurrence equations
- eigenvalues and eigenvectors
- diagonalisation of a square matrix

Key Terms, Notations and Formulae

- coupled system of first-order recurrence equations: $\mathbf{x}_t = A\mathbf{x}_{t-1}$
- eigenvalues are solutions, λ, to $|A - \lambda I| = 0$
- an eigenvector corresponding to eigenvalue λ is a non-zero $\mathbf{x}$ such that

$$(A - \lambda I)\mathbf{x} = \mathbf{0}$$

- if λ_1, λ_2 are eigenvalues of a 2×2 matrix A and $\mathbf{v}_1, \mathbf{v}_2$ corresponding eigenvectors (which are not multiples of each other), then $P = (\mathbf{v}_1\ \mathbf{v}_2)$ is invertible and $P^{-1}AP = D$ where

$$D = \begin{pmatrix} \lambda_1 & 0 \\ 0 & \lambda_2 \end{pmatrix}$$

Exercises

29.1 Sequences x_t and y_t satisfy the system of recurrence equations

$$\begin{aligned} x_t &= -3x_{t-1} - 4y_{t-1}, \\ y_t &= 2x_{t-1} + 3y_{t-1}. \end{aligned}$$

Show that $x_t - x_{t-2} = 0$ and hence find the sequences if $x_0 = 2$ and $y_0 = 1$.

29.2 Find the eigenvalues of the matrix

$$A = \begin{pmatrix} 7 & 1 \\ 3 & 5 \end{pmatrix}$$

and, for each eigenvalue, find a corresponding eigenvector.

29.3 The matrix A is

$$A = \begin{pmatrix} 2 & -1 \\ 0 & 1 \end{pmatrix}.$$

Without determining the characteristic polynomial of A, verify that the vectors

$$\begin{pmatrix} 1 \\ 1 \end{pmatrix} \quad \text{and} \quad \begin{pmatrix} 1 \\ 0 \end{pmatrix}$$

are eigenvectors of A, and determine the eigenvalues of A.

29.4 Find the eigenvalues of the matrix

$$A = \begin{pmatrix} 5 & -1 \\ -3 & 7 \end{pmatrix}$$

and, for each eigenvalue, find a corresponding eigenvector. Hence find an invertible matrix P and a diagonal matrix D such that $P^{-1}AP = D$.

Problems

29.1 Sequences x_t and y_t satisfy $x_0 = y_0 = 1$ and

$$x_t = 0.85x_{t-1} + 0.10y_{t-1},$$
$$y_t = 0.15x_{t-1} + 0.90y_{t-1}.$$

Show that x_t satisfies

$$x_t - 1.75x_{t-1} + 0.75x_{t-1} = 0.$$

Hence find x_t and y_t.

29.2 Find the eigenvalues of the matrix

$$A = \begin{pmatrix} -1 & 1 \\ 2 & 3 \end{pmatrix}.$$

29.3 Find the eigenvalues of the matrix

$$A = \begin{pmatrix} 1 & 2 \\ 2 & 4 \end{pmatrix}$$

and, for each eigenvalue, find a corresponding eigenvector.

29.4 Find the eigenvalues of the matrix

$$A = \begin{pmatrix} -5 & -4 \\ 2 & 1 \end{pmatrix}$$

and, for each eigenvalue, find a corresponding eigenvector. Hence find an invertible matrix P and a diagonal matrix D such that $P^{-1}AP = D$.

30 Applications of Diagonalisation

30.1 Using Diagonalisation: Powers of Matrices

For a positive integer n, the nth power of a matrix A is simply

$$A^n = \underbrace{AAA \cdots A}_{n \text{ times}}.$$

For example, $A^2 = AA$ and $A^4 = AAAA$.

Example Suppose

$$A = \begin{pmatrix} 2 & 4 \\ 3 & 3 \end{pmatrix}$$

(a matrix we worked with in the previous chapter). We have

$$A^2 = \begin{pmatrix} 2 & 4 \\ 3 & 3 \end{pmatrix} \begin{pmatrix} 2 & 4 \\ 3 & 3 \end{pmatrix} = \begin{pmatrix} 16 & 20 \\ 15 & 21 \end{pmatrix}$$

and

$$A^3 = AAA = AA^2 = \begin{pmatrix} 2 & 4 \\ 3 & 3 \end{pmatrix} \begin{pmatrix} 16 & 20 \\ 15 & 21 \end{pmatrix} = \begin{pmatrix} 92 & 124 \\ 93 & 123 \end{pmatrix}.$$

■

How might we determine other powers of A? Could there be a formula, an expression involving n, for the power A^n?

It can be useful, as we shall see shortly, to determine A^n for a general integer n. Diagonalisation helps here. If we can write $P^{-1}AP = D$, then $A = PDP^{-1}$ and

$$\begin{aligned} A^n &= \underbrace{AAA \cdots A}_{n \text{ times}} \\ &= \underbrace{(PDP^{-1})(PDP^{-1})(PDP^{-1}) \cdots (PDP^{-1})}_{n \text{ times}} \\ &= PD(P^{-1}P)D(P^{-1}P)D(P^{-1}P) \cdots D(P^{-1}P)DP^{-1} \\ &= P\underbrace{DDD \cdots D}_{n \text{ times}}P^{-1} \\ &= PD^nP^{-1} \end{aligned}$$

as $P^{-1}P = I$. The product PD^nP^{-1} is easy to compute since D^n is simply the diagonal matrix with entries equal to the nth power of those of D.

Example Let's consider again the matrix

$$A = \begin{pmatrix} 2 & 4 \\ 3 & 3 \end{pmatrix}.$$

We have seen, in the previous chapter, that A can be diagonalised. Explicitly, if

$$P = \begin{pmatrix} 1 & 4 \\ 1 & -3 \end{pmatrix},$$

then P is invertible and $P^{-1}AP = D$, where

$$D = \begin{pmatrix} 6 & 0 \\ 0 & -1 \end{pmatrix}.$$

It follows that

$$\begin{aligned} A^n &= PD^nP^{-1} \\ &= \begin{pmatrix} 1 & 4 \\ 1 & -3 \end{pmatrix} \begin{pmatrix} 6^n & 0 \\ 0 & (-1)^n \end{pmatrix} \left(-\frac{1}{7} \begin{pmatrix} -3 & -4 \\ -1 & 1 \end{pmatrix} \right) \\ &= \frac{1}{7} \begin{pmatrix} 1 & 4 \\ 1 & -3 \end{pmatrix} \begin{pmatrix} 3(6^n) & 4(6^n) \\ (-1)^n & -(-1)^n \end{pmatrix} \\ &= \frac{1}{7} \begin{pmatrix} 3(6^n) + 4(-1)^n & 4(6^n) - 4(-1)^n \\ 3(6^n) - 3(-1)^n & 4(6^n) + 3(-1)^n \end{pmatrix}. \end{aligned}$$

You can easily check, by taking $n = 2$ and $n = 3$, that this general formula for A^n agrees with what we saw above for A^2 and A^3. ■

30.2 Solving Recurrence Systems Using Matrix Powers

Consider the general system of recurrence equations $\mathbf{x}_t = A\mathbf{x}_{t-1}$. Then, taking $t = 1$, $\mathbf{x}_1 = A\mathbf{x}_0$ and, taking $t = 2$, $\mathbf{x}_2 = A\mathbf{x}_1 = A(A\mathbf{x}_0) = A^2\mathbf{x}_0$ and, in general, it can be seen that $\mathbf{x}_t = A^t\mathbf{x}_0$. So, if we know a formula for the power A^t, we can solve the system.

Example Consider the system

$$\begin{pmatrix} x_t \\ y_t \end{pmatrix} = \begin{pmatrix} 2 & 4 \\ 3 & 3 \end{pmatrix} \begin{pmatrix} x_{t-1} \\ y_{t-1} \end{pmatrix},$$

where $x_0 = 5$ and $y_0 = -2$, from the previous chapter. As just noted, the solution is $\mathbf{x}_t = A^t\mathbf{x}_0$ and we calculated the general powers of A above, so we have

$$\mathbf{x}_t = A^t\mathbf{x}_0 = \frac{1}{7} \begin{pmatrix} 3(6^t) + 4(-1)^t & 4(6^t) - 4(-1)^t \\ 3(6^t) - 3(-1)^t & 4(6^t) + 3(-1)^t \end{pmatrix} \begin{pmatrix} 5 \\ -2 \end{pmatrix} = \begin{pmatrix} 6^t + 4(-1)^t \\ 6^t - 3(-1)^t \end{pmatrix}.$$

Therefore, the solution is

$$x_t = 6^t + 4(-1)^t \quad \text{and} \quad y_t = 6^t - 3(-1)^t,$$

as before. ■

30.3 Solving Recurrence Systems Using Change of Variable

We can use diagonalisation in another way as the key to a general method for solving systems of recurrence equations. Given a system $\mathbf{x}_t = A\mathbf{x}_{t-1}$, in which A is diagonalisable, we perform a *change of variable*, as follows. Suppose that $P^{-1}AP = D$, where D is diagonal. We introduce two new sequences u_t, v_t related to x_t, y_t. The relationship is given by $\mathbf{x}_t = P\mathbf{u}_t$, where

$$\mathbf{u}_t = \begin{pmatrix} u_t \\ v_t \end{pmatrix}.$$

Equivalently, the new sequences are related to the original ones by $\mathbf{u}_t = P^{-1}\mathbf{x}_t$.

Now, $\mathbf{x}_t = P\mathbf{u}_t$ and $A\mathbf{x}_{t-1} = A(P\mathbf{u}_{t-1}) = AP\mathbf{u}_{t-1}$. Therefore, $\mathbf{x}_t = A\mathbf{x}_{t-1}$ tells us that $P\mathbf{u}_t = AP\mathbf{u}_{t-1}$, so

$$\mathbf{u}_t = P^{-1}AP\mathbf{u}_{t-1} = D\mathbf{u}_{t-1}.$$

Since D is diagonal, this is very easy to solve for $\mathbf{u}_t$. To find $\mathbf{x}_t$, we then use the fact that $\mathbf{x}_t = P\mathbf{u}_t$.

Example Consider again the system

$$\begin{pmatrix} x_t \\ y_t \end{pmatrix} = A \begin{pmatrix} x_{t-1} \\ y_{t-1} \end{pmatrix} = \begin{pmatrix} 2 & 4 \\ 3 & 3 \end{pmatrix} \begin{pmatrix} x_{t-1} \\ y_{t-1} \end{pmatrix},$$

where $x_0 = 5$ and $y_0 = -2$.

We have seen that if

$$P = \begin{pmatrix} 1 & 4 \\ 1 & -3 \end{pmatrix},$$

then $P^{-1}AP = D$, where D is the diagonal matrix

$$D = \begin{pmatrix} 6 & 0 \\ 0 & -1 \end{pmatrix}.$$

Let u_t, v_t be given by $\mathbf{x}_t = P\mathbf{u}_t$, where

$$\mathbf{u}_t = \begin{pmatrix} u_t \\ v_t \end{pmatrix}.$$

Then, as described above, we have $\mathbf{u}_t = D\mathbf{u}_{t-1}$, which means

$$\begin{pmatrix} u_t \\ v_t \end{pmatrix} = \begin{pmatrix} 6 & 0 \\ 0 & -1 \end{pmatrix} \begin{pmatrix} u_{t-1} \\ v_{t-1} \end{pmatrix},$$

so the (uncoupled) equations for u_t and v_t are

$$u_t = 6u_{t-1} \quad \text{and} \quad v_t = -v_{t-1}.$$

We see immediately that the solutions of these are

$$u_t = 6^t u_0 \quad \text{and} \quad v_t = (-1)^t v_0.$$

Now, $\mathbf{u}_t = P^{-1}\mathbf{x}_t$, so taking $t = 0$ we have

$$\begin{pmatrix} u_0 \\ v_0 \end{pmatrix} = P^{-1}\begin{pmatrix} x_0 \\ y_0 \end{pmatrix} = -\frac{1}{7}\begin{pmatrix} -3 & -4 \\ -1 & 1 \end{pmatrix}\begin{pmatrix} 5 \\ -2 \end{pmatrix} = \begin{pmatrix} 1 \\ 1 \end{pmatrix}$$

and so $u_t = 6^t$ and $v_t = (-1)^t$. Then, using $\mathbf{x}_t = P\mathbf{u}_t$, we have

$$\begin{pmatrix} x_t \\ y_t \end{pmatrix} = \begin{pmatrix} 1 & 4 \\ 1 & -3 \end{pmatrix}\begin{pmatrix} u_t \\ v_t \end{pmatrix} = \begin{pmatrix} 1 & 4 \\ 1 & -3 \end{pmatrix}\begin{pmatrix} 6^t \\ (-1)^t \end{pmatrix} = \begin{pmatrix} 6^t + 4(-1)^t \\ 6^t - 3(-1)^t \end{pmatrix}.$$

That is

$$x_t = 6^t + 4(-1)^t \quad \text{and} \quad y_t = 6^t - 3(-1)^t,$$

as before. ■

30.4 A Population Model

At the start of the previous chapter, we introduced a coupled system of recurrence equations that model the population within a city and its suburbs. We had

$$x_t = 0.85x_{t-1} + 0.10y_{t-1},$$
$$y_t = 0.15x_{t-1} + 0.90y_{t-1},$$

where x_t and y_t are the number of people living in the city and suburbs, respectively, in year t. Initially, there were a million in each. So, in millions, $x_0 = y_0 = 1$. We would like to explore what happens to the population in the long run. This system is $\mathbf{x}_t = A\mathbf{x}_{t-1}$, where

$$A = \begin{pmatrix} 0.85 & 0.10 \\ 0.15 & 0.90 \end{pmatrix}.$$

We diagonalise A in the usual way. Its characteristic polynomial is

$$\begin{aligned} |A - \lambda I| &= \begin{pmatrix} 0.85 - \lambda & 0.10 \\ 0.15 & 0.90 - \lambda \end{pmatrix} \\ &= (0.85 - \lambda)(0.9 - \lambda) - 0.015 \\ &= \lambda^2 - 1.75\lambda + 0.75 \\ &= (\lambda - 1)(\lambda - 0.75). \end{aligned}$$

So the eigenvalues (its roots) are 1 and 0.75. To find an eigenvector corresponding to eigenvalue 1, we solve the system $(A - I)\mathbf{x} = \mathbf{0}$, which is

$$\begin{pmatrix} -0.15 & 0.10 \\ 0.15 & -0.10 \end{pmatrix} \begin{pmatrix} x_1 \\ x_2 \end{pmatrix} = \begin{pmatrix} 0 \\ 0 \end{pmatrix},$$

equivalent to the single equation $x_1 = 2x_2/3$. So an eigenvector is $\begin{pmatrix} 2 \\ 3 \end{pmatrix}$. For eigenvalue 0.75, we solve

$$\begin{pmatrix} 0.10 & 0.10 \\ 0.15 & 0.15 \end{pmatrix} \begin{pmatrix} x_1 \\ x_2 \end{pmatrix} = \begin{pmatrix} 0 \\ 0 \end{pmatrix},$$

equivalent to the single equation $x_1 = -x_2$. So an eigenvector is $\begin{pmatrix} -1 \\ 1 \end{pmatrix}$. Let

$$P = \begin{pmatrix} 2 & -1 \\ 3 & 1 \end{pmatrix},$$

the matrix with these eigenvectors as its columns. Then

$$P^{-1}AP = D = \begin{pmatrix} 1 & 0 \\ 0 & 0.75 \end{pmatrix}.$$

We can now use either the method involving powers of the matrix A, or the change of variable method. The former tells us that

$$\begin{aligned} \mathbf{x}_t &= A^t \mathbf{x}_0 \\ &= PD^tP^{-1} \begin{pmatrix} 1 \\ 1 \end{pmatrix} \\ &= \begin{pmatrix} 2 & -1 \\ 3 & 1 \end{pmatrix} \begin{pmatrix} 1 & 0 \\ 0 & 0.75^t \end{pmatrix} \frac{1}{5} \begin{pmatrix} 1 & 1 \\ -3 & 2 \end{pmatrix} \begin{pmatrix} 1 \\ 1 \end{pmatrix} \\ &= \begin{pmatrix} 2 & -1 \\ 3 & 1 \end{pmatrix} \begin{pmatrix} 1 & 0 \\ 0 & 0.75^t \end{pmatrix} \begin{pmatrix} 0.4 \\ -0.2 \end{pmatrix} \\ &= \begin{pmatrix} 2 & -1 \\ 3 & 1 \end{pmatrix} \begin{pmatrix} 0.4 \\ -0.2(0.75)^t \end{pmatrix} \\ &= \begin{pmatrix} 0.8 + 0.2(0.75)^t \\ 1.2 - 0.2(0.75)^t \end{pmatrix}. \end{aligned}$$

So, in millions, the city population is $x_t = 0.8 + 0.2(0.75)^t$ and the suburban population is $y_t = 1.2 - 0.2(0.75)^t$. In the long run, as t tends to infinity, these approach the limiting values of 0.8 and 1.2, respectively. The city population decreases and tends towards 800 000 and the suburban population increases and tends towards 1 200 000.

30.5 Systems of Differential Equations

A change of variable method can also be used to deal with certain systems of *differential* equations. We consider a *linear system* of first-order differential equations for two functions $f(t), g(t)$ of the following form:

$$\begin{aligned} f'(t) &= a_{11}f(t) + a_{12}g(t), \\ g'(t) &= a_{21}f(t) + a_{22}g(t). \end{aligned}$$

If we write

$$\mathbf{f}(t) = \begin{pmatrix} f(t) \\ g(t) \end{pmatrix},$$

then this can be expressed as $\mathbf{f}'(t) = A\mathbf{f}(t)$, where

$$A = \begin{pmatrix} a_{11} & a_{12} \\ a_{21} & a_{22} \end{pmatrix}.$$

We can, just as for recurrence equations, use diagonalisation and a change of variable, so that the system can be re-expressed as an *uncoupled* system involving related functions. Suppose that A can be diagonalised, with $P^{-1}AP = D$, where P is invertible and D is diagonal. This would mean that

$$D = \begin{pmatrix} \lambda_1 & 0 \\ 0 & \lambda_2 \end{pmatrix} \quad \text{and} \quad P = (\mathbf{v}_1 \ \mathbf{v}_2),$$

where λ_1, λ_2 are eigenvalues of A and $\mathbf{v}_1, \mathbf{v}_2$ are corresponding eigenvectors. Let us now introduce two new functions, $F(t)$ and $G(t)$, by setting

$$\begin{pmatrix} f(t) \\ g(t) \end{pmatrix} = P \begin{pmatrix} F(t) \\ G(t) \end{pmatrix}.$$

We can write this as $\mathbf{f}(t) = P\mathbf{F}(t)$, where

$$\mathbf{f}(t) = \begin{pmatrix} f(t) \\ g(t) \end{pmatrix}.$$

An equivalent formulation is $\mathbf{F}(t) = P^{-1}\mathbf{f}(t)$.

The entries of the matrix

$$P = \begin{pmatrix} p_{11} & p_{12} \\ p_{21} & p_{22} \end{pmatrix}$$

are fixed numbers that do not change, so

$$\begin{aligned} \mathbf{f}'(t) = (P\mathbf{F}(t))' &= \begin{pmatrix} p_{11}F(t) + p_{12}G(t) \\ p_{21}F(t) + p_{22}G(t) \end{pmatrix}' \\ &= \begin{pmatrix} p_{11}F'(t) + p_{12}G'(t) \\ p_{21}F'(t) + p_{22}G'(t) \end{pmatrix} \\ &= P \begin{pmatrix} F'(t) \\ G'(t) \end{pmatrix} = P\mathbf{F}'(t). \end{aligned}$$

So the system $\mathbf{f}'(t) = A\mathbf{f}(t)$ is equivalent to

$$P\mathbf{F}'(t) = A\mathbf{f}(t) = A(P\mathbf{F}(t)) = AP\mathbf{F}(t),$$

and hence

$$\mathbf{F}'(t) = P^{-1}(AP\mathbf{F}(t)) = P^{-1}AP\mathbf{F}(t) = D\mathbf{F}(t).$$

The system $\mathbf{F}'(t) = D\mathbf{F}(t)$ is easy to solve: it is uncoupled. Once we have found $\mathbf{F}(t)$, we can then find $\mathbf{f}(t)$. We illustrate with an example.

Example We find the functions $f(t), g(t)$ satisfying $f(0) = 9$, $g(0) = 2$ and

$$\begin{aligned} f'(t) &= 2f(t) + 4g(t), \\ g'(t) &= 3f(t) + 3g(t). \end{aligned}$$

This is $\mathbf{f}'(t) = A\mathbf{f}(t)$, where

$$A = \begin{pmatrix} 2 & 4 \\ 3 & 3 \end{pmatrix}$$

is the matrix we have used throughout our examples in this and the previous chapter. We know that A is diagonalisable and that $P^{-1}AP = D$, where

$$P = \begin{pmatrix} 1 & 4 \\ 1 & -3 \end{pmatrix} \quad \text{and} \quad D = \begin{pmatrix} 6 & 0 \\ 0 & -1 \end{pmatrix}.$$

As proposed, we introduce two functions $F(t), G(t)$ defined by $\mathbf{f}(t) = P\mathbf{F}(t)$ (where $\mathbf{F}(t)$ is, as above, the column vector with entries $F(t)$ and $G(t)$). Then, as explained in the analysis just given, $\mathbf{F}'(t) = D\mathbf{F}(t)$, which means

$$F'(t) = 6F(t) \quad \text{and} \quad G'(t) = -G(t).$$

Now, we use the fact that if a function $h(t)$ satisfies $h'(t) = ah(t)$ for a constant a, then $h(t) = e^{at}h(0)$. (This is easily seen: the equation is a first-order separable equation.) So

$$F(t) = e^{6t}F(0) \quad \text{and} \quad G(t) = e^{-t}G(0).$$

Now, $\mathbf{F}(t) = P^{-1}\mathbf{f}(t)$, so taking $t = 0$ we have

$$\begin{pmatrix} F(0) \\ G(0) \end{pmatrix} = P^{-1}\begin{pmatrix} f(0) \\ g(0) \end{pmatrix} = -\frac{1}{7}\begin{pmatrix} -3 & -4 \\ -1 & 1 \end{pmatrix}\begin{pmatrix} 9 \\ 2 \end{pmatrix} = \begin{pmatrix} 5 \\ 1 \end{pmatrix}$$

and so $F(t) = 5e^{6t}$ and $G(t) = e^{-t}$. Then, using $\mathbf{f}(t) = P\mathbf{F}(t)$, we have

$$\begin{pmatrix} f(t) \\ g(t) \end{pmatrix} = \begin{pmatrix} 1 & 4 \\ 1 & -3 \end{pmatrix}\begin{pmatrix} 5e^{6t} \\ e^{-t} \end{pmatrix} = \begin{pmatrix} 5e^{6t} + 4e^{-t} \\ 5e^{6t} - 3e^{-t} \end{pmatrix}$$

and we have solved the system: the solution is

$$f(t) = 5e^{6t} + 4e^{-t} \quad \text{and} \quad g(t) = 5e^{6t} - 3e^{-t}.$$

Worked Examples

Example 30.1 Suppose that

$$A = \begin{pmatrix} 1 & 4 \\ 1/2 & 0 \end{pmatrix}.$$

Find an expression for the nth power, A^n.

Solution

The characteristic polynomial $|A - \lambda I|$ is $\lambda^2 - \lambda - 2 = (\lambda - 2)(\lambda + 1)$. (Check this!) So the eigenvalues are -1 and 2. Eigenvectors for -1 are given by

$$\begin{aligned} 2x_1 + 4x_2 &= 0, \\ \frac{1}{2}x_1 + x_2 &= 0, \end{aligned}$$

so we may take as eigenvector $\begin{pmatrix} 2 \\ -1 \end{pmatrix}$. Eigenvectors for 2 are given by

$$\begin{aligned} -x_1 + 4x_2 &= 0, \\ \frac{1}{2}x_1 - 2x_2 &= 0, \end{aligned}$$

so we may take $\begin{pmatrix} 4 \\ 1 \end{pmatrix}$. Let P be the matrix

$$P = \begin{pmatrix} 2 & 4 \\ -1 & 1 \end{pmatrix} \quad \text{and its inverse is} \quad P^{-1} = \frac{1}{6}\begin{pmatrix} 1 & -4 \\ 1 & 2 \end{pmatrix}.$$

We have $P^{-1}AP = D = \begin{pmatrix} -1 & 0 \\ 0 & 2 \end{pmatrix}$. The nth power of the matrix A is given by

$$\begin{aligned} A^n &= PD^nP^{-1} \\ &= \frac{1}{6}\begin{pmatrix} 2 & 4 \\ -1 & 1 \end{pmatrix}\begin{pmatrix} (-1)^n & 0 \\ 0 & 2^n \end{pmatrix}\begin{pmatrix} 1 & -4 \\ 1 & 2 \end{pmatrix} \\ &= \frac{1}{6}\begin{pmatrix} 2(-1)^n + 4(2^n) & -8(-1)^n + 8(2^n) \\ -(-1)^n + 2^n & 4(-1)^n + 2(2^n) \end{pmatrix}. \end{aligned}$$

Example 30.2 Sequences x_t and y_t satisfy $x_0 = 1$, $y_0 = -1$ and

$$\begin{aligned} x_t &= 5x_{t-1} - y_{t-1}, \\ y_t &= -3x_{t-1} + 7y_{t-1}. \end{aligned}$$

Find explicit expressions, in terms of t, for x_t and y_t. Describe the behaviour of the sequences as t tends to infinity. How does the ratio x_t/y_t behave?

Solution

Note that we solved the same problem by reduction to a second-order system in Example 29.1. Now we use the methods described in this chapter. This system is $\mathbf{x}_t = A\mathbf{x}_{t-1}$, where

$$A = \begin{pmatrix} 5 & -1 \\ -3 & 7 \end{pmatrix}$$

is the matrix studied in Exercise 29.4. There, we found that A can be diagonalised as follows. If we take

$$P = \begin{pmatrix} 1 & 1 \\ 1 & -3 \end{pmatrix} \quad \text{and} \quad D = \begin{pmatrix} 4 & 0 \\ 0 & 8 \end{pmatrix},$$

then P is invertible and $P^{-1}AP = D$.

We will use the change of variable method to solve this problem (though we could, alternatively, use powers of matrices). We consider

$$\mathbf{x}_t = \begin{pmatrix} x_t \\ y_t \end{pmatrix} \quad \text{and} \quad \mathbf{u}_t = \begin{pmatrix} u_t \\ v_t \end{pmatrix},$$

where $\mathbf{x}_t$ and $\mathbf{u}_t$ are related by $\mathbf{x}_t = P\mathbf{u}_t$. This means that $\mathbf{x}_{t-1} = P\mathbf{u}_{t-1}$ and, substituting this into $\mathbf{x}_t = A\mathbf{x}_{t-1}$, we get $P\mathbf{u}_t = AP\mathbf{u}_{t-1}$, or $\mathbf{u}_t = P^{-1}AP\mathbf{u}_{t-1}$, which means $\mathbf{u}_t = D\mathbf{u}_{t-1}$, as $P^{-1}AP = D$. Using this, we have

$$\begin{pmatrix} u_t \\ v_t \end{pmatrix} = \begin{pmatrix} 4 & 0 \\ 0 & 8 \end{pmatrix} \begin{pmatrix} u_{t-1} \\ v_{t-1} \end{pmatrix} \quad \text{so} \quad u_t = 4u_{t-1} \quad \text{and} \quad v_t = 8v_{t-1},$$

and this pair of recurrence equations can easily be solved to get

$$u_t = 4^t u_0 \quad \text{and} \quad v_t = 8^t v_0.$$

Now, $\mathbf{u}_t = P^{-1}\mathbf{x}_t$, so taking $t = 0$ we have

$$\begin{pmatrix} u_0 \\ v_0 \end{pmatrix} = P^{-1} \begin{pmatrix} x_0 \\ y_0 \end{pmatrix} = \frac{1}{-4} \begin{pmatrix} -3 & -1 \\ -1 & 1 \end{pmatrix} \begin{pmatrix} 1 \\ -1 \end{pmatrix} = \begin{pmatrix} 1/2 \\ 1/2 \end{pmatrix}$$

and so $u_t = 4^t/2$ and $v_t = 8^t/2$. Then, using $\mathbf{x}_t = P\mathbf{u}_t$, we have

$$\begin{pmatrix} x_t \\ y_t \end{pmatrix} = P \begin{pmatrix} u_t \\ v_t \end{pmatrix} = \begin{pmatrix} 1 & 1 \\ 1 & -3 \end{pmatrix} \begin{pmatrix} 4^t/2 \\ 8^t/2 \end{pmatrix} = \frac{1}{2} \begin{pmatrix} 4^t + 8^t \\ 4^t - 3(8^t) \end{pmatrix}.$$

There is an alternative way to find the numbers u_0 and v_0, which avoids calculating P^{-1}. We could proceed as follows. From

$$\begin{pmatrix} u_t \\ v_t \end{pmatrix} = \begin{pmatrix} 4^t u_0 \\ 8^t v_0 \end{pmatrix},$$

and using $\mathbf{x}_t = P\mathbf{u}_t$, we have

$$\begin{pmatrix} x_t \\ y_t \end{pmatrix} = \begin{pmatrix} 1 & 1 \\ 1 & -3 \end{pmatrix} \begin{pmatrix} 4^t u_0 \\ 8^t v_0 \end{pmatrix} = \begin{pmatrix} 4^t u_0 + 8^t v_0 \\ 4^t u_0 - 3(8^t) v_0 \end{pmatrix},$$

which means that

$$x_t = 4^t u_0 + 8^t v_0 \quad \text{and} \quad y_t = 4^t u_0 - 3(8^t) v_0.$$

Then with the initial conditions $x_0 = 1$ and $y_0 = -1$, we get the equations

$$1 = u_0 + v_0 \quad \text{and} \quad -1 = u_0 - 3v_0,$$

which are easily solved to get $u_0 = 1/2$ and $v_0 = 1/2$. Consequently, we find that

$$x_t = \frac{4^t + 8^t}{2} \quad \text{and} \quad y_t = \frac{4^t - 3(8^t)}{2}$$

is the solution, as before.

We can then see that since

$$x_t = \frac{4^t + 8^t}{2} = \frac{8^t}{2}\left[\left(\frac{1}{2}\right)^t + 1\right],$$

$x_t \to \infty$ as $t \to \infty$, whereas since

$$y_t = \frac{4^t - 3(8^t)}{2} = \frac{8^t}{2}\left[\left(\frac{1}{2}\right)^t - 3\right],$$

$y_t \to -\infty$ as $t \to \infty$. Lastly, we can see that

$$\frac{x_t}{y_t} = \frac{\left(\frac{1}{2}\right)^t + 1}{\left(\frac{1}{2}\right)^t - 3} \to -\frac{1}{3},$$

as $t \to \infty$.

Example 30.3 Find the eigenvalues of the matrix

$$A = \begin{pmatrix} 11 & -6 \\ 18 & -10 \end{pmatrix},$$

and find an eigenvector corresponding to each eigenvalue. Hence find an invertible matrix P and a diagonal matrix D such that $P^{-1}AP = D$.

Use your result to find the functions $f(t)$ and $g(t)$ that satisfy the differential equations

$$f'(t) = 11f(t) - 6g(t),$$
$$g'(t) = 18f(t) - 10g(t)$$

and the initial conditions $f(0) = 2$ and $g(0) = 1$.

Solution

To find the eigenvalues of the matrix, we solve the equation

$$\begin{vmatrix} 11-\lambda & -6 \\ 18 & -10-\lambda \end{vmatrix} = 0.$$

This gives us

$$(11 - \lambda)(-10 - \lambda) + 108 = 0,$$

which, multiplying out the brackets, gives us the quadratic equation

$$\lambda^2 - \lambda - 2 = 0 \quad \text{and hence} \quad (\lambda + 1)(\lambda - 2) = 0,$$

so the eigenvalues are -1 and 2. To find the corresponding eigenvectors we seek a non-zero vector $\mathbf{x}$, which is a solution to the equation $(A - \lambda I)\mathbf{x} = \mathbf{0}$.

For $\lambda = -1$, we solve

$$\begin{pmatrix} 12 & -6 \\ 18 & -9 \end{pmatrix} \begin{pmatrix} x \\ y \end{pmatrix} = \mathbf{0},$$

which is equivalent to $y = 2x$, so an eigenvector is $\begin{pmatrix} 1 \\ 2 \end{pmatrix}$.

For $\lambda = 2$, we solve

$$\begin{pmatrix} 9 & -6 \\ 18 & -12 \end{pmatrix} \begin{pmatrix} x \\ y \end{pmatrix} = \mathbf{0},$$

which is equivalent to $y = \frac{3}{2}x$, so an eigenvector is $\begin{pmatrix} 2 \\ 3 \end{pmatrix}$.

So, if we take

$$P = \begin{pmatrix} 1 & 2 \\ 2 & 3 \end{pmatrix} \quad \text{and} \quad D = \begin{pmatrix} -1 & 0 \\ 0 & 2 \end{pmatrix},$$

then we have an invertible matrix P and a diagonal matrix D such that $P^{-1}AP = D$.

The given system of differential equations can be written as $\mathbf{f}'(t) = A\mathbf{f}(t)$, where

$$\mathbf{f}(t) = \begin{pmatrix} f(t) \\ g(t) \end{pmatrix}.$$

So, using our answer above, we can define two new functions $F(t)$ and $G(t)$ by

$$\mathbf{F}(t) = \begin{pmatrix} F(t) \\ G(t) \end{pmatrix} = P^{-1}\mathbf{f}(t)$$

or, equivalently, $\mathbf{f}(t) = P\mathbf{F}(t)$. Substituting this into $\mathbf{f}'(t) = A\mathbf{f}(t)$, we then get $P\mathbf{F}'(t) = AP\mathbf{F}(t)$, which means $\mathbf{F}'(t) = P^{-1}AP\mathbf{F}(t) = D\mathbf{F}(t)$. Using this, we then have

$$\begin{pmatrix} F'(t) \\ G'(t) \end{pmatrix} = \begin{pmatrix} -1 & 0 \\ 0 & 2 \end{pmatrix} \begin{pmatrix} F(t) \\ G(t) \end{pmatrix},$$

so

$$F'(t) = -F(t) \quad \text{and} \quad G'(t) = 2G(t),$$

and this pair of differential equations can easily be solved to get

$$F(t) = e^{-t}F(0) \quad \text{and} \quad G(t) = e^{2t}G(0).$$

We could, at this stage, use $\mathbf{F}(0) = P^{-1}\mathbf{f}(0)$ to determine $F(0)$ and $G(0)$ from the initial values $f(0)$ and $g(0)$. Alternatively, we can proceed as follows, determining these values slightly later in the calculation.

Using $\mathbf{f}(t) = P\mathbf{F}(t)$, we have

$$\begin{pmatrix} f(t) \\ g(t) \end{pmatrix} = \begin{pmatrix} 1 & 2 \\ 2 & 3 \end{pmatrix} \begin{pmatrix} e^{-t}F(0) \\ e^{2t}G(0) \end{pmatrix} = \begin{pmatrix} e^{-t}F(0) + 2e^{2t}G(0) \\ 2e^{-t}F(0) + 3e^{2t}G(0) \end{pmatrix},$$

which means that

$$f(t) = e^{-t}F(0) + 2e^{2t}G(0) \quad \text{and} \quad g(t) = 2e^{-t}F(0) + 3e^{2t}G(0)$$

is the general solution to our system of differential equations. Then, with the initial conditions $f(0) = 2$ and $g(0) = 1$, we get the equations

$$2 = F(0) + 2G(0) \qquad \text{and} \qquad 1 = 2F(0) + 3G(0),$$

which are easily solved to get $F(0) = -4$ and $G(0) = 3$. So

$$f(t) = -4e^{-t} + 6e^{2t} \quad \text{and} \quad g(t) = -8e^{-t} + 9e^{2t}$$

is our particular solution.

Main Topics

- the powers of a matrix and calculating these via diagonalisation
- solving systems of recurrence equations using matrix powers
- solving systems of recurrences using diagonalisation and change of variable
- solving systems of first-order differential equations using diagonalisation and change of variable

Key Terms, Notations and Formulae

- the nth power of a matrix A is $A^n = \underbrace{AA A \cdots A}_{n \text{ times}}$
- solving recurrence systems using powers of matrices: $\mathbf{x}_t = A\mathbf{x}_{t-1}$ has solution $\mathbf{x}_t = A^t\mathbf{x}_0$
- if A can be diagonalised as $P^{-1}AP = D$, then $A^t = PD^tP^{-1}$
- solving recurrence systems using change of variable: if $\mathbf{x}_t = A\mathbf{x}_{t-1}$ and if A can be diagonalised as $P^{-1}AP = D$, then $\mathbf{x}_t = P\mathbf{u}_t$ and $\mathbf{u}_t = D\mathbf{u}_{t-1}$ (which is easily solved)
- solving systems of first-order differential equations using change of variable: if $\mathbf{f}'(t) = A\mathbf{f}(t)$ and if A can be diagonalised as $P^{-1}AP = D$, then $\mathbf{f}(t) = P\mathbf{F}(t)$ and $\mathbf{F}'(t) = D\mathbf{F}(t)$ (which is easily solved)

Exercises

30.1 Suppose that

$$A = \begin{pmatrix} 7 & 1 \\ 3 & 5 \end{pmatrix}.$$

Find an expression for the nth power, A^n, of A, simplifying your answer as far as possible.

30.2 Find the eigenvalues of the matrix

$$A = \begin{pmatrix} -3 & -4 \\ 2 & 3 \end{pmatrix},$$

and find an eigenvector corresponding to each eigenvalue. Hence find an invertible matrix P and a diagonal matrix D such that $P^{-1}AP = D$.

Use your result to find the sequences x_t and y_t that satisfy the system of recurrence equations

$$x_t = -3x_{t-1} - 4y_{t-1},$$
$$y_t = 2x_{t-1} + 3y_{t-1}$$

and the initial conditions $x_0 = 2$ and $y_0 = 1$.

Describe how these sequences behave as t increases.

30.3 Find the eigenvalues of the matrix

$$A = \begin{pmatrix} 1 & 2 \\ 0 & -1 \end{pmatrix},$$

and find an eigenvector corresponding to each eigenvalue. Hence find an invertible matrix P and a diagonal matrix D such that $P^{-1}AP = D$.

Hence, or otherwise, find the functions $f(t)$ and $g(t)$ which satisfy the system of differential equations

$$f'(t) = f(t) + 2g(t) \quad \text{and} \quad g'(t) = -g(t),$$

and for which $f(0) = 1$ and $g(0) = 1$.

Problems

30.1 For the matrix

$$A = \begin{pmatrix} 2 & -3 \\ 1 & -2 \end{pmatrix},$$

find an invertible matrix P and a diagonal matrix D such that $P^{-1}AP = D$.

Use your answer to find a general expression for the nth power of A, A^n. Simplify your answer as far as possible for the cases when n is even and when n is odd.

30.2 Find the sequences x_t and y_t which are such that $x_0 = 1$, $y_0 = 4$ and

$$x_t = -5x_{t-1} - 4y_{t-1},$$
$$y_t = 2x_{t-1} + y_{t-1}.$$

30.3 Find an invertible matrix P and a diagonal matrix D such that $P^{-1}AP = D$, where A is the matrix

$$\begin{pmatrix} 7 & 1 \\ 3 & 5 \end{pmatrix}.$$

Hence, or otherwise, find the functions $f(t)$ and $g(t)$ which are such that $f(0) = 2$, $g(0) = 1$ and

$$f'(t) = 7f(t) + g(t),$$
$$g'(t) = 3f(t) + 5g(t).$$

Appendix A Solutions to Exercises

Chapter 1

Exercise

1.1 The supply and demand functions are

$$q^S(p) = 3p - 5 \quad \text{and} \quad q^D(p) = 145 - 2p.$$

By the standard method we obtain the equilibrium values $p^* = 30$ and $q^* = 85$.

When the excise tax is imposed, the equilibrium values q^T and p^T are given by

$$q^T = 3(p^T - T) - 5 = 145 - 2p^T.$$

It follows that $5p^T = 150 + 3T$ and $p^T = 30 + (3/5)T$. The corresponding quantity sold is $q^T = 145 - 2p^T = 85 - (6/5)T$ thousand bottles.

1.2 The supply and demand functions are

$$q^S(p) = \frac{p-2}{3} \quad \text{and} \quad q^D(p) = 6 - p.$$

By the standard method we obtain the equilibrium values $p^* = 5$ and $q^* = 1$.

When the excise tax is imposed, the equilibrium values q^T and p^T are given by

$$q^T = \frac{p^T - T - 2}{3} = 6 - p^T.$$

It follows that $4p^T = 20 + T$ and $p^T = 5 + T/4$. The corresponding quantity is $q^T = 6 - p^T = 1 - T/4$.

The amount of money the government obtains from taxing this good is given by

$$Tq^T = T\left(1 - \frac{T}{4}\right) = T - \frac{T^2}{4},$$

as this is just the tax per unit multiplied by the number of units sold.

1.3 We solve $240 - 4p = 8p$ to find that the equilibrium price is $p^* = 20$. It follows that the equilibrium quantity is $q^* = 8(20) = 160$.

When the tax is imposed, the price then satisfies

$$240 - 4p = 8p(1 - r).$$

Note that it is only the supply side that changes and that it changes to reflect the fact that, as far as suppliers are concerned, the effective price is $p(1-r)$. Solving, we find

$$p = \frac{60}{3-2r}.$$

Then

$$q = 8p(1-r) = \frac{480(1-r)}{3-2r}.$$

(We could, instead, use the demand equation to determine q, leading to the same result.)

Chapter 2

Exercise

2.1 We have $A \cap B = \{2, 5\}$, since this is the set of objects that belong to *both* A and B. Then

$$(A \cap B) \cup C = \{2, 5\} \cup \{1, 7, 8, 11\} = \{1, 2, 5, 7, 8, 11\}.$$

Also

$$A \cup C = \{1, 2, 3, 5, 7, 8, 11\} \quad \text{and} \quad B \cup C = \{1, 2, 4, 5, 7, 8, 11\}$$

and so

$$(A \cup C) \cap (B \cup C) = \{1, 2, 3, 5, 7, 8, 11\} \cap \{1, 2, 4, 5, 7, 8, 11\} = \{1, 2, 5, 7, 8, 11\}.$$

We see here that $(A \cap B) \cup C = (A \cup C) \cap (B \cup C)$. In fact, this is always true, not just for these three particular sets.

2.2 If $y = f(x) = 3x + 2$ then we may solve this for x by noting that $x = (y-2)/3$. It follows that $f^{-1}(y) = (y-2)/3$ or, equivalently, $f^{-1}(x) = (x-2)/3$.

2.3 The composition fg is

$$(fg)(x) = f(g(x)) = \sqrt{g(x)} = \sqrt{x^2+1}.$$

The composition gf is

$$(gf)(x) = g((f(x)) = g(\sqrt{x}) = (\sqrt{x})^2 + 1 = x + 1.$$

2.4 We can see that $h = fg$ where $g(x) = 1 + x^2$ and $f(x) = 1/x$.

2.5 The graph $y = x^2 + 4x + 3$ is a quadratic, with a positive x^2 term, and hence it has the parabolic 'U' shape. To locate its position, we find where it crosses the axes. It crosses the y-axis when $x = 0$, and hence at $(0, 3)$. To find where it crosses the x-axis (if at all), we need to solve $y = 0$; that is $x^2 + 4x + 3 = 0$. There are at least three ways we can do this.

We could spot that this factorises as $(x+3)(x+1) = 0$, so that the solutions are $x = -3$ and $x = -1$.

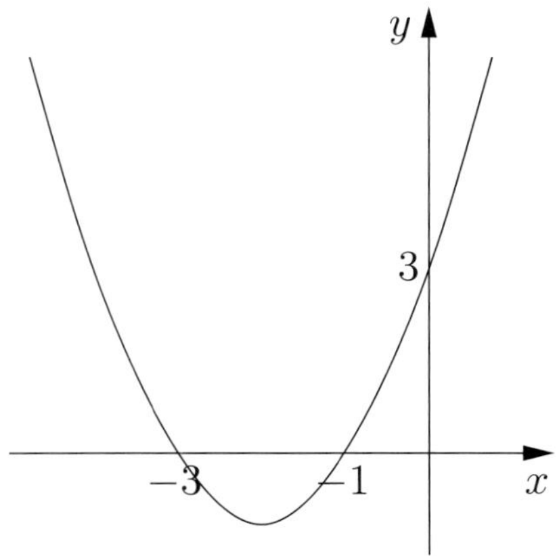

Figure A.1 The sketch for Exercise 2.5

Alternatively, we can use the formula for the solutions of a quadratic equation, with $a = 1, b = 4, c = 3$. This gives

$$\frac{-b \pm \sqrt{b^2 - 4ac}}{2a} = (-4 \pm \sqrt{4})/2 = -3, -1.$$

Or we could complete the square: $x^2 + 4x + 3 = (x + 2)^2 - 1$, so $x^2 + 4x + 3 = 0$ if and only if $x + 2 = \pm 1$, which gives $x = -3, -1$.

With this information, we can sketch the curve as illustrated in Figure A.1.

Chapter 3

Exercise

3.1 The recurrence is of the form $x_n = ax_{n-1} + b$ for $n \geq 1$, where $a = 1/3$ and $b = 2$. We have the time-independent solution $x^* = b/(1 - a)$, so $x^* = 2/(2/3) = 3$. The general formula for the solution is $x_n = x^* + (x_0 - x^*)a^n$. So

$$x_n = 3 + (4 - 3)\left(\frac{1}{3}\right)^n = 3 + \left(\frac{1}{3}\right)^n.$$

As $n \to \infty$, $(1/3)^n$ tends to 0 and decreases, so the sequence decreases to 3.

3.2 Let y_N be the number of fish N years from now.

We have $y_0 = 10\,000$ and, for $n \geq 1$:

$$y_n = 0.95y_{n-1} + 1000.$$

This is of the form $y_n = ay_{n-1} + b$ for $n \geq 1$, where $a = 0.95$ and $b = 1000$.

We have the time-independent solution

$$y^* = \frac{b}{1 - a} = \frac{1000}{1 - 0.95} = 20\,000.$$

The general formula for the solution is $y_N = y^* + (y_0 - y^*)(0.95)^N$. So

$$y_N = 20\,000 - 10\,000(0.95)^N.$$

This increases with N and tends to 20 000 in the long run.

Chapter 4

Exercise

4.1 Let y_n be the amount of money after the nth deposit. Then $y_0 = 0$ and we have, for $n \geq 1$:

$$y_n = 1.04y_{n-1} + A.$$

That is, we have the recurrence $y_n = ay_{n-1} + b$ for $n \geq 1$, where $a = 1.04$ and $b = A$. We have the time-independent solution

$$y^* = \frac{b}{1-a} = \frac{A}{1-1.04} = -25A.$$

The general formula for the solution is $y_N = y^* + (y_0 - y^*)a^N$. So

$$y_N = -25A + 25A(1.04)^N.$$

4.2 Let y_n be the amount of money owed after the nth repayment.

The amount owed after the nth payment is the $(n-1)$th amount, y_{n-1}, plus interest (which is $0.02y_{n-1}$), minus the payment D; so $y_n = ay_{n-1} + b$ for $n \geq 1$, where $a = 1.02$ and $b = -D$. In particular, $y_1 = 10\,000(1.02) - D$.

We have $y_0 = 10\,000$. The time-independent solution is

$$y^* = \frac{b}{1-a} = -D/(1-1.02) = 50D.$$

The general formula for the solution is $y_n = y^* + (y_0 - y^*)a^n$. So

$$y_n = 50D + (10\,000 - 50D)(1.02)^n.$$

After 25 repayments, the balance is zero and hence

$$0 = 50D + (10\,000 - 50D)(1.02)^{25}.$$

So

$$50D((1.02)^{25} - 1) = 10\,000$$

and hence

$$D = \frac{200(1.02)^{25}}{1.02^{25} - 1} = \frac{200(1.64)}{1.64 - 1} = \frac{200(164)}{64} = 512.5.$$

4.3 Let y_t be the balance of the prize fund after t full years. Then we have, for $t \geq 1$:

$$y_t = (1.05)y_{t-1} - P$$

and suppose that $y_0 = D$, which is what we are trying to find. Thus, we have the recurrence $y_n = ay_{n-1} + b$ for $n \geq 1$, where $a = 1.05$ and $b = -P$. We have the time-independent solution

$$y^* = \frac{b}{1-a} = \frac{-P}{1-1.05} = 20P.$$

The general formula for the solution is $y_N = y^* + (y_0 - y^*)a^N$. So

$$y_N = 20P + (D - 20P)(1.05)^N.$$

We need to have $y_N \geq 0$, so that the prize can be awarded the required number of times. So, with a bit of work, we see that

$$D \geq 20P\left(1 - \frac{1}{(1.05)^N}\right)$$

and this quantity is the minimum required donation.

Chapter 5

Exercise

5.1 The equilibrium price is given by $0.05p - 4 = 20 - 0.15p$, so $p = 120$. The inverse demand function is obtained by solving the equation $q = 20 - 0.15p$ for p, so

$$p^D(q) = \frac{400}{3} - \frac{20}{3}q.$$

Now, using the given equations, we have

$$p_t = \frac{400}{3} - \frac{20}{3}q_t \quad \text{and} \quad q_t = 0.05p_{t-1} - 4$$

and, eliminating q_t, we get

$$p_t = \frac{400}{3} - \frac{20}{3}(0.05p_{t-1} - 4) = 160 - \frac{1}{3}p_{t-1}.$$

This has time-independent solution

$$p^* = \frac{160}{1-(-1/3)} = 120,$$

which is the equilibrium price. The solution for p_t is

$$p_t = 120 + (p_0 - 120)\left(-\frac{1}{3}\right)^t.$$

5.2 The equilibrium price is given by $p - 4 = 20 - p$, so $p = 12$. The inverse demand function is $p^D(q) = 20 - q$ and so, using the cobweb model, we have

$$p_t = 20 - q_t \quad \text{and} \quad q_t = p_{t-1} - 4$$

and, eliminating q_t, we get

$$p_t = 20 - (p_{t-1} - 4) = 24 - p_{t-1}.$$

This has time-independent solution

$$p^* = \frac{24}{1-(-1)} = 12,$$

which is the equilibrium price. Given $p_0 = 10$, the solution for p_t is therefore

$$p_t = 12 + (10 - 12)(-1)^t = 12 - 2(-1)^t.$$

The behaviour of p_t is determined by the behaviour of the term $(-1)^t$. This oscillates constantly between 1 and -1 as t tends to infinity; thus, p_t oscillates constantly between 10 and 14. When this happens, we say that the market equilibrium is *cyclic*.

Chapter 6

Exercise

6.1 By the composite function rule

$$\left((3x+7)^{15}\right)' = 15(3x+7)^{14}(3) = 45(3x+7)^{14}.$$

We have $\sqrt{x^4+1} = (x^4+1)^{1/2}$. By the composite function rule

$$\left((x^4+1)^{1/2}\right)' = \frac{1}{2}(x^4+1)^{-1/2}(4x^3) = \frac{2x^3}{\sqrt{x^4+1}}.$$

6.2 The derivative of $f(x) = \sqrt{x}$ is $f'(x) = 1/(2\sqrt{x})$. Therefore, by the approximation

$$\Delta f \simeq f'(x)\Delta x,$$

we have

$$f(101) - f(100) \simeq f'(100)(1) = \frac{1}{2\sqrt{100}} = \frac{1}{20}.$$

The approximate change is therefore $1/20 = 0.05$. (Incidentally, the exact change is $\sqrt{101} - \sqrt{100}$, which is 0.04987562, so the approximation is good.)

Generally, if n is large and x increases from n to $n+1$, then the change is given approximately by

$$f(n+1) - f(n) \simeq f'(n)(1) = \frac{1}{2\sqrt{n}},$$

as required.

Chapter 7

Exercise

7.1 If we call the larger principal P, then the balance is $Pe^{0.02t}$ after t years. The smaller principal (being 80% of the larger principal) will then be $0.8P$ and its balance will be $0.8Pe^{0.05t}$ after t years. The two balances are the same when

$$Pe^{0.02t} = 0.8Pe^{0.05t}, \quad \text{that is} \quad e^{-0.03t} = 0.8.$$

Using the definition of the logarithm, this gives us

$$-0.03t = \ln(0.8),$$

and so t is approximately 7.438 years. Thus, approximately, the balances will be the same after 7 years and 5 months.

7.2 We have $\ln g(x) = \ln x^x = x \ln x$. We now differentiate both sides of the equation $\ln g(x) = x \ln x$, using the composite function rule for the left-hand side and the product rule for the right-hand side:

$$\left(\frac{1}{g(x)}\right) g'(x) = 1 + \ln x.$$

So $g'(x) = (1 + \ln x)x^x$.

7.3 Let

$$g(x) = \frac{1}{x} \ln\left(\frac{1}{x}\right) \quad \text{and} \quad h(x) = \left(1 - \frac{1}{x}\right) \ln\left(1 - \frac{1}{x}\right).$$

We have

$$g'(x) = -\frac{1}{x^2} \ln\left(\frac{1}{x}\right) + \frac{1}{x}\left(\frac{1}{1/x}\right)\left(-\frac{1}{x^2}\right) = -\frac{1}{x^2} \ln\left(\frac{1}{x}\right) - \frac{1}{x^2}$$

and

$$h'(x) = \frac{1}{x^2} \ln\left(1 - \frac{1}{x}\right) + \left(1 - \frac{1}{x}\right)\left(\frac{1}{1 - 1/x}\right)\left(\frac{1}{x^2}\right) = \frac{1}{x^2} \ln\left(1 - \frac{1}{x}\right) + \frac{1}{x^2}.$$

So

$$f'(x) = g'(x) + h'(x) = \frac{1}{x^2}\left(\ln\left(1 - \frac{1}{x}\right) - \ln\left(\frac{1}{x}\right)\right).$$

This simplifies to the required expression because

$$\begin{aligned} \ln\left(1 - \frac{1}{x}\right) - \ln\left(\frac{1}{x}\right) &= \ln\left(\frac{x-1}{x}\right) - \ln\left(\frac{1}{x}\right) \\ &= \ln(x-1) - \ln x - (\ln 1 - \ln x) \\ &= \ln(x-1). \end{aligned}$$

Chapter 8

Exercise

8.1 The derivative is

$$f'(x) = 6x^2 - 18x = 6x(x-3).$$

The solutions to $f'(x) = 0$ are 0 and 3 and these are therefore the critical points.

We now determine the nature of these critical points. Let's examine the sign of $f'(x)$ in their vicinity.

- For $x = 0$, we have $f'(x) = 6x(x-3)$, which is positive for $x < 0$ (since it is then the product of two negative numbers). For x just greater than 0, $x > 0$ and $x-3 < 0$, so that $f'(x) < 0$. (*Note:* we are interested only in the signs of $f'(x)$ *just* to either side of the critical point, in its immediate vicinity.) Thus, at $x = 0, f'$ changes sign from positive to negative and hence $x = 0$ is a maximum point.
- For $x = 3$, when x is just less than 3, $6x(x-3) < 0$ and when $x > 3$, $6x(x-3) > 0$; thus, since f' changes from negative to positive around the point, $x = 3$ is a minimum point.

Alternatively, we could use the second derivative. We have

$$f''(x) = (6x^2 - 18x)' = 12x - 18.$$

Since $f''(0) = 18 < 0$, $x = 0$ is a maximum point. Since $f''(3) = 18 > 0$, $x = 3$ is a minimum point.

Now let us sketch the curve $y = f(x)$. We saw that it has a maximum point at $x = 0$ and a minimum point at $x = 3$. The corresponding values of $f(x)$ are $f(0) = 1$ and $f(3) = -26$.

The curve crosses the y-axis when $y = f(0) = 1$. It crosses the x-axis when $2x^3 - 9x^2 + 1 = 0$. Now, this is not an easy equation to solve! However, we can get some idea of the points where it crosses the x-axis by considering the shape of the curve. Note that as $x \to \infty, f(x) \to \infty$ and as $x \to -\infty, f(x) \to -\infty$. Also, we have $f(0) > 0$ and $f(3) < 0$. These observations imply that the graph must cross the x-axis somewhere to the left of 0 (since it must move from negative y-values to a positive y-value), it must cross again somewhere between 0 and 3 (since $f(0) > 0$ and $f(3) < 0$) and it must cross again at some point greater than 3 (because $f(x) \to \infty$ as $x \to \infty$ and hence must be positive from some point).

We therefore have the sketch in Figure A.2 (in which we have omitted the tricky x-axis crossings).

8.2 We first calculate the derivative, using the product rule to get

$$f'(x) = e^{-x^2} - (2x)xe^{-x^2} = e^{-x^2}(1 - 2x^2).$$

There are two solutions of $f'(x) = 0$, namely $x = 1/\sqrt{2}$ and $x = -1/\sqrt{2}$. (Note that e^{-x^2} is never equal to 0.) In other words, these values of x give the critical points.

Let's examine the sign of $f'(x)$ as x goes from just less than $-1/\sqrt{2}$ to just greater than $-1/\sqrt{2}$. For $x < -1/\sqrt{2}$, $1 - 2x^2 < 0$ and so $f'(x) < 0$, while for x just greater

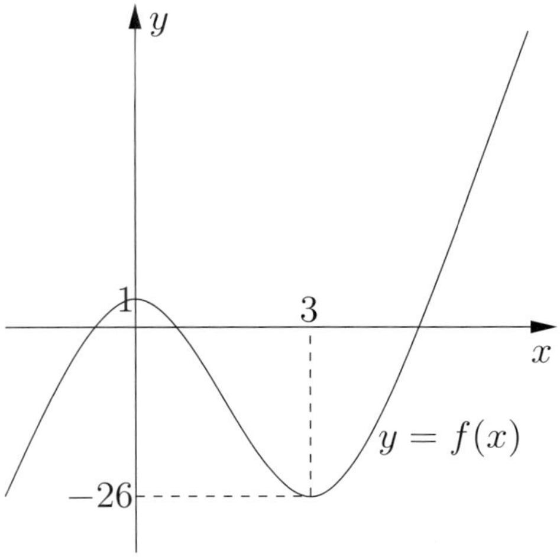

Figure A.2 The sketch for Exercise 8.1

than $-1/\sqrt{2}$, $1-2x^2 > 0$ and $f'(x) > 0$. It follows that $-1/\sqrt{2}$ is a minimum point. In a similar way, one can check – and you should do this – that for x just less than $1/\sqrt{2}$, the derivative is positive and for x just greater than $1/\sqrt{2}$, the derivative is negative, so that we may deduce $1/\sqrt{2}$ is a maximum point.

Now for the second-derivative test. We may calculate $f''(x)$, using the product rule, to get

$$f''(x) = -2xe^{-x^2}(1 - 2x^2) - 4xe^{-x^2}.$$

For the second-derivative test, we need to determine, at each critical point, whether the second derivative is positive or negative. It is its sign that matters, not its value. Notice that the first term in the expression for f'' will be 0 when we take x to be one of the critical points (because each of these satisfies $1 - 2x^2 = 0$). Note also that e^{-x^2} is always positive. So we can see that $f''(-1/\sqrt{2}) > 0$, so this point is a minimum; and $f''(1/\sqrt{2}) < 0$, so this point is a maximum.

The question asked us to take both approaches to determining the nature of critical points, but of course, in practice, we would only need to take one approach.

8.3 We have

$$f'(x) = (2x + 10)e^{-x} - (x^2 + 10x - 10)e^{-x} = (20 - 8x - x^2)e^{-x}.$$

Critical points are given by $f'(x) = 0$. This is $x^2 + 8x - 20 = 0$ (noting that $e^{-x} \neq 0$), which is $(x + 10)(x - 2) = 0$. So the critical points are $x = -10$ and $x = 2$.

We can check their nature by examining the behaviour of the sign of the derivative around each point, or we can use the second derivative test.

With the second-derivative approach, we have

$$f''(x) = -e^{-x}(20 - 8x - x^2) - (2x + 8)e^{-x}.$$

When $x = -10$, we have

$$f''(-10) = e^{10}(0) + 12e^{10} > 0,$$

so this is a minimum point and, when $x = 2$:

$$f''(2) = e^{-2}(0) - 12e^{-2} < 0,$$

so this is a maximum point.

If using the sign of the derivative, we examine how the sign of the derivative changes around the critical points. At $x = -10$, this changes from $-$ to $+$ so this point is a minimum and at $x = 2$, this changes from $+$ to $-$ so this point is a maximum.

Chapter 9

Exercise

9.1 As $q \geq 0$, the demand equation gives us

$$q = \sqrt{31}(4 + p^3)^{-1/2},$$

and so, differentiating this with the composite function rule, we get

$$q' = \sqrt{31}\left(-\frac{1}{2}\right)(4 + p^3)^{-3/2}(3p^2) = -\frac{3\sqrt{31}p^2}{2(4 + p^3)^{3/2}}$$

and so the elasticity of demand is

$$\varepsilon(p) = -\frac{q'p}{q} = \frac{3\sqrt{31}p^3/[2(4 + p^3)^{3/2}]}{\sqrt{31}/(4 + p^3)^{1/2}} = \frac{3}{2}\left(\frac{p^3}{4 + p^3}\right).$$

Indeed, as $p \geq 0$, we have

$$\frac{p^3}{4 + p^3} < 1,$$

and so $\varepsilon(p) < 3/2$ (i.e., the elasticity of demand never exceeds $3/2$).

9.2 (a) The inverse demand function is

$$p^D(q) = \frac{300 - q}{20} = 15 - 0.05q.$$

(b) The profit function is

$$\Pi(q) = qp^D(q) - C(q) = q(15 - 0.05q) - (q + 0.02q^2) = 14q - 0.07q^2.$$

(c) We have $\Pi'(q) = 14 - 0.14q$, so $q = 100$ is a critical point. The second derivative of Π is $\Pi''(q) = -0.14$, which is negative, so the critical point is a maximum. The value of the profit there is $\Pi(100) = 1400 - 700 = 700$, whereas $\Pi(0) = 0$ and $\Pi(200) = 0$. Since the maximum profit in the interval $[0, 200]$ must be either at a maximum point or an end-point, it follows that the maximum profit is 700, obtained when $q = 100$.
(d) The price when $q = 100$ is $p^D(100) = 15 - (0.05)(100) = 10$.

Chapter 10

Exercise

10.1 The cost function is $C(q) = 50\,000 + 25q + 0.001q^2$. The fixed cost is obtained by setting $q = 0$, giving $FC = 50\,000$. The marginal cost is $MC = 25 + 0.002q$. The marginal cost is 25.2 if the output is 100, but it rises to 45 if the output is 10 000.

10.2 (a) The fixed cost is $C(0) = 4$.
(b) Suppose the firm is in production; that is $0 < q \leq 5$, where 5 is the given upper limit on its capacity. Then $p = C'(q)$ and the profit function is $\Pi(q) = qC'(q) - C(q)$. Hence

$$\Pi(q) = q(3q^2 - 6q + 5) - (q^3 - 3q^2 + 5q + 4) = 2q^3 - 3q^2 - 4.$$

(c) The startup point is the value q_s such that $\Pi(q_s) = \Pi(0) = -C(0)$. Solving the equation $2q^3 - 3q^2 - 4 = -4$, or $q^2(2q - 3) = 0$, it follows that $q_s = 3/2$. The corresponding value of p is $p_s = C'(3/2) = 11/4$.
(d) The breakeven point q_b satisfies $\Pi(q_b) = 0$. We therefore need to solve the equation $2q^3 - 3q^2 - 4 = 0$. Trial and error reveals the factorisation

$$2q^3 - 3q^2 - 4 = (q - 2)(2q^2 + q + 2).$$

The quadratic equation $2q^2 + q + 2 = 0$ has no solutions, since $1^2 - 4(2)(2) < 0$, so 2 is the only solution and the breakeven point is $q_b = 2$.
(e) The startup point is $q_s = 3/2$, and the corresponding going price is $p_s = 11/4$. Thus the firm supplies nothing when $p < 11/4$, and for $p \geq 11/4$ the amount q supplied is related to p by

$$p = C'(q) = 3q^2 - 6q + 5.$$

But this can hold only when q is in the feasible range $q \leq 5$ and $q = 5$ corresponds to a price $p = C'(5) = 50$. So, when $p > 50$, the firm will produce its full capacity, 5. Thus, the firm's supply set is the union of three pieces:

- the points $(0, p)$ for $0 < p < 11/4$;
- the points (q, p) such that $p = 3q^2 - 6q + 5$ for $11/4 \leq p \leq 50$;
- the points $(5, p)$ for $p > 50$.

Chapter 11

Exercise

11.1 We have $f(x, y) = x^3y - xy^{-1}$, so

$$\frac{\partial f}{\partial x} = 3x^2y - y^{-1} = 3x^2y - \frac{1}{y} \quad \text{and} \quad \frac{\partial f}{\partial y} = x^3 - x(-y^{-2}) = x^3 + \frac{x}{y^2}.$$

11.2 We have

$$\frac{\partial f}{\partial x} = y\ln\left(\frac{x}{y}\right) + xy\left(\frac{y}{x}\right)\left(\frac{1}{y}\right) = y\ln\left(\frac{x}{y}\right) + y$$

and

$$\frac{\partial f}{\partial y} = x \ln\left(\frac{x}{y}\right) + xy\left(\frac{y}{x}\right)\left(-\frac{x}{y^2}\right) = x \ln\left(\frac{x}{y}\right) - x.$$

So we can see that

$$x\frac{\partial f}{\partial x} + y\frac{\partial f}{\partial y} = x\left(y \ln\left(\frac{x}{y}\right) + y\right) + y\left(x \ln\left(\frac{x}{y}\right) - x\right) = 2xy \ln\left(\frac{x}{y}\right) = 2f(x,y).$$

11.3 By the chain rule:

$$\begin{aligned}\frac{dF}{dt} &= \frac{\partial f}{\partial x}\frac{dx}{dt} + \frac{\partial f}{\partial y}\frac{dy}{dt}\\ &= 2xy(3) + x^2(2t)\\ &= 6(2+3t)(t^2+1) + 2t(2+3t)^2\\ &= 36t^3 + 36t^2 + 26t + 12.\end{aligned}$$

Chapter 12

Exercise

12.1 The equation defining y implicitly as a function of x is of the form $g(x,y) = 1$, where $g(x,y) = x^2y^3 - 6x^3y^2 + 2xy$. According to the formula

$$\frac{dy}{dx} = -\frac{\partial g/\partial x}{\partial g/\partial y}$$

and so, with

$$\frac{\partial g}{\partial x} = 2xy^3 - 18x^2y^2 + 2y \quad \text{and} \quad \frac{\partial g}{\partial y} = 3x^2y^2 - 12x^3y + 2x,$$

we get

$$\frac{dy}{dx} = -\frac{2xy^3 - 18x^2y^2 + 2y}{3x^2y^2 - 12x^3y + 2x}.$$

We want to calculate this derivative when $x = 1/2$. Clearly we first need to find the corresponding value of y. Putting $x = 1/2$ into the defining equation $x^2y^3 - 6x^3y^2 + 2xy = 1$, we obtain

$$\tfrac{1}{4}y^3 - \tfrac{3}{4}y^2 + y = 1, \quad \text{or} \quad y^3 - 3y^2 + 4y - 4 = 0.$$

This equation factorises as

$$(y-2)(y^2 - y + 2) = 0.$$

(Check this!) The quadratic $y^2 - y + 2$ has negative discriminant and so has no zeroes. It follows that when $x = 1/2$, $y = 2$. Substituting these values into the expression for dy/dx, we see that the derivative when $x = 1/2$ is 6.

12.2 To show that the function is homogeneous of degree 5, we have

$$\begin{aligned}f(cx,cy) &= ((cx)^4+(cy)^4)^{1/2}\left((cx)^2(cy)+(cx)^3\right)\\ &= (c^4)^{1/2}(x^4+y^4)^{1/2}c^3(x^2y+x^3)\\ &= c^5f(x,y).\end{aligned}$$

Now, we have

$$\frac{\partial f}{\partial x} = 2x^3(x^4+y^4)^{-1/2}(x^2y+x^3)+(x^4+y^4)^{1/2}(2xy+3x^2)$$

and

$$\frac{\partial f}{\partial y} = 2y^3(x^4+y^4)^{-1/2}(x^2y+x^3)+x^2(x^4+y^4)^{1/2}.$$

It follows that

$$\begin{aligned}x\frac{\partial f}{\partial x}+y\frac{\partial f}{\partial y} &= (2x^4+2y^4)(x^4+y^4)^{-1/2}(x^2y+x^3)+(x^4+y^4)^{1/2}(2x^2y+3x^3+x^2y)\\ &= 2(x^4+y^4)(x^4+y^4)^{-1/2}(x^2y+x^3)+(x^4+y^4)^{1/2}(3x^2y+3x^3)\\ &= 2(x^4+y^4)^{1/2}(x^2y+x^3)+3(x^4+y^4)^{1/2}(x^2y+x^3)\\ &= 5(x^4+y^4)^{1/2}(x^2y+x^3) = 5f(x,y),\end{aligned}$$

as predicted by Euler's Theorem.

12.3 We have

$$f(cx,cy) = c^{a+1}xy^a\sin\left(\frac{x}{y}\right)+c^{b-1}\frac{x^b}{y}.$$

For the function to be homogeneous, we must have $a+1=b-1$, so that $b=a+2$. The degree of homogeneity is $a+1$.

We can now replace b by $a+2$ before calculating the partial derivatives, or calculate them in terms of both a and b and then substitute for b in terms of a. Either way, we obtain

$$\frac{\partial f}{\partial x} = y^a\sin\left(\frac{x}{y}\right)+xy^{a-1}\cos\left(\frac{x}{y}\right)+(a+2)\frac{x^{a+1}}{y}$$

and

$$\frac{\partial f}{\partial y} = axy^{a-1}\sin\left(\frac{x}{y}\right)-x^2y^{a-2}\cos\left(\frac{x}{y}\right)-\frac{x^{a+2}}{y^2}.$$

To verify Euler's Theorem, we need to carefully check that

$$x\frac{\partial f}{\partial x}+y\frac{\partial f}{\partial y} = (a+1)f.$$

The details are omitted here.

Chapter 13

Exercise

13.1 The partial derivatives are

$$f_1(x, y) = 2x - 2y \quad \text{and} \quad f_2(x, y) = -2x + 4y - 2.$$

For critical points, we solve $f_1(x, y) = 0$ and $f_2(x, y) = 0$, which means

$$2x - 2y = 0 \quad \text{and} \quad -2x + 4y - 2 = 0.$$

The first equation gives us $y = x$ and then the second, with this information, tells us that $x = y = 1$. So there is one critical point, $(1, 1)$.

The second derivatives are

$$f_{11}(x, y) = 2, \quad f_{12}(x, y) = -2 \quad \text{and} \quad f_{22}(x, y) = 4.$$

So, using $D = f_{11}f_{22} - f_{12}^2$, we have

$$D = (2)(4) - (-2)^2 = 4 > 0 \quad \text{and} \quad f_{11}(1, 1) = 2 > 0,$$

which means that the critical point is a minimum.

13.2 The partial derivatives are

$$f_1(x, y) = -\frac{4}{x^2} - \frac{2}{y} \quad \text{and} \quad f_2(x, y) = \frac{2x}{y^2} + 1.$$

We solve $f_1(x, y) = 0$ and $f_2(x, y) = 0$, which means

$$-\frac{4}{x^2} - \frac{2}{y} = 0 \quad \text{and} \quad \frac{2x}{y^2} + 1 = 0.$$

The first equation gives us $y = -(1/2)x^2$ and the second $2x/(-x^2/2)^2 = -1$, so $x^3 = -8$ and $x = -2$. Then, we have $y = -x^2/2 = -2$ and so there is one critical point, $(-2, -2)$. Of course, there are other valid ways of solving these equations.

The second derivatives are

$$f_{11}(x, y) = 8/x^3, \quad f_{12}(x, y) = 2/y^2 \quad \text{and} \quad f_{22}(x, y) = -4x/y^3.$$

At $(-2, -2)$ we have $f_{11} = -1$, $f_{12} = 1/2$ and $f_{22} = -1$, so that

$$D = f_{11}f_{22} - f_{12}^2 = (-1)(-1) - (1/2)^2 > 0 \quad \text{and} \quad f_{11} = -1 < 0,$$

which means that this is a maximum point.

13.3 The partial derivatives are

$$f_1(x, y) = -4x + 2y^2 \quad \text{and} \quad f_2(x, y) = 2y - 4y^3 + 4xy.$$

We solve $f_1(x, y) = 0$ and $f_2(x, y) = 0$, which means

$$-4x + 2y^2 = 0 \quad \text{and} \quad 2y - 4y^3 + 4xy = 0.$$

The first equation gives us $x = y^2/2$ and the second $2y - 4y^3 + 2y^3 = 0$, which is $2y(1-y^2) = 0$, so that $y = 0, 1, -1$. Then, using $x = y^2/2$ again, we get $x = 0, 1/2, 1/2$, respectively. So there are three critical points: $(0, 0), (1/2, 1), (1/2, -1)$.

The second derivatives are

$$f_{11}(x, y) = -4, \quad f_{12}(x, y) = 4y \quad \text{and} \quad f_{22}(x, y) = 2 - 12y^2 + 4x.$$

So, using $D = f_{11}f_{22} - f_{12}^2$, we have

- at $(0, 0)$, $D = (-4)(2) - (0)^2 < 0$, so it's a saddle point;
- at $(1/2, 1)$, $D = (-4)(-8) - (4)^2 > 0$ and $f_{11} = -4 < 0$, so it's a maximum;
- at $(1/2, -1)$, $D = (-4)(-8) - (-4)^2 > 0$ and $f_{11} = -4 < 0$, so it's a maximum.

13.4 The profit is given by

$$\Pi(x, y) = xp^X + yp^Y - (2x + y).$$

We want to get this as a function of x and y, so we need to find p^X and p^Y in terms of x and y.

We have

$$-p^X + p^Y = 6x - 13 \quad \text{and} \quad p^X - 4p^Y = 6y - 26,$$

so adding these equations gives

$$-3p^Y = 6x + 6y - 39, \quad \text{or} \quad p^Y = 13 - 2x - 2y.$$

Then, using the first equation again, this gives us

$$p^X = p^Y - 6x + 13 = (13 - 2x - 2y) - 6x + 13 = 26 - 8x - 2y.$$

The profit is therefore

$$\Pi(x, y) = x(26 - 8x - 2y) + y(13 - 2x - 2y) - (2x + y) = 24x - 8x^2 - 4xy + 12y - 2y^2.$$

Solving for the critical points, we need

$$\Pi_1(x, y) = 24 - 16x - 4y = 0 \quad \text{and} \quad \Pi_2(x, y) = -4x + 12 - 4y = 0,$$

which gives us the solution $(x, y) = (1, 2)$.

The second derivatives are

$$\Pi_{11}(x, y) = -16, \quad \Pi_{12}(x, y) = -4 \quad \text{and} \quad \Pi_{22}(x, y) = -4.$$

So, at $(1, 2)$, since

$$\Pi_{22}\Pi_{22} - (\Pi_{12})^2 = (-16)(-4) - (-4)^2 = 48 > 0 \quad \text{and} \quad \Pi_{11} = -16 < 0,$$

this point maximises profit.

Chapter 14

Exercise

14.1 We have

$$\mathbf{x} + \mathbf{y} = (1 + 3, 2 + 2, 3 + 1) = (4, 4, 4)$$

and

$$\mathbf{x}.\mathbf{y} = (1 \times 3) + (2 \times 2) + (3 \times 1) = 10.$$

14.2 Consider first Ann. As c increases, her indifference curves $x_1 + x_2 = c$ move in a north-easterly direction. It is clear that the 'last' one which intersects the budget set will be the one which meets it at the extreme point $(0, 15)$ (Figure A.3a). This, then, is the optimal bundle for Ann.

The analogous sketches for Bill and Chas are shown in Figures A.3b and c. It can be seen that the optimal bundle for Bill is $(10, 0)$ and the optimal bundle for Chas is $(0, 15)$.

Note that the method used in Section 14.5 is not applicable in cases like this where the optimal bundle is at an extreme point of the budget set.

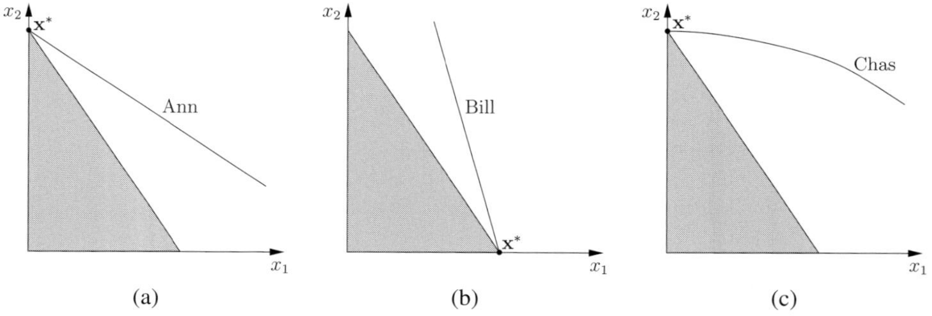

Figure A.3 The optimal bundles for Ann, Bill and Chas.

14.3 Susan's budget constraint is $2x_1 + x_2 \leq 4$ and her indifference curves $u(x_1, x_2) = x_1^2 x_2^2 = c$ have the typical 'Cobb–Douglas' form. This means that her optimal bundle (x_1^*, x_2^*) of flour and sugar occurs when her indifference curve is tangent to the budget line.

The slope of an indifference curve is given by

$$\frac{dx_2}{dx_1} = -\frac{\partial u/\partial x_1}{\partial u/\partial x_2} = -\frac{2x_1x_2^2}{2x_1^2x_2} = -\frac{x_2}{x_1}$$

and the slope of the budget line is -2, so at the optimal bundle we must have

$$-\frac{x_2^*}{x_1^*} = -2, \quad \text{or} \quad x_2^* = 2x_1^*.$$

Since the optimal bundle lies on the budget line, it also satisfies the equation $2x_1^*+x_2^* = 4$ and so, solving these equations, we get $x_1^* = 1$ and $x_2^* = 2$. That is, Susan's optimal bundle is $(1, 2)$.

Chapter 15

Exercise

15.1 We know that, since A is a 2×3 matrix and B is a 3×2 matrix, the product AB can be formed and it will be a 2×2 matrix. There are therefore four entries to compute. The top-left entry is computed by forming the dot product of

$$\text{the first row } \begin{pmatrix} 1 & 2 & 3 \end{pmatrix} \text{ of } A \text{ and the first column } \begin{pmatrix} 2 \\ 0 \\ 1 \end{pmatrix} \text{ of } B$$

and is therefore

$$(1 \times 2) + (2 \times 0) + (3 \times 1) = 5.$$

The other entries are computed in a similar way. For example, the bottom-right entry of AB (indicated in bold in the following equation) is the dot product of the row and column indicated in bold:

$$AB = \begin{pmatrix} 1 & 2 & 3 \\ \mathbf{3} & \mathbf{2} & \mathbf{5} \end{pmatrix} \begin{pmatrix} 2 & \mathbf{1} \\ 0 & \mathbf{-1} \\ 1 & \mathbf{2} \end{pmatrix} = \begin{pmatrix} 5 & 5 \\ 11 & \mathbf{11} \end{pmatrix}.$$

15.2 With $Y = (500, 10\,000, 1000)$, we have

$$YR = \begin{pmatrix} 500 & 10\,000 & 1000 \end{pmatrix} \begin{pmatrix} 1.05 & 0.95 \\ 1.05 & 1.05 \\ 1.37 & 1.42 \end{pmatrix} = \begin{pmatrix} 12\,395 & 12\,395 \end{pmatrix}$$

so it is *riskless*, since the guaranteed return is 12 395 in each state.

With $Z = \begin{pmatrix} 1000 & -2000 & 1000 \end{pmatrix}$, the cost of Z is $1000 - 2000 + 1000 = 0$ and

$$ZR = \begin{pmatrix} 1000 & -2000 & 1000 \end{pmatrix} \begin{pmatrix} 1.05 & 0.95 \\ 1.05 & 1.05 \\ 1.37 & 1.42 \end{pmatrix} = \begin{pmatrix} 320 & 270 \end{pmatrix}.$$

Hence, Z is an arbitrage portfolio, since it costs nothing, $(ZR)_i \geq 0$ in all states (so no money is ever lost) and $(ZR)_i > 0$ in at least one state (so, in at least one state, a profit is made).

Chapter 16

Exercise

16.1 We reduce the augmented matrix as follows:

$$\left(\begin{array}{ccc|c} 1 & 1 & 1 & 6 \\ 2 & 4 & 1 & 5 \\ 2 & 3 & 1 & 6 \end{array}\right) \to \left(\begin{array}{ccc|c} 1 & 1 & 1 & 6 \\ 0 & 2 & -1 & -7 \\ 2 & 3 & 1 & 6 \end{array}\right) \to \left(\begin{array}{ccc|c} 1 & 1 & 1 & 6 \\ 0 & 1 & -1/2 & -7/2 \\ 0 & 1 & -1 & -6 \end{array}\right)$$

$$\to \left(\begin{array}{ccc|c} 1 & 1 & 1 & 6 \\ 0 & 1 & -1/2 & -7/2 \\ 0 & 0 & -1/2 & -5/2 \end{array}\right) \to \left(\begin{array}{ccc|c} 1 & 1 & 1 & 6 \\ 0 & 1 & -1/2 & -7/2 \\ 0 & 0 & 1 & 5 \end{array}\right).$$

You should check which elementary row operations have been used at each step. For example, going from the first to the second matrix, we have subtracted twice the first row from the second.

The final augmented matrix represents the system

$$\begin{aligned} x_1 + x_2 + x_3 &= 6, \\ x_2 - \frac{1}{2}x_3 &= -\frac{7}{2}, \\ x_3 &= 5. \end{aligned}$$

As we have seen, the solution of this system can be found by 'working backwards':

$$x_3 = 5, \quad x_2 = \frac{1}{2}x_3 - \frac{7}{2} = -1 \quad \text{and} \quad x_1 = 6 - x_2 - x_3 = 2.$$

16.2 We have

$$f'(x) = -\frac{a}{x^2} + b + 2cx.$$

The given information therefore implies that

$$\begin{aligned} a + b + c &= 8, \\ \frac{a}{2} + 2b + 4c &= 2, \\ -a + b + 2c &= -9. \end{aligned}$$

Therefore

$$\begin{aligned} a + b + c &= 8, \\ a + 4b + 8c &= 4, \\ a - b - 2c &= 9. \end{aligned}$$

Using elementary row operations to reduce the augmented matrix, we have

$$\left(\begin{array}{ccc|c} 1 & 1 & 1 & 8 \\ 1 & 4 & 8 & 4 \\ 1 & -1 & -2 & 9 \end{array}\right) \to \left(\begin{array}{ccc|c} 1 & 1 & 1 & 8 \\ 0 & 3 & 7 & -4 \\ 0 & -2 & -3 & 1 \end{array}\right) \to \left(\begin{array}{ccc|c} 1 & 1 & 1 & 8 \\ 0 & 3 & 7 & -4 \\ 0 & 0 & 5 & -5 \end{array}\right)$$

$$\to \left(\begin{array}{ccc|c} 1 & 1 & 1 & 8 \\ 0 & 1 & 7/3 & -4/3 \\ 0 & 0 & 1 & -1 \end{array}\right).$$

So, by back-substitution, $c = -1$; $b + (7/3)c = -4/3$, meaning $b = 1$; and $a + b + c = 8$, meaning $a = 8$.

16.3 We have

$$f'(x) = 3ax^2 - \frac{b}{x^2} + \frac{c}{x} \quad \text{and} \quad f''(x) = 6ax + \frac{2b}{x^3} - \frac{c}{x^2},$$

so the given information means that

$$\begin{aligned} a + b + c &= 0, \\ 3a - b + c &= 20, \\ 6a + 2b - c &= 40. \end{aligned}$$

Using elementary row operations to reduce the augmented matrix, we have

$$\left(\begin{array}{ccc|c} 1 & 1 & 1 & 0 \\ 3 & -1 & 1 & 20 \\ 6 & 2 & -1 & 40 \end{array}\right) \to \left(\begin{array}{ccc|c} 1 & 1 & 1 & 0 \\ 0 & -4 & -2 & 20 \\ 0 & -4 & -7 & 40 \end{array}\right) \to \left(\begin{array}{ccc|c} 1 & 1 & 1 & 0 \\ 0 & 2 & 1 & -10 \\ 0 & -4 & -7 & 40 \end{array}\right)$$

$$\to \left(\begin{array}{ccc|c} 1 & 1 & 1 & 0 \\ 0 & 2 & 1 & -10 \\ 0 & 0 & 1 & -4 \end{array}\right) \to \left(\begin{array}{ccc|c} 1 & 1 & 1 & 0 \\ 0 & 1 & 1/2 & -5 \\ 0 & 0 & 1 & -4 \end{array}\right).$$

So, from the row-echelon form, we have $c = -4$, $b = -5 - (1/2)c = -3$ and $a = -b - c = 7$.

16.4 The equations are equivalent to

$$\begin{aligned} x_1 + x_2 \qquad\quad &= 70, \\ x_1 + 5x_2 + 7x_3 &= 294, \\ 6x_1 - 2x_2 + 4x_3 &= 44. \end{aligned}$$

Using elementary row operations to reduce the augmented matrix, we have

$$\left(\begin{array}{ccc|c} 1 & 1 & 0 & 70 \\ 1 & 5 & 7 & 294 \\ 6 & -2 & 4 & 44 \end{array}\right) \to \left(\begin{array}{ccc|c} 1 & 1 & 0 & 70 \\ 0 & 4 & 7 & 224 \\ 0 & -8 & 4 & -376 \end{array}\right) \to \left(\begin{array}{ccc|c} 1 & 1 & 0 & 70 \\ 0 & 4 & 7 & 224 \\ 0 & 0 & 18 & 72 \end{array}\right),$$

so the equilibrium prices are $x_3 = 72/18 = 4$, $4x_2 = 224 - 7x_3 = 196$, so that $x_2 = 49$, and $x_1 = 70 - x_2 = 21$.

Chapter 17

Exercise

17.1 The augmented matrix is

$$\left(\begin{array}{ccc|c} 0 & 1 & 2 & 7 \\ 3 & -1 & -1 & 5 \\ 6 & 2 & 6 & 38 \end{array}\right),$$

which we reduce as follows:

$$\left(\begin{array}{ccc|c} 0 & 1 & 2 & 7 \\ 3 & -1 & -1 & 5 \\ 6 & 2 & 6 & 38 \end{array}\right) \to \left(\begin{array}{ccc|c} 3 & -1 & -1 & 5 \\ 0 & 1 & 2 & 7 \\ 6 & 2 & 6 & 38 \end{array}\right) \to \left(\begin{array}{ccc|c} 3 & -1 & -1 & 5 \\ 0 & 1 & 2 & 7 \\ 0 & 4 & 8 & 28 \end{array}\right)$$

$$\to \left(\begin{array}{ccc|c} 3 & -1 & -1 & 5 \\ 0 & 1 & 2 & 7 \\ 0 & 1 & 2 & 7 \end{array}\right) \to \left(\begin{array}{ccc|c} 1 & -1/3 & -1/3 & 5/3 \\ 0 & 1 & 2 & 7 \\ 0 & 0 & 0 & 0 \end{array}\right).$$

The system is therefore equivalent to

$$x_1 - \frac{1}{3}x_2 - \frac{1}{3}x_3 = \frac{5}{3},$$
$$x_2 + 2x_3 = 7.$$

(Since there are two non-zero rows, the rank is 2.) To find the general solution, let $x_3 = s$. Then $x_2 = 7 - 2s$ and

$$x_1 = \frac{5}{3} + \frac{1}{3}x_2 + \frac{1}{3}x_3 = \frac{5}{3} + \frac{1}{3}(7 - 2s) + \frac{1}{3}s = 4 - \frac{1}{3}s.$$

The general solution is therefore

$$\mathbf{x} = \begin{pmatrix} x_1 \\ x_2 \\ x_3 \end{pmatrix} = \begin{pmatrix} 4 - \frac{1}{3}s \\ 7 - 2s \\ s \end{pmatrix} = \begin{pmatrix} 4 \\ 7 \\ 0 \end{pmatrix} + \begin{pmatrix} -\frac{1}{3}s \\ -2s \\ s \end{pmatrix} = \begin{pmatrix} 4 \\ 7 \\ 0 \end{pmatrix} + s\begin{pmatrix} -\frac{1}{3} \\ -2 \\ 1 \end{pmatrix},$$

which, taking

$$\mathbf{v} = \begin{pmatrix} 4 \\ 7 \\ 0 \end{pmatrix}, \quad \mathbf{u} = \begin{pmatrix} -\frac{1}{3} \\ -2 \\ 1 \end{pmatrix},$$

is of the required form.

17.2 We start with the augmented matrix, and perform row operations as follows:

$$\left(\begin{array}{cccc|c} 1 & 1 & 1 & 1 & 4 \\ 1 & 3 & -1 & 5 & 8 \\ 2 & -1 & 3 & -1 & 3 \\ 2 & -3 & a & -5 & b \end{array}\right) \to \left(\begin{array}{cccc|c} 1 & 1 & 1 & 1 & 4 \\ 0 & 2 & -2 & 4 & 4 \\ 0 & -3 & 1 & -3 & -5 \\ 0 & -5 & a-2 & -7 & b-8 \end{array}\right)$$

$$\to \left(\begin{array}{cccc|c} 1 & 1 & 1 & 1 & 4 \\ 0 & 1 & -1 & 2 & 2 \\ 0 & 0 & -2 & 3 & 1 \\ 0 & 0 & a-7 & 3 & b+2 \end{array}\right) \to \left(\begin{array}{cccc|c} 1 & 1 & 1 & 1 & 4 \\ 0 & 1 & -1 & 2 & 2 \\ 0 & 0 & -2 & 3 & 1 \\ 0 & 0 & 0 & 3a-15 & a+2b-3 \end{array}\right).$$

Of course, it is possible to perform different row operations and end up with a slightly different reduced matrix. Now we need to realise the following: the system will have infinitely many solutions if and only if the last row is the all-zero row. For, if $3a-15 \neq 0$, there will be exactly one solution; and if $3a-15=0$ and $a+2b-3 \neq 0$, there will be no solutions (the system being inconsistent in this case). So, for infinitely many solutions, we must have $3a-15=0$ and $a+2b-3=0$, which means $a=5$ and $b=-1$. In this case, the final reduced matrix is

$$\left(\begin{array}{cccc|c} 1 & 1 & 1 & 1 & 0 \\ 0 & 1 & -1 & 2 & 0 \\ 0 & 0 & -2 & 3 & 0 \\ 0 & 0 & 0 & 0 & 0 \end{array}\right).$$

The last row gives no information, so we simply set $w=t$, which can be any real number. Then the third row reveals that

$$-2z+3w=1,$$

so

$$z=\frac{3}{2}t-\frac{1}{2}.$$

The second row tells us

$$y=2-2w+z=2-2t+\frac{3}{2}t-\frac{1}{2}=\frac{3}{2}-\frac{1}{2}t.$$

The first row then yields

$$x=4-w-z-y=4-t-\left(\frac{3}{2}t-\frac{1}{2}\right)-\left(\frac{3}{2}-\frac{1}{2}t\right)=3-2t.$$

17.3 We start with the augmented matrix and perform row operations as follows:

$$\left(\begin{array}{ccc|c} 1 & 1 & 1 & 4+a \\ 2 & 1 & 0 & 3+a \\ 1 & -1 & -2 & -4 \end{array}\right) \to \left(\begin{array}{ccc|c} 1 & 1 & 1 & 4+a \\ 0 & -1 & -2 & -5-a \\ 0 & -2 & -3 & -8-a \end{array}\right) \to \left(\begin{array}{ccc|c} 1 & 1 & 1 & 4+a \\ 0 & 1 & 2 & 5+a \\ 0 & 0 & 1 & 2+a \end{array}\right).$$

So, from the row-echelon form, we have

$$\begin{aligned} z &= 2 + a, \\ y &= 5 + a - 2(2 + a) = 1 - a, \\ x &= 4 + a - (1 - a) - (2 + a) = 1 + a. \end{aligned}$$

For x, y, z to all be positive, we must have $2+a, 1-a, 1+a > 0$, which means $a > -2$, $a < 1$ and $a > -1$. So we need $-1 < a < 1$.

17.4 A vector of state prices is

$$\mathbf{p} = \begin{pmatrix} p_1 \\ p_2 \\ p_3 \end{pmatrix} \text{ such that } p_1,\ p_2,\ p_3 > 0 \text{ and } R\mathbf{p} = \mathbf{u}, \text{ where } \mathbf{u} = \begin{pmatrix} 1 \\ 1 \\ 1 \end{pmatrix}.$$

Consider the system $R\mathbf{p} = \mathbf{u}$. We start with the augmented matrix and perform row operations as follows:

$$\left(\begin{array}{ccc|c} 1.05 & 0.95 & 1.0 & 1 \\ 1.05 & 1.05 & 1.05 & 1 \\ 1.20 & 1.26 & 1.23 & 1 \end{array}\right) \to \left(\begin{array}{ccc|c} 1 & 1 & 1 & 20/21 \\ 1.05 & 0.95 & 1.0 & 1 \\ 1.20 & 1.26 & 1.23 & 1 \end{array}\right)$$

$$\to \left(\begin{array}{ccc|c} 1 & 1 & 1 & 20/21 \\ 0 & 0.1 & 0.05 & 0 \\ 0 & 0.06 & 0.03 & -1/7 \end{array}\right) \to \left(\begin{array}{ccc|c} 1 & 1 & 1 & 20/21 \\ 0 & 1 & 0.5 & 0 \\ 0 & 0 & 0 & -1/7 \end{array}\right).$$

The system is inconsistent: state prices do not exist for the given matrix R and so arbitrage portfolios do exist.

To find an arbitrage portfolio which guarantees a return of at least \$200 in each state, you can set $X = \begin{pmatrix} x & -2000 & 2000 - x \end{pmatrix}$ and look at the inequalities $(XR)_j \geq 200$. From these you can conclude that $x \leq 666.66$ will work. For example, take $X = \begin{pmatrix} 600 & -2000 & 1400 \end{pmatrix}$ so that $XR = \begin{pmatrix} 210 & 234 & 222 \end{pmatrix}$.

Or, you might look at R and make an educated guess, noticing that, the last asset has the best predicted returns. So, trying $Y = \begin{pmatrix} 0 & -2000 & 2000 \end{pmatrix}$ we get $YR = \begin{pmatrix} 300 & 420 & 360 \end{pmatrix}$.

Chapter 18

Exercise

18.1 In order to determine if the matrix is invertible and, if so, to determine the inverse, we form the matrix

$$(A \mid I) = \left(\begin{array}{ccc|ccc} 1 & 2 & 4 & 1 & 0 & 0 \\ 1 & 3 & 6 & 0 & 1 & 0 \\ -1 & 0 & 1 & 0 & 0 & 1 \end{array}\right).$$

Then we carry out elementary row operations:

$$\left(\begin{array}{ccc|ccc} 1 & 2 & 4 & 1 & 0 & 0 \\ 1 & 3 & 6 & 0 & 1 & 0 \\ -1 & 0 & 1 & 0 & 0 & 1 \end{array}\right) \to \left(\begin{array}{ccc|ccc} 1 & 2 & 4 & 1 & 0 & 0 \\ 0 & 1 & 2 & -1 & 1 & 0 \\ 0 & 2 & 5 & 1 & 0 & 1 \end{array}\right)$$

$$\to \left(\begin{array}{ccc|ccc} 1 & 2 & 4 & 1 & 0 & 0 \\ 0 & 1 & 2 & -1 & 1 & 0 \\ 0 & 0 & 1 & 3 & -2 & 1 \end{array}\right) \to \left(\begin{array}{ccc|ccc} 1 & 2 & 0 & -11 & 8 & -4 \\ 0 & 1 & 0 & -7 & 5 & -2 \\ 0 & 0 & 1 & 3 & -2 & 1 \end{array}\right)$$

$$\to \left(\begin{array}{ccc|ccc} 1 & 0 & 0 & 3 & -2 & 0 \\ 0 & 1 & 0 & -7 & 5 & -2 \\ 0 & 0 & 1 & 3 & -2 & 1 \end{array}\right).$$

This is now in the form $(I \mid B)$, so we deduce that A is invertible and that

$$A^{-1} = \begin{pmatrix} 3 & -2 & 0 \\ -7 & 5 & -2 \\ 3 & -2 & 1 \end{pmatrix}.$$

It is very easy to make mistakes when row-reducing a matrix, so the next thing you should do is *check* that $AA^{-1} = I$.

18.2 We begin with A, by row-reducing $(A \mid I)$:

$$\left(\begin{array}{ccc|ccc} 1 & 2 & -1 & 1 & 0 & 0 \\ 0 & 1 & 2 & 0 & 1 & 0 \\ 3 & 8 & 1 & 0 & 0 & 1 \end{array}\right) \to \left(\begin{array}{ccc|ccc} 1 & 2 & -1 & 1 & 0 & 0 \\ 0 & 1 & 2 & 0 & 1 & 0 \\ 0 & 2 & 4 & -3 & 0 & 1 \end{array}\right).$$

The next step, $R_3 - 2R_2$, will yield a row of zeros in the first half, and therefore the matrix A is not invertible.

For the matrix B, we row-reduce $(B \mid I)$:

$$\left(\begin{array}{ccc|ccc} -1 & 2 & 1 & 1 & 0 & 0 \\ 0 & 1 & 2 & 0 & 1 & 0 \\ 3 & 1 & 4 & 0 & 0 & 1 \end{array}\right) \to \left(\begin{array}{ccc|ccc} 1 & -2 & -1 & -1 & 0 & 0 \\ 0 & 1 & 2 & 0 & 1 & 0 \\ 3 & 1 & 4 & 0 & 0 & 1 \end{array}\right)$$

$$\to \left(\begin{array}{ccc|ccc} 1 & -2 & -1 & -1 & 0 & 0 \\ 0 & 1 & 2 & 0 & 1 & 0 \\ 0 & 7 & 7 & 3 & 0 & 1 \end{array}\right) \to \left(\begin{array}{ccc|ccc} 1 & -2 & -1 & -1 & 0 & 0 \\ 0 & 1 & 2 & 0 & 1 & 0 \\ 0 & 0 & -7 & 3 & -7 & 1 \end{array}\right)$$

$$\to \left(\begin{array}{ccc|ccc} 1 & -2 & -1 & -1 & 0 & 0 \\ 0 & 1 & 2 & 0 & 1 & 0 \\ 0 & 0 & 1 & -\frac{3}{7} & 1 & -\frac{1}{7} \end{array}\right) \to \left(\begin{array}{ccc|ccc} 1 & -2 & 0 & -\frac{10}{7} & 1 & -\frac{1}{7} \\ 0 & 1 & 0 & \frac{6}{7} & -1 & \frac{2}{7} \\ 0 & 0 & 1 & -\frac{3}{7} & 1 & -\frac{1}{7} \end{array}\right)$$

$$\to \left(\begin{array}{ccc|ccc} 1 & 0 & 0 & \frac{2}{7} & -1 & \frac{3}{7} \\ 0 & 1 & 0 & \frac{6}{7} & -1 & \frac{2}{7} \\ 0 & 0 & 1 & -\frac{3}{7} & 1 & -\frac{1}{7} \end{array}\right).$$

This final matrix shows that

$$B^{-1} = \frac{1}{7}\begin{pmatrix} 2 & -7 & 3 \\ 6 & -7 & 2 \\ -3 & 7 & -1 \end{pmatrix}.$$

We can check this is correct by noting that

$$B^{-1}B = \frac{1}{7}\begin{pmatrix} 2 & -7 & 3 \\ 6 & -7 & 2 \\ -3 & 7 & -1 \end{pmatrix}\begin{pmatrix} -1 & 2 & 1 \\ 0 & 1 & 2 \\ 3 & 1 & 4 \end{pmatrix} = \frac{1}{7}\begin{pmatrix} 7 & 0 & 0 \\ 0 & 7 & 0 \\ 0 & 0 & 7 \end{pmatrix} = I.$$

18.3 Using the formula given in Section 18.4, with $a = 1$, $b = 2$, $c = 1$ and $d = 3$, we have $ad - bc = 1$ and

$$A^{-1} = \frac{1}{ad-bc}\begin{pmatrix} d & -b \\ -c & a \end{pmatrix} = \begin{pmatrix} 3 & -2 \\ -1 & 1 \end{pmatrix}.$$

Chapter 19

Exercise

19.1 Let x_1 be the amount of coal production required (in dollars), x_2 the amount of electricity and x_3 the amount of transportation. Then, by considering in turn the total demand for each of the three commodities:

$$\begin{aligned} x_1 &= 50\,000 + 0.65x_2 + 0.55x_3, \\ x_2 &= 25\,000 + 0.25x_1 + 0.05x_2 + 0.10x_3, \\ x_3 &= 0.25x_1 + 0.05x_2. \end{aligned}$$

That is

$$\mathbf{x} = \begin{pmatrix} 50\,000 \\ 25\,000 \\ 0 \end{pmatrix} + \begin{pmatrix} 0 & 0.65 & 0.55 \\ 0.25 & 0.05 & 0.10 \\ 0.25 & 0.05 & 0 \end{pmatrix}\mathbf{x} = \mathbf{d} + A\mathbf{x}.$$

As usual, $(I - A)\mathbf{x} = \mathbf{d}$ and $\mathbf{x} = (I - A)^{-1}\mathbf{d}$. The inverse of $(I - A)$ can be calculated by row operations. For the sake of complete exposition, we give below the computation, although in practice matrix inversions such as this are done on a computer.

Keeping numbers to six decimal places where necessary and accepting that this may introduce some rounding errors, we see that

$$\left(\begin{array}{ccc|ccc} 1 & -0.65 & -0.55 & 1 & 0 & 0 \\ -0.25 & 0.95 & -0.10 & 0 & 1 & 0 \\ -0.25 & -0.05 & 1 & 0 & 0 & 1 \end{array}\right)$$

$$\to \left(\begin{array}{ccc|ccc} 1 & -0.65 & -0.55 & 1 & 0 & 0 \\ 0 & 0.7875 & -0.2375 & 0.25 & 1 & 0 \\ 0 & -0.2125 & 0.8625 & 0.25 & 0 & 1 \end{array}\right)$$

$$\to \left(\begin{array}{ccc|ccc} 1 & -0.65 & -0.55 & 1 & 0 & 0 \\ 0 & 1 & -0.301587 & 0.317460 & 1.269841 & 0 \\ 0 & -0.2125 & 0.8625 & 0.25 & 0 & 1 \end{array}\right)$$

$$\to \left(\begin{array}{ccc|ccc} 1 & -0.65 & -0.55 & 1 & 0 & 0 \\ 0 & 1 & -0.301587 & 0.317460 & 1.269841 & 0 \\ 0 & 0 & 0.798413 & 0.317460 & 0.269841 & 1 \end{array}\right)$$

$$\to \left(\begin{array}{ccc|ccc} 1 & -0.65 & -0.55 & 1 & 0 & 0 \\ 0 & 1 & -0.301587 & 0.317460 & 1.269841 & 0 \\ 0 & 0 & 1 & 0.397614 & 0.337972 & 1.252485 \end{array}\right)$$

$$\to \left(\begin{array}{ccc|ccc} 1 & -0.65 & 0 & 1.218688 & 0.185885 & 0.688867 \\ 0 & 1 & 0 & 0.437375 & 1.371769 & 0.377733 \\ 0 & 0 & 1 & 0.397614 & 0.337972 & 1.252485 \end{array}\right)$$

$$\to \left(\begin{array}{ccc|ccc} 1 & 0 & 0 & 1.502982 & 1.0775349 & 0.934393 \\ 0 & 1 & 0 & 0.437375 & 1.371769 & 0.377733 \\ 0 & 0 & 1 & 0.397614 & 0.337972 & 1.252485 \end{array}\right).$$

Therefore, we have

$$(I - A)^{-1} = \begin{pmatrix} 1.502982 & 1.077535 & 0.934393 \\ 0.437376 & 1.371769 & 0.377733 \\ 0.397614 & 0.337972 & 1.252485 \end{pmatrix}$$

and so the production schedule is

$$\mathbf{x} = (I - A)^{-1}\mathbf{d} = \begin{pmatrix} 102\,087 \\ 56\,163 \\ 28\,330 \end{pmatrix}$$

to the nearest dollar.

19.2 Let x_1, x_2, x_3 and x_4 be the amount of coal, iron, electricity and railways measured in dollars. Using the information given, we then have

$$\begin{aligned} x_1 &= 8000 + 0.6x_2 + 0.5x_3 + 0.5x_4, \\ x_2 &= 1000 + 0.1x_1 + 0.2x_4, \\ x_3 &= 4000 + 0.2x_1 + 0.01x_2 + 0.02x_3 + 0.25x_4, \\ x_4 &= 6000 + 0.5x_1 + 0.2x_2 + 0.01x_3. \end{aligned}$$

In matrix form, the system is

$$\mathbf{x} = \begin{pmatrix} x_1 \\ x_2 \\ x_3 \\ x_4 \end{pmatrix} = \begin{pmatrix} 8000 \\ 1000 \\ 4000 \\ 6000 \end{pmatrix} + \begin{pmatrix} 0 & 0.6 & 0.5 & 0.5 \\ 0.1 & 0 & 0 & 0.2 \\ 0.2 & 0.01 & 0.02 & 0.25 \\ 0.5 & 0.2 & 0.01 & 0 \end{pmatrix} \begin{pmatrix} x_1 \\ x_2 \\ x_3 \\ x_4 \end{pmatrix} = \mathbf{d} + A\mathbf{x},$$

and so we need to solve the equation $(I - A)\mathbf{x} = \mathbf{d}$, that is

$$\begin{pmatrix} 1 & -0.6 & -0.5 & -0.5 \\ -0.1 & 1 & 0 & -0.2 \\ -0.2 & -0.01 & 0.98 & -0.25 \\ -0.5 & -0.2 & -0.01 & 1 \end{pmatrix} \begin{pmatrix} x_1 \\ x_2 \\ x_3 \\ x_4 \end{pmatrix} = \begin{pmatrix} 8000 \\ 1000 \\ 4000 \\ 6000 \end{pmatrix}.$$

We are not asked to solve these equations, but we could if we were prepared to put in the effort.

Chapter 20

Exercise

20.1 We have

$$\begin{vmatrix} 5 & 7 & -2 \\ 10 & 15 & -2 \\ 5 & 8 & -1 \end{vmatrix} = \begin{vmatrix} 5 & 7 & -2 \\ 0 & 1 & 2 \\ 0 & 1 & 1 \end{vmatrix} = \begin{vmatrix} 5 & 7 & -2 \\ 0 & 1 & 2 \\ 0 & 0 & -1 \end{vmatrix} = 5(1)(-1) = -5.$$

20.2 We have

$$\begin{vmatrix} 1 & 2 & -1 & 4 \\ -1 & 3 & 0 & 2 \\ 2 & 1 & 1 & 2 \\ 1 & 4 & 1 & 3 \end{vmatrix} = \begin{vmatrix} 1 & 2 & -1 & 4 \\ 0 & 5 & -1 & 6 \\ 0 & -3 & 3 & -6 \\ 0 & 2 & 2 & -1 \end{vmatrix} = \begin{vmatrix} 1 & 2 & -1 & 4 \\ 0 & 5 & -1 & 6 \\ 0 & 0 & \frac{12}{5} & -\frac{12}{5} \\ 0 & 0 & \frac{12}{5} & -\frac{17}{5} \end{vmatrix}$$

$$= \begin{vmatrix} 1 & 2 & -1 & 4 \\ 0 & 5 & -1 & 6 \\ 0 & 0 & \frac{12}{5} & -\frac{12}{5} \\ 0 & 0 & 0 & -1 \end{vmatrix} = 1(5)\left(\frac{12}{5}\right)(-1) = -12.$$

20.3 First, we find its determinant. We can use the formula from Section 20.1, or row operations. We have $\det A = 7$ and, since this is non-zero, the inverse does exist. We now calculate the relevant 2×2 determinants:

$$A_{11} = \begin{vmatrix} 1 & 2 \\ 1 & 4 \end{vmatrix} = 2, \qquad A_{12} = \begin{vmatrix} 0 & 2 \\ 3 & 4 \end{vmatrix} = -6, \qquad A_{13} = \begin{vmatrix} 0 & 1 \\ 3 & 1 \end{vmatrix} = -3,$$

$$A_{21} = \begin{vmatrix} 2 & 1 \\ 1 & 4 \end{vmatrix} = 7, \qquad A_{22} = \begin{vmatrix} -1 & 1 \\ 3 & 4 \end{vmatrix} = -7, \qquad A_{23} = \begin{vmatrix} -1 & 2 \\ 3 & 1 \end{vmatrix} = -7,$$

$$A_{31} = \begin{vmatrix} 2 & 1 \\ 1 & 2 \end{vmatrix} = 3, \qquad A_{32} = \begin{vmatrix} -1 & 1 \\ 0 & 2 \end{vmatrix} = -2, \qquad A_{33} = \begin{vmatrix} -1 & 2 \\ 0 & 1 \end{vmatrix} = -1.$$

Then, the inverse is

$$A^{-1} = \frac{1}{\det A}\text{adj}(A) = \frac{1}{7}\begin{pmatrix} +A_{11} & -A_{21} & +A_{31} \\ -A_{12} & +A_{22} & -A_{32} \\ +A_{13} & -A_{23} & +A_{33} \end{pmatrix} = \frac{1}{7}\begin{pmatrix} 2 & -7 & 3 \\ 6 & -7 & 2 \\ -3 & 7 & -1 \end{pmatrix}.$$

Chapter 21

Exercise

21.1 The Lagrangean is

$$L = \frac{2y}{y+2} + \frac{x}{x+1} - \lambda(x + y - 120),$$

and the equations to solve are

$$\begin{aligned} \frac{\partial L}{\partial x} &= \frac{1}{(x+1)^2} - \lambda = 0, \\ \frac{\partial L}{\partial y} &= \frac{4}{(y+2)^2} - \lambda = 0, \\ x + y - 120 &= 0. \end{aligned}$$

From the first two equations, eliminating λ, we have

$$\lambda = \frac{1}{(x+1)^2} = \frac{4}{(y+2)^2}$$

so, given that $x, y > 0$, $y + 2 = 2(x + 1)$ and hence $y = 2x$. Then the third equation gives $3x = 120$, so $x = 40$ and $y = 80$.

21.2 (i) The cost is $vk + wl$ and we have to minimise this subject to the constraint $k^{1/4}l^{1/12} = Q$, so the Lagrangean is

$$L = vk + wl - \lambda(k^{1/4}l^{1/12} - Q).$$

The equations to solve are

$$\begin{aligned} \frac{\partial L}{\partial k} &= v - \frac{\lambda}{4}k^{-3/4}l^{1/12} = 0, \\ \frac{\partial L}{\partial l} &= w - \frac{\lambda}{12}k^{1/4}l^{-11/12} = 0, \\ k^{1/4}l^{1/12} - Q &= 0. \end{aligned}$$

From the first two equations, eliminating λ, we have

$$\lambda = 4vk^{3/4}l^{-1/12} = 12wk^{-1/4}l^{11/12},$$

so that $k = 3(w/v)l$. Then the third equation gives

$$\left(\frac{3w}{v}l\right)^{1/4} l^{1/12} = Q, \quad \text{and so} \quad l^{1/3} = \left(\frac{v}{3w}\right)^{1/4} Q,$$

which means that

$$l = \left(\frac{v}{3w}\right)^{3/4} Q^3 \quad \text{and} \quad k = \frac{3w}{v}\left(\frac{v}{3w}\right)^{3/4} Q^3 = \left(\frac{3w}{v}\right)^{1/4} Q^3.$$

(ii) Using part (i), the cost function is given by

$$C(Q) = v\left(\frac{3w}{v}\right)^{1/4} Q^3 + w\left(\frac{v}{3w}\right)^{3/4} Q^3 = \frac{4}{3^{3/4}} v^{3/4} w^{1/4} Q^3,$$

and so the profit function, $\Pi(Q) = pQ - C(Q)$, is

$$\Pi(Q) = pQ - \frac{4}{3^{3/4}} v^{3/4} w^{1/4} Q^3.$$

(iii) The condition $\Pi'(Q) = 0$ yields

$$p = \frac{12}{3^{3/4}} v^{3/4} w^{1/4} Q^2,$$

and so we have

$$Q^2 = \frac{3^{3/4}}{12} \frac{p}{v^{3/4} w^{1/4}}, \quad \text{so that} \quad Q = \frac{3^{3/8}}{12^{1/2}} \frac{p^{1/2}}{v^{3/8} w^{1/8}} = \left(\frac{p}{12}\right)^{1/2} \left(\frac{27}{v^3 w}\right)^{1/8}.$$

Note that, as $Q > 0$, we have

$$\Pi''(Q) = -\frac{24}{3^{3/4}} v^{3/4} w^{1/4} Q < 0,$$

and so this value of Q does indeed give a maximum.

Chapter 22

Exercise

22.1 The budget constraint is $x + 16y = 1280$ and so the Lagrangean is

$$L = x^{3/4} y^{1/4} - \lambda(x + 16y - 1280).$$

The first-order conditions are

$$\frac{\partial L}{\partial x} = \frac{3}{4} x^{-1/4} y^{1/4} - \lambda = 0,$$
$$\frac{\partial L}{\partial y} = \frac{1}{4} x^{3/4} y^{-3/4} - 16\lambda = 0,$$
$$x + 16y - 1280 = 0.$$

From the first two equations, we obtain two expressions for λ:

$$\lambda = \frac{3}{4} x^{-1/4} y^{1/4} = \frac{1}{64} x^{3/4} y^{-3/4},$$

which means that $y = x/48$. Now the third equation gives

$$x + 16\left(\frac{x}{48}\right) = 1280,$$

so $x = 960$ and $y = 960/48 = 20$.

Now, with the optimal values of x and y, the value of λ can be calculated using either of the two expressions we obtained above. From the first of these, we have

$$\lambda = \frac{3}{4}\left(960^{-1/4}\right)\left(20^{1/4}\right) = \frac{3}{4}\left(\frac{20}{960}\right)^{1/4} = \frac{3}{4}\left(48^{-1/4}\right).$$

If the income rises by \$2, then the maximum utility will increase by approximately 2λ, which is $\frac{3}{2}(48^{-1/4})$.

22.2 To maximise $u(x, y)$ subject to the budget constraint $px + qy = M$, we use the Lagrangean

$$L = \frac{xy}{x+y} - \lambda(px + qy - M).$$

We solve the equations

$$\begin{aligned}
\frac{\partial L}{\partial x} &= \frac{y(x+y) - xy}{(x+y)^2} - p\lambda = \frac{y^2}{(x+y)^2} - p\lambda = 0,\\
\frac{\partial L}{\partial y} &= \frac{x(x+y) - xy}{(x+y)^2} - q\lambda = \frac{x^2}{(x+y)^2} - q\lambda = 0,\\
& \qquad\qquad\qquad\qquad\qquad px + qy - M = 0.
\end{aligned}$$

Eliminating λ from the first two equations, we see that

$$\frac{y^2}{p} = \frac{x^2}{q}, \quad \text{so that} \quad y = \sqrt{\frac{p}{q}}\,x$$

as $x, y > 0$. Now the third equation gives

$$px + q\left(\sqrt{\frac{p}{q}}\,x\right) = M,$$

which means that the solutions are

$$x^* = \frac{M}{p + \sqrt{pq}} \quad \text{and} \quad y^* = \sqrt{\frac{p}{q}}\frac{M}{p + \sqrt{pq}} = \frac{M}{q + \sqrt{pq}}.$$

Now, with these, we have

$$\begin{aligned}
x^* + y^* &= \frac{M/\sqrt{p}}{\sqrt{p} + \sqrt{q}} + \frac{M/\sqrt{q}}{\sqrt{q} + \sqrt{p}} = \frac{M}{\sqrt{p} + \sqrt{q}}\left(\frac{1}{\sqrt{p}} + \frac{1}{\sqrt{q}}\right)\\
&= \frac{M}{\sqrt{p} + \sqrt{q}}\left(\frac{\sqrt{q} + \sqrt{p}}{\sqrt{p}\sqrt{q}}\right) = \frac{M}{\sqrt{pq}},
\end{aligned}$$

as required.

The corresponding value of the Lagrange multiplier is

$$\lambda^* = \frac{1}{p}\frac{(y^*)^2}{(x^* + y^*)^2} = \frac{1}{p}\left(\frac{M/\sqrt{q}}{\sqrt{q} + \sqrt{p}}\right)^2\left(\frac{\sqrt{pq}}{M}\right)^2 = \frac{1}{(\sqrt{q} + \sqrt{p})^2}.$$

We also have the maximum achievable utility

$$V = \frac{x^*y^*}{x^* + y^*} = \left(\frac{M/\sqrt{p}}{\sqrt{p}+\sqrt{q}}\right)\left(\frac{M/\sqrt{q}}{\sqrt{q}+\sqrt{p}}\right)\left(\frac{\sqrt{pq}}{M}\right) = \frac{M}{(\sqrt{p}+\sqrt{q})^2},$$

and so we can see that

$$\frac{\partial V}{\partial M} = \frac{1}{(\sqrt{p}+\sqrt{q})^2} = \lambda^*,$$

as expected.

Chapter 23

Exercise

23.1 The auxiliary equation is

$$z^2 - z - 6 = (z-3)(z+2) = 0,$$

so for some constants A and B the general solution is

$$y_t = A3^t + B(-2)^t.$$

23.2 The auxiliary equation is

$$z^2 - 5z + 4 = (z-4)(z-1) = 0,$$

so for some constants A and B the general solution is

$$y_t = A(1^t) + B(4^t) = A + B4^t.$$

Using the initial conditions, we have $y_0 = A + B = 1$ and $y_1 = A + 4B = 10$, from which it follows that $B = 3$ and $A = -2$. The solution is therefore $y_t = -2 + 3(4^t)$.

23.3 The auxiliary equation is

$$z^2 - z + \frac{1}{4} = \left(z - \frac{1}{2}\right)^2 = 0$$

so, because this has just one solution, the general solution is

$$y_t = (Ct + D)\left(\frac{1}{2}\right)^t$$

for some constants C and D. Using the initial conditions, we have $y_0 = D = 2$ and $y_1 = (C+D)/2 = 5$, from which it follows that $D = 2$ and $C = 8$. The solution is therefore

$$y_t = (8t + 2)\left(\frac{1}{2}\right)^t.$$

23.4 Substituting the expressions for the demand and supply functions:

$$p_t - p_{t-1} = \frac{7}{12}\left[1 - p_{t-1} - p_{t-1}\right] - \frac{1}{12}\left[1 - p_{t-2} - p_{t-2}\right].$$

Rearranging, we obtain the recurrence equation

$$p_t + \frac{1}{6}p_{t-1} - \frac{1}{6}p_{t-2} = \frac{1}{2}.$$

The auxiliary equation of the associated homogeneous equation is

$$z^2 + \frac{1}{6}z - \frac{1}{6} = \left(z - \frac{1}{3}\right)\left(z + \frac{1}{2}\right) = 0$$

and the constant solution is $p^* = 1/2$. Hence the general solution is

$$p_t = \frac{1}{2} + A\left(\frac{1}{3}\right)^t + B\left(-\frac{1}{2}\right)^t$$

for some constants A and B. Using the initial conditions, we have

$$\frac{1}{2} + A + B = \frac{5}{2} \quad \text{and} \quad \frac{1}{2} + \frac{1}{3}A - \frac{1}{2}B = \frac{1}{3},$$

from which it follows that $A = 1$ and $B = 1$. The solution is therefore

$$p_t = \frac{1}{2} + \left(\frac{1}{3}\right)^t + \left(-\frac{1}{2}\right)^t.$$

Chapter 24

Exercise

24.1 Here the dominant term is $(-1/2)^t$, since we can write

$$y_t = 3 + \left(-\frac{1}{2}\right)^t \left[1 - \left(\frac{2}{3}\right)^t\right],$$

and $1 - \left(\frac{2}{3}\right)^t \to 1$ as $t \to \infty$. It follows that y_t oscillates decreasingly to 3 as $t \to \infty$.

24.2 Assuming the usual equilibrium conditions $Y_t = Q_t$ and $Y_t = C_t + I_t$, we have

$$\begin{aligned} Y_t &= C_t + I_t \\ &= 10 + \frac{1}{4}Y_{t-1} + 20 + \frac{1}{2}(Q_{t-1} - Q_{t-2}) \\ &= 30 + \frac{1}{4}Y_{t-1} + \frac{1}{2}(Y_{t-1} - Y_{t-2}) \\ &= 30 + \frac{3}{4}Y_{t-1} - \frac{1}{2}Y_{t-2}. \end{aligned}$$

Thus we obtain

$$Y_t - \frac{3}{4}Y_{t-1} + \frac{1}{2}Y_{t-2} = 30.$$

24.3 As $(\frac{3}{4})^2 < 4(\frac{1}{2})$, the auxiliary equation has no solutions. This means that the general solution will have the form

$$Y_t = Y^* + r^t\,[E\cos(\theta t) + F\sin(\theta t)]$$

where $r = \sqrt{1/2} < 1$ and

$$Y^* = \frac{30}{1 - \frac{3}{4} + \frac{1}{2}} = 40.$$

Thus, in the long run, Y_t will oscillate decreasingly to 40 (unless, of course, $E = F = 0$, in which case Y_t is constant at 40 for all t).

Chapter 25

Exercise

25.1 We know that the function f is an anti-derivative of $t^2 + 2\sin t$. Integrating this, by the standard integrals and the rules just seen, we have

$$\int (t^2 + 2\sin t)\,dt = \frac{t^3}{3} - 2\cos t + c,$$

where c is a constant of integration. But we know more about f: we know that $f(0) = 1$, so we know that

$$f(t) = \frac{t^3}{3} - 2\cos t + c,$$

where the constant c is such that $f(0) = 1$. Now, substituting $t = 0$ into the expression for f, we have

$$f(0) = 0 - 2\cos 0 + c = -2 + c,$$

and for this to equal 1, c must be 3. Therefore, the function f is given by

$$f(t) = \frac{t^3}{3} - 2\cos t + 3.$$

25.2 The integral $\int t^2\,dt$ is $t^3/3 + c$, so

$$\int_0^2 t^2\,dt = \left[\frac{t^3}{3}\right]_0^2 = \frac{2^3}{3} - 0 = \frac{8}{3}.$$

The integral $\int e^t\,dt$ is $e^t + c$, so

$$\int_{-1}^{1} e^t\,dt = \left[e^t\right]_{-1}^{1} = e^1 - e^{-1} = e - \frac{1}{e}.$$

25.3 To find the equilibrium quantity, q^*, we solve

$$\frac{100}{q+5} = \frac{q+45}{5},$$

which is equivalent to

$$(q + 45)(q + 5) = 500, \quad q^2 + 50q - 275 = 0, \quad (q + 55)(q - 5) = 0.$$

Since q^* cannot be negative, $q^* = 5$. (Alternatively, we could use the formula for the solutions of a quadratic equation.) The equilibrium price is $p^* = (q^* + 45)/5 = 10$.

The consumer surplus is then

$$\begin{aligned} CS &= \int_0^5 \frac{100}{q+5}\,dq - (5)(10) \\ &= \Big[100\ln(q+5)\Big]_0^5 - 50 \\ &= (100\ln 10 - 100\ln 5) - 50 \\ &= 100\ln 2 - 50. \end{aligned}$$

Chapter 26

Exercise

26.1 We make the substitution $u = 2x^2 + 2$. We have $du = 4x\,dx$, so the integral reduces to

$$\int u^{1/2}\left(\frac{1}{4}\right)du = \frac{1}{4}\left(\frac{2}{3}u^{3/2}\right) + c = \frac{1}{6}(2x^2+2)^{3/2} + c.$$

26.2 (a) With $u = x - 1$, we have $du = dx$ and so, on noting that $x = u + 1$, the integral becomes

$$\begin{aligned} \int x\sqrt{x-1}\,dx &= \int (u+1)u^{1/2}\,du \\ &= \int (u^{3/2} + u^{1/2})du \\ &= \frac{2}{5}u^{5/2} + \frac{2}{3}u^{3/2} + c \\ &= \frac{2}{5}(x-1)^{5/2} + \frac{2}{3}(x-1)^{3/2} + c. \end{aligned}$$

(b) With $u = \sqrt{x-1}$, we have

$$\frac{du}{dx} = \frac{1}{2}\frac{1}{\sqrt{x-1}} = \frac{1}{2u},$$

and so $dx = 2u\,du$. Also $u = \sqrt{x-1}$ and so $u^2 = x - 1$, from which we obtain $x = u^2 + 1$. So the integral is

$$\begin{aligned}\int x\sqrt{x-1}\,dx &= \int (u^2+1)u(2u)\,du \\ &= 2\int (u^4+u^2)\,du \\ &= 2\left(\frac{u^5}{5}+\frac{u^3}{3}\right)+c \\ &= \frac{2}{5}\left(\sqrt{x-1}\right)^5+\frac{2}{3}\left(\sqrt{x-1}\right)^3+c.\end{aligned}$$

This is (of course!) the same as the answer obtained using the first substitution. But which is easier? Well, the details of the substitution are easier for the first, but the actual integration of the transformed integral is easier for the second (because it does not involve fractional powers). On balance, probably the first substitution is easier. But both are correct! Do not think there is necessarily only one way to solve a problem.

26.3 We can approach this integral by making the substitution $u = 1+\sqrt{x}$. We then have

$$du = \frac{1}{2\sqrt{x}}\,dx, \text{ so that } dx = 2\sqrt{x}\,du = 2(u-1)\,du$$

and the integral becomes

$$\begin{aligned}\int \frac{\sqrt{x}}{1+\sqrt{x}}\,dx &= \int \frac{2(u-1)^2}{u}\,du \\ &= \int\left(2u-4+\frac{2}{u}\right)du \\ &= u^2-4u+2\ln u+c \\ &= (1+\sqrt{x})^2-4(1+\sqrt{x})+2\ln(1+\sqrt{x})+c.\end{aligned}$$

The substitution $u = \sqrt{x}$ also works. With this choice, we then have

$$du = \frac{1}{2\sqrt{x}}\,dx, \text{ so that } dx = 2\sqrt{x}\,du = 2u\,du$$

and the integral becomes

$$\begin{aligned}\int \frac{\sqrt{x}}{1+\sqrt{x}}\,dx &= \int \frac{2u^2}{1+u}\,du = \int \frac{2u(1+u-1)}{1+u}\,du \\ &= \int\left(2u-\frac{2u}{1+u}\right)du = \int\left(2u-\frac{2(1+u-1)}{1+u}\right)du\end{aligned}$$

$$= \int \left(2u - 2 + \frac{2}{1+u}\right) du$$
$$= u^2 - 2u + 2\ln(1+u) + c$$
$$= x - 2\sqrt{x} + 2\ln(1+\sqrt{x}) + c.$$

This is the same as the first answer, of course, as it should be. For the first answer can be simplified to $x - 2\sqrt{x} - 3 + 2\ln(1+\sqrt{x}) + c$ and this is the same as the second (with a different constant c).

26.4 We use the substitution $u = x^2+4x+5$, so that $du = (2x+4)dx$ and the integral is

$$\int_0^1 \frac{x+2}{x^2+4x+5}\,dx = \int_5^{10} \frac{1}{u}\left(\frac{1}{2}\right) du = \frac{1}{2}\Big[\ln u\Big]_5^{10} = \frac{1}{2}\Big(\ln 10 - \ln 5\Big) = \frac{1}{2}\ln 2.$$

26.5 We note that $x^2 + 4x + 3 = (x+1)(x+3)$, so the method of partial fractions is applicable. We write

$$\frac{1}{x^2+4x+3} = \frac{A_1}{x+1} + \frac{A_2}{x+3},$$

and multiply through by $(x+1)(x+3)$ to obtain

$$1 = A_1(x+3) + A_2(x+1).$$

Taking $x = -1$ gives $2A_1 = 1$, so $A_1 = 1/2$. Taking $x = -3$ gives $-2A_2 = 1$, so $A_2 = -1/2$. Thus, the integral is

$$\int \frac{dx}{x^2+4x+3} = \int \left(\frac{1/2}{x+1} - \frac{1/2}{x+3}\right) dx = \frac{1}{2}\ln(x+1) - \frac{1}{2}\ln(x+3) + c,$$

provided x is in a range where the logarithms are defined; that is $x > -1$.

26.6 We use the substitution $u = e^x$, so that $du = e^x\,dx = u\,dx$ and, as $e^{2x} = (e^x)^2 = u^2$, the integral is

$$\int \frac{e^x}{e^{2x}-1}\,dx = \int \frac{u}{u^2-1}\left(\frac{1}{u}\right) du = \int \frac{du}{u^2-1}.$$

Now, $u^2 - 1 = (u-1)(u+1)$, so the method of partial fractions is applicable. We write

$$\frac{1}{u^2-1} = \frac{A_1}{u-1} + \frac{A_2}{u+1}$$

and multiply through by $(u-1)(u+1)$ to obtain

$$1 = A_1(u+1) + A_2(u-1).$$

Taking $u = 1$ gives $2A_1 = 1$, so $A_1 = 1/2$. Taking $x = -1$ gives $-2A_2 = 1$, so $A_2 = -1/2$. Thus, the integral is

$$\int \frac{du}{u^2-1} = \int \left(\frac{1/2}{u-1} - \frac{1/2}{u+1} \right) du = \frac{1}{2}\ln(u-1) - \frac{1}{2}\ln(u+1) + c,$$

provided u is in a range where the logarithms are defined; that is $u > 1$. This leads us to the final answer:

$$\int \frac{e^x}{e^{2x}-1}\, dx = \frac{1}{2}\ln(e^x-1) - \frac{1}{2}\ln(e^x+1) + c,$$

as long as $x > 0$.

26.7 We use the substitution $w = 1 + x^2$, so that $dw = 2x\,dx$ and the integral is

$$\int \frac{\ln(1+x^2)}{x^3}\, dx = \int \frac{\ln(1+x^2)}{x^4}\, x\,dx = \int \frac{\ln w}{(w-1)^2} \left(\frac{1}{2}\right) dw = \frac{1}{2} \int \frac{\ln w}{(w-1)^2}\, du.$$

Now we use integration by parts, taking $u'(t) = (w-2)^{-2}$ and $v(t) = \ln w$. This gives

$$\begin{aligned}\int (w-1)^{-2} \ln w\, dw &= -(w-1)^{-1} \ln w - \int \left(-(w-1)^{-1} \right) \frac{1}{w}\, dw \\ &= -\frac{\ln w}{w-1} + \int \frac{dw}{w(w-1)}.\end{aligned}$$

Here the method of partial fractions is applicable and we write

$$\frac{1}{w(w-1)} = \frac{A_1}{w} + \frac{A_2}{w-1}$$

and multiply through by $w(w-1)$ to obtain

$$1 = A_1(w-1) + A_2 w.$$

Taking $w = 0$ gives $-A_1 = 1$, so $A_1 = -1$. Taking $w = 1$ gives $A_2 = 1$. Thus, the integral is

$$\int \frac{dw}{w(w-1)} = \int \left(\frac{-1}{w} + \frac{1}{w-1} \right) dx = -\ln w + \ln(w-1) + c,$$

provided w is in a range where the logarithms are defined; that is $w > 1$. Putting all this together, we get

$$\begin{aligned}\int \frac{\ln(1+x^2)}{x^3}\, dx &= \frac{1}{2} \left(-\frac{\ln w}{w-1} - \ln w + \ln(w-1) \right) + c \\ &= -\frac{\ln(1+x^2)}{2x^2} - \frac{1}{2}\ln(1+x^2) + \frac{1}{2}\ln(x^2) + c,\end{aligned}$$

as long as $x \neq 0$.

Chapter 27

Exercise

27.1 The equation is separable. Separating and integrating, we obtain

$$\int \frac{1}{y^2}\,dy = \int -\frac{t^3}{\sqrt{1+t^4}}\,dt.$$

We can work out the integrals (using the substitution $r = 1 + t^4$ in the second one) and obtain

$$-\frac{1}{y} = -\frac{1}{2}\sqrt{1+t^4} + c,$$

where c is some constant. So, if we let $k = -2c$, we find that

$$y = \frac{2}{\sqrt{1+t^4}+k},$$

and this is the *general solution*: each value of the constant k gives a function $y(t)$ which satisfies the differential equation.

27.2 We have

$$-\frac{p}{q}\frac{dq}{dp} = \frac{p^2}{p^2+5p+6},$$

where $q = q^D(p)$ is the demand function. This equation is separable and so, separating and integrating, we obtain

$$\int \frac{dq}{q} = \int -\frac{p}{p^2+5p+6}\,dp.$$

We can work out the integrals, using the method of partial fractions, to see that

$$-\frac{p}{p^2+5p+6} = -\frac{p}{(p+2)(p+3)} = \frac{2}{p+2} - \frac{3}{p+3}$$

for the second one, and obtain

$$\ln q = 2\ln(p+2) - 3\ln(p+3) + c,$$

which is valid since q and p are positive. Taking the exponential of both sides:

$$q = \frac{(p+2)^2 e^c}{(p+3)^3} = K\frac{(p+2)^2}{(p+3)^3},$$

where $K = e^c$ is some positive constant. Given that $q(1) = 9$, we have $9K/64 = 9$, so $K = 64$. Therefore, the demand function is

$$q^D(p) = 64\frac{(p+2)^2}{(p+3)^3}.$$

27.3 The equation, written in standard form, is

$$\frac{dA}{dt} - rA(t) = t,$$

in which, in the standard terminology, $P(t) = -r$ and $Q(t) = t$. The integrating factor is

$$\mu(t) = e^{\int P(t)\,dt} = e^{\int -r\,dt} = e^{-rt}.$$

So

$$A(t)e^{-rt} = \int te^{-rt}\,dt = -\frac{t}{r}e^{-rt} + \int \frac{1}{r}e^{-rt}\,dt = -\frac{t}{r}e^{-rt} - \frac{1}{r^2}e^{-rt} + c,$$

where we have used integration by parts. Thus

$$A(t) = -\frac{t}{r} - \frac{1}{r^2} + ce^{rt}.$$

Now, given that the balance at time $t = 0$ is P, we have

$$P = -\frac{1}{r^2} + c,$$

so $c = P + 1/r^2$ and

$$A(t) = Pe^{rt} + \frac{e^{rt} - (1 + rt)}{r^2}.$$

27.4 The given differential equation can be written as

$$\frac{dv}{dt} - v(t) = e^{-2t},$$

which is in standard linear form

$$\frac{dv}{dt} + P(t)v = Q(t),$$

with $P(t) = -1$ and $Q(t) = e^{-2t}$. An integrating factor is

$$e^{\int P(t)\,dt} = e^{\int -t\,dt} = e^{-t}.$$

So, as

$$\frac{d}{dt}(v\mu(t)) = \mu(t)Q(t),$$

we have

$$ve^{-t} = \int e^{-t}e^{-2t}\,dt = \int e^{-3t}\,dt = \frac{e^{-3t}}{-3} + c$$

for some arbitrary constant c and so

$$v(t) = -\frac{1}{3}e^{-2t} + c\,e^{t}$$

is the general solution. Given that $v(0) = 2/3$, we can use this to get $c = 1$, so that

$$v(t) = -\frac{1}{3}e^{-2t} + e^{t}$$

is the particular solution we seek.

Chapter 28

Exercise

28.1 First we find the solution of the corresponding homogeneous equation. The auxiliary equation is $z^2 - 2z - 3 = 0$, which is $(z-3)(z+1) = 0$. This has two solutions, 3 and -1, so the general solution to the homogeneous equation is

$$Ae^{3x} + Be^{-x},$$

where A and B are constants. In order to find a particular solution, noting that the right-hand side of the equation is linear, we try $f^*(x) = ax + b$. Then, f^* is a solution if and only if

$$0 - 2a - 3(ax + b) = 1 + x,$$

or equivalently

$$-3ax - (2a + 3b) = 1 + x,$$

from which it follows that

$$-3a = 1 \quad \text{and} \quad -2a - 3b = 1.$$

Solving, we find $a = -1/3$ and $b = -1/9$, so a particular solution is

$$f^*(x) = -\frac{x}{3} - \frac{1}{9}.$$

The general solution to the differential equation is therefore

$$f(x) = -\frac{x}{3} - \frac{1}{9} + Ae^{3x} + Be^{-x}.$$

To determine the solution satisfying the given initial conditions, we note that

$$f'(x) = -\frac{1}{3} + 3Ae^{3x} - Be^{-x}.$$

So the initial conditions tell us that

$$-\frac{1}{9} + A + B = 1 \quad \text{and} \quad -\frac{1}{3} + 3A - B = 2,$$

from which we obtain $A = 31/36$ and $B = 1/4$. The required solution is therefore

$$f(x) = -\frac{x}{3} - \frac{1}{9} + \frac{31}{36}e^{3x} + \frac{1}{4}e^{-x}.$$

28.2 Let $y^*(t) = a\sin t + b\cos t$. Then

$$\frac{dy^*}{dt} = a\cos t - b\sin t \quad \text{and} \quad \frac{d^2y^*}{dt^2} = -a\sin t - b\cos t.$$

If y^* is a solution of the equation, then

$$(-a\sin t - b\cos t) - 5(a\cos t - b\sin t) + 6(a\sin t + b\cos t) = 18\sin t,$$

so

$$(5a + 5b)\sin t - (5a - 5b)\cos t = 18\sin t.$$

We must have $5a+5b = 18$ and $-5a+5b = 0$, so $a = b = 9/5$. A particular solution is therefore

$$y^*(t) = \frac{9}{5}\sin t + \frac{9}{5}\cos t.$$

To find the general solution, we note that the corresponding homogeneous equation has $z^2 - 5z + 6 = 0$ as its auxiliary equation. This is $(z-2)(z-3) = 0$ and so we have two solutions, 2 and 3, which means that the general solution to the homogeneous equation is

$$Ae^{2t} + Be^{3t},$$

where A and B are constants. The general solution is therefore

$$y(t) = \frac{9}{5}\sin t + \frac{9}{5}\cos t + Ae^{2t} + Be^{3t}.$$

The initial condition $y(0) = 0$ means that $9/5 + A + B = 0$. The derivative of y is

$$y'(t) = \frac{9}{5}\cos t - \frac{9}{5}\sin t + 2Ae^{2t} + 3Be^{3t},$$

so, since $y'(0) = 1$, we must have $9/5 + 2A + 3B = 1$. Solving these equations for A and B, we find $A = -23/5$ and $B = 14/5$. Therefore the required solution is

$$y(t) = \frac{9}{5}\sin t + \frac{9}{5}\cos t - \frac{23}{5}e^{2t} + \frac{14}{5}e^{3t}.$$

Chapter 29

Exercise

29.1 We have, from the first equation, and using the second equation with $t-1$ in place of t:

$$x_t = -3x_{t-1} - 4y_{t-1} = -3x_{t-1} - 4(2x_{t-2} + 3y_{t-2}) = -3x_{t-1} - 8x_{t-2} - 12y_{t-2}.$$

Now, to get an equation just involving x_t, x_{t-1} and x_{t-2}, we express y_{t-2} in terms of the sequence x_t. From the first of the two original equations, we have $4y_{t-1} = -x_t - 3x_{t-1}$. It follows (on replacing t by $t-1$) that

$$y_{t-2} = \frac{1}{4}(-x_{t-1} - 3x_{t-2}).$$

Therefore

$$\begin{aligned} x_t &= -3x_{t-1} - 8x_{t-2} - 12y_{t-2} \\ &= -3x_{t-1} - 8x_{t-2} - 3(-x_{t-1} - 3x_{t-2}) \\ &= x_{t-2} \end{aligned}$$

and we have the following second-order recurrence equation for x_t:

$$x_t - x_{t-2} = 0.$$

The auxiliary equation $z^2 - 1 = 0$ has solutions 1 and -1. So, for some constants A and B:

$$x_t = A + B(-1)^t.$$

Now, as already observed, $4y_{t-1} = -x_t - 3x_{t-1}$, so (replacing $t-1$ by t and, correspondingly, t by $t+1$):

$$\begin{aligned} y_t &= \frac{1}{4}(-x_{t+1} - 3x_t) \\ &= \frac{1}{4}\left[-\left[A + B(-1)^{t+1}\right] - 3\left[A + B(-1)^t\right]\right] \\ &= \frac{1}{4}\left[-\left[A - B(-1)^t\right] - 3A - 3B(-1)^t\right] \\ &= \frac{1}{4}\left[-4A - 2B(-1)^t\right] \\ &= -A - \frac{B}{2}(-1)^t. \end{aligned}$$

Now we use the initial conditions $x_0 = 2$, $y_0 = 1$. The first gives $A + B = 2$ and the second $-A - (B/2) = 1$. Solving these, we obtain $A = -4$ and $B = 6$. The solutions are therefore

$$x_t = -4 + 6(-1)^t \quad \text{and} \quad y_t = 4 - 3(-1)^t.$$

29.2 The eigenvalues of A are found by solving $|A - \lambda I| = 0$, which is

$$\lambda^2 - 12\lambda + 32 = 0 \qquad \text{or} \qquad (\lambda - 4)(\lambda - 8) = 0,$$

so the eigenvalues are 4 and 8. For an eigenvector corresponding to 4, we solve $(A - 4I)\mathbf{x} = \mathbf{0}$, which is

$$\begin{pmatrix} 3 & 1 \\ 3 & 1 \end{pmatrix}\begin{pmatrix} x \\ y \end{pmatrix} = \begin{pmatrix} 0 \\ 0 \end{pmatrix},$$

and this system is equivalent to the single equation $3x + y = 0$, a solution to which is $x = 1$, $y = -3$. So, an eigenvector corresponding to 4 is

$$\begin{pmatrix} 1 \\ -3 \end{pmatrix},$$

or any non-zero multiple of this. Similarly, to find an eigenvector corresponding to eigenvalue 8, we solve $(A - 8I)\mathbf{x} = \mathbf{0}$, which is

$$\begin{pmatrix} -1 & 1 \\ 3 & -3 \end{pmatrix}\begin{pmatrix} x \\ y \end{pmatrix} = \begin{pmatrix} 0 \\ 0 \end{pmatrix},$$

and this system is equivalent to the single equation $-x + y = 0$, a solution to which is $x = 1$, $y = 1$. So, an eigenvector corresponding to 8 is

$$\begin{pmatrix} 1 \\ 1 \end{pmatrix},$$

or, again, any non-zero multiple of this.

29.3 Normally, we determine the eigenvalues of a matrix (by finding the roots of the characteristic polynomial) and then, once we know the eigenvalues, we can determine corresponding eigenvectors. But here we can proceed differently (and more directly), because we are given the eigenvectors in the question.

Recall that what it means for vector $\mathbf{x}$ to be an eigenvector corresponding to eigenvalue λ is that $\mathbf{x}$ is non-zero and $A\mathbf{x} = \lambda\mathbf{x}$. Thus, since

$$\begin{pmatrix} 2 & -1 \\ 0 & 1 \end{pmatrix} \begin{pmatrix} 1 \\ 1 \end{pmatrix} = \begin{pmatrix} 1 \\ 1 \end{pmatrix} = 1 \begin{pmatrix} 1 \\ 1 \end{pmatrix},$$

it follows that 1 is an eigenvalue and $\begin{pmatrix} 1 \\ 1 \end{pmatrix}$ a corresponding eigenvector.

Similarly, since

$$\begin{pmatrix} 2 & -1 \\ 0 & 1 \end{pmatrix} \begin{pmatrix} 1 \\ 0 \end{pmatrix} = \begin{pmatrix} 2 \\ 0 \end{pmatrix} = 2 \begin{pmatrix} 1 \\ 0 \end{pmatrix},$$

it follows that 2 is an eigenvalue and $\begin{pmatrix} 1 \\ 0 \end{pmatrix}$ is a corresponding eigenvector.

So these vectors are indeed eigenvectors, and the eigenvalues are 1 and 2.

29.4 To find the eigenvalues of this matrix, we solve the equation $|A - \lambda I| = 0$. This is

$$\begin{vmatrix} 5-\lambda & -1 \\ -3 & 7-\lambda \end{vmatrix} = (5-\lambda)(7-\lambda) - 3 = 0,$$

which, multiplying out the brackets, gives us the quadratic equation

$$\lambda^2 - 12\lambda + 32 = 0,$$

which becomes

$$(\lambda - 4)(\lambda - 8) = 0,$$

and so the eigenvalues are 4 and 8. To find the corresponding eigenvectors we seek a non-zero vector $\mathbf{x}$, which is a solution to the equation $(A - \lambda I)\mathbf{x} = \mathbf{0}$.

For $\lambda = 4$, we solve

$$\begin{pmatrix} 1 & -1 \\ -3 & 3 \end{pmatrix} \begin{pmatrix} x \\ y \end{pmatrix} = \mathbf{0},$$

which is equivalent to $x = y$, so an eigenvector is

$$\begin{pmatrix} 1 \\ 1 \end{pmatrix}$$

(or any non-zero multiple of this).

For $\lambda = 8$, we solve

$$\begin{pmatrix} -3 & -1 \\ -3 & -1 \end{pmatrix} \begin{pmatrix} x \\ y \end{pmatrix} = \mathbf{0},$$

which becomes $y = -3x$, so an eigenvector is

$$\begin{pmatrix} 1 \\ -3 \end{pmatrix}$$

(or any non-zero multiple of this).

It follows from the standard theory that if we let

$$P = \begin{pmatrix} 1 & 1 \\ 1 & -3 \end{pmatrix} \quad \text{and} \quad D = \begin{pmatrix} 4 & 0 \\ 0 & 8 \end{pmatrix},$$

then $P^{-1}AP = D$.

Of course, this is only one of the many possible pairs of matrices that we could choose for P and D; others are possible depending on what eigenvectors we choose to form the columns of P and the order in which we choose to place them. For instance, another answer is that $P^{-1}AP = D$, when

$$P = \begin{pmatrix} 1 & 1 \\ -3 & 1 \end{pmatrix} \quad \text{and} \quad D = \begin{pmatrix} 8 & 0 \\ 0 & 4 \end{pmatrix}.$$

Chapter 30

Exercise

30.1 The usual approach to diagonalisation (see Exercise 29.2) shows that if

$$P = \begin{pmatrix} 1 & 1 \\ -3 & 1 \end{pmatrix} \quad \text{and} \quad D = \begin{pmatrix} 4 & 0 \\ 0 & 8 \end{pmatrix},$$

then $P^{-1}AP = D$. Now we use the fact that, since $P^{-1}AP = D$, we have $A = PDP^{-1}$. So $A^n = PD^nP^{-1}$ and hence, since

$$P^{-1} = \frac{1}{4}\begin{pmatrix} 1 & -1 \\ 3 & 1 \end{pmatrix},$$

we have

$$A^n = \begin{pmatrix} 1 & 1 \\ -3 & 1 \end{pmatrix}\begin{pmatrix} 4 & 0 \\ 0 & 8 \end{pmatrix}^n \frac{1}{4}\begin{pmatrix} 1 & -1 \\ 3 & 1 \end{pmatrix} = \frac{1}{4}\begin{pmatrix} 1 & 1 \\ -3 & 1 \end{pmatrix}\begin{pmatrix} 4^n & 0 \\ 0 & 8^n \end{pmatrix}\begin{pmatrix} 1 & -1 \\ 3 & 1 \end{pmatrix}.$$

On calculating this matrix product (details omitted), we obtain

$$A^n = \frac{1}{4}\begin{pmatrix} 4^n + 3(8^n) & -4^n + 8^n \\ -3(4^n) + 3(8^n) & 3(4^n) + 8^n \end{pmatrix}.$$

30.2 To find the eigenvalues of this matrix, we solve the equation

$$\begin{vmatrix} -3-\lambda & -4 \\ 2 & 3-\lambda \end{vmatrix} = 0.$$

This gives us

$$(\lambda^2 - 9) + 8 = 0,$$

which is the quadratic equation $\lambda^2 = 1$, and so the eigenvalues are -1 and 1. To find the corresponding eigenvectors we seek a non-zero vector $\mathbf{x}$, which is a solution to the equation $(A - \lambda I)\mathbf{x} = \mathbf{0}$.

For $\lambda = -1$, we solve

$$\begin{pmatrix} -2 & -4 \\ 2 & 4 \end{pmatrix} \begin{pmatrix} x \\ y \end{pmatrix} = \mathbf{0},$$

which is equivalent to $x = -2y$, so an eigenvector is $\begin{pmatrix} -2 \\ 1 \end{pmatrix}$.

For $\lambda = 1$, we solve

$$\begin{pmatrix} -4 & -4 \\ 2 & 2 \end{pmatrix} \begin{pmatrix} x \\ y \end{pmatrix} = \mathbf{0},$$

which is equivalent to $x = -y$, so an eigenvector is $\begin{pmatrix} -1 \\ 1 \end{pmatrix}$.

So, if we take

$$P = \begin{pmatrix} -2 & -1 \\ 1 & 1 \end{pmatrix} \quad \text{and} \quad D = \begin{pmatrix} -1 & 0 \\ 0 & 1 \end{pmatrix},$$

then we have an invertible matrix P and a diagonal matrix D, such that $P^{-1}AP = D$.

Using the normal change of variable approach, we let $\mathbf{x}_t = P\mathbf{u}_t$, where

$$\mathbf{x}_t = \begin{pmatrix} x_t \\ y_t \end{pmatrix} \quad \text{and} \quad \mathbf{u}_t = \begin{pmatrix} u_t \\ v_t \end{pmatrix},$$

so that the given system of recurrence equations can be written as $\mathbf{x}_t = A\mathbf{x}_{t-1}$ and, equivalently, $\mathbf{u}_t = D\mathbf{u}_{t-1}$. Using this, we have

$$\begin{pmatrix} u_t \\ v_t \end{pmatrix} = \begin{pmatrix} -1 & 0 \\ 0 & 1 \end{pmatrix} \begin{pmatrix} u_{t-1} \\ v_{t-1} \end{pmatrix},$$

so

$$u_t = -u_{t-1} \quad \text{and} \quad v_t = v_{t-1},$$

and this pair of recurrence equations can easily be solved to yield

$$u_t = (-1)^t u_0 \qquad \text{and} \qquad v_t = (1)^t v_0 = v_0.$$

Now, $\mathbf{u}_t = P^{-1}\mathbf{x}_t$, so taking $t = 0$ we have

$$\begin{pmatrix} u_0 \\ v_0 \end{pmatrix} = P^{-1} \begin{pmatrix} x_0 \\ y_0 \end{pmatrix} = \frac{1}{-1} \begin{pmatrix} 1 & 1 \\ -1 & -2 \end{pmatrix} \begin{pmatrix} 2 \\ 1 \end{pmatrix} = \begin{pmatrix} -3 \\ 4 \end{pmatrix}$$

and so $u_t = -3(-1)^t$ and $v_t = 4$. Then, using $\mathbf{x}_t = P\mathbf{u}_t$, we have

$$\begin{pmatrix} x_t \\ y_t \end{pmatrix} = \begin{pmatrix} -2 & -1 \\ 1 & 1 \end{pmatrix} \begin{pmatrix} -3(-1)^t \\ 4 \end{pmatrix} = \begin{pmatrix} 6(-1)^t - 4 \\ -3(-1)^t + 4 \end{pmatrix},$$

which means that

$$x_t = 6(-1)^t - 4 \quad \text{and} \quad y_t = -3(-1)^t + 4$$

are the required solutions to our coupled system of recurrence equations.

An alternative approach to finding A and B that does not involve using the inverse of P can be taken as follows. Using $\mathbf{x}_t = P\mathbf{u}_t$, we have

$$\begin{pmatrix} x_t \\ y_t \end{pmatrix} = \begin{pmatrix} -2 & -1 \\ 1 & 1 \end{pmatrix} \begin{pmatrix} (-1)^t u_0 \\ v_0 \end{pmatrix} = \begin{pmatrix} -2(-1)^t u_0 - v_0 \\ (-1)^t u_0 + v_0 \end{pmatrix}.$$

Then, using the initial conditions $x_0 = 2$ and $y_0 = 1$, we can solve the equations

$$2 = -2u_0 - v_0 \qquad \text{and} \qquad 1 = u_0 + v_0$$

by adding them to get $u_0 = -3$ and, in turn, this gives us $v_0 = 4$. Consequently, we find that

$$x_t = 6(-1)^t - 4 \quad \text{and} \quad y_t = -3(-1)^t + 4$$

as before.

Depending on whether t is even or odd, we see that x_t will be 2 or -10 and y_t will be 1 or 7. Thus, as t increases, x_t oscillates constantly between 2 and -10 and y_t oscillates constantly between 1 and 7.

30.3 To find the eigenvalues of this matrix, we solve the equation

$$\begin{vmatrix} 1-\lambda & 2 \\ 0 & -1-\lambda \end{vmatrix} = 0.$$

This gives us

$$(1-\lambda)(-1-\lambda) = 0,$$

and so the eigenvalues are 1 and -1. To find the corresponding eigenvectors we seek a non-zero vector $\mathbf{x}$, which is a solution to the equation $(A - \lambda I)\mathbf{x} = \mathbf{0}$.

For $\lambda = 1$, we solve

$$\begin{pmatrix} 0 & 2 \\ 0 & -2 \end{pmatrix} \begin{pmatrix} x \\ y \end{pmatrix} = \mathbf{0},$$

which is equivalent to $2y = 0$, so an eigenvector is $\begin{pmatrix} 1 \\ 0 \end{pmatrix}$.

For $\lambda = -1$, we solve

$$\begin{pmatrix} 2 & 2 \\ 0 & 0 \end{pmatrix} \begin{pmatrix} x \\ y \end{pmatrix} = \mathbf{0},$$

which is equivalent to $x + y = 0$, so an eigenvector is $\begin{pmatrix} 1 \\ -1 \end{pmatrix}$.

So, if we take

$$P = \begin{pmatrix} 1 & 1 \\ 0 & -1 \end{pmatrix} \quad \text{and} \quad D = \begin{pmatrix} 1 & 0 \\ 0 & -1 \end{pmatrix},$$

then we have an invertible matrix P and a diagonal matrix D such that $P^{-1}AP = D$.

Using the normal change of variable approach, we let $\mathbf{f}(t) = P\mathbf{F}(t)$, where

$$\mathbf{f}(t) = \begin{pmatrix} f(t) \\ g(t) \end{pmatrix} \quad \text{and} \quad \mathbf{F}(t) = \begin{pmatrix} F(t) \\ G(t) \end{pmatrix},$$

so that the given system of recurrence equations can be written as $\mathbf{f}'(t) = A\mathbf{f}(t)$ and, equivalently, $\mathbf{F}'(t) = D\mathbf{F}(t)$. Using this, we then have

$$\begin{pmatrix} F'(t) \\ G'(t) \end{pmatrix} = \begin{pmatrix} 1 & 0 \\ 0 & -1 \end{pmatrix} \begin{pmatrix} F(t) \\ G(t) \end{pmatrix},$$

so

$$F'(t) = F(t) \quad \text{and} \quad G'(t) = -G(t),$$

and this pair of differential equations can easily be solved to get

$$F(t) = e^t F(0) \qquad \text{and} \qquad G(t) = e^{-t}G(0).$$

We could, at this stage, use $\mathbf{F}(0) = P^{-1}\mathbf{f}(0)$ to determine $F(0)$ and $G(0)$ from the initial values $f(0)$ and $g(0)$. Alternatively, we can proceed as follows.

Using $\mathbf{f}(t) = P\mathbf{F}(t)$, we have

$$\begin{pmatrix} f(t) \\ g(t) \end{pmatrix} = \begin{pmatrix} 1 & 1 \\ 0 & -1 \end{pmatrix} \begin{pmatrix} e^t F(0) \\ e^{-t}G(0) \end{pmatrix} = \begin{pmatrix} e^t F(0) + e^{-t}G(0) \\ -e^{-t}G(0) \end{pmatrix},$$

which means that

$$f(t) = e^t F(0) + e^{-t}G(0) \quad \text{and} \quad g(t) = -e^{-t}G(0)$$

is the general solution to our system of differential equations. Then, with the initial conditions $f(0) = 1$ and $g(0) = 1$, we get the equations

$$1 = F(0) + G(0) \qquad \text{and} \qquad 1 = -G(0),$$

which are easily solved to get $F(0) = 2$ and $G(0) = -1$. So

$$f(t) = 2e^t - e^{-t} \qquad \text{and} \qquad g(t) = e^{-t}$$

is the required solution.

Alternatively, we can follow the 'otherwise' route and not use diagonalisation. We can reduce these coupled differential equations to a second-order differential equation. This is easily done by taking the first equation and writing it as

$$2g(t) = f'(t) - f(t),$$

so that we can differentiate it to get

$$2g'(t) = f''(t) - f'(t),$$

and then, by noting that the second equation gives us $2g'(t) = -2g(t)$, we have

$$f''(t) - f'(t) = -\left(f'(t) - f(t)\right)$$

and hence

$$f''(t) - f(t) = 0.$$

This is a standard second-order differential equation of the type we know how to solve. The auxiliary equation is $m^2 - 1 = 0$, with solutions $m = \pm 1$, and these two distinct real solutions imply that the general solution is

$$f(t) = Ae^t + Be^{-t},$$

for constants A and B. We can then find the general solution for $g(t)$ by noting that

$$2g(t) = f'(t) - f(t) = \left(Ae^t - Be^{-t}\right) - \left(Ae^t + Be^{-t}\right) = -2Be^{-t}$$

and so

$$g(t) = -Be^{-t}.$$

Then, as we did above, we can use these general solutions and the given initial conditions to find the required particular solutions.

Appendix B Answers to Problems

1.1 $x = 1,\ y = 2.$

1.2 $x = 4,\ y = 1.$

1.3 E consists of one point, $(27, 13/2)$. $q^S(p) = 6p - 12$, $q^D(p) = 40 - 2p$, $p^S(q) = \frac{1}{6}q + 2$, $p^D(q) = 20 - \frac{1}{2}q$.

1.4 $p^T = \frac{13}{2} + \frac{3}{4}T.$

1.5 $27T - \frac{3}{2}T^2$, 13.125.

1.6 $p^T = \frac{3}{4} + \frac{3}{4}T$, $q^T = 5 - 3T.$

2.1 Only the first and third functions have inverses. For the first, $f^{-1}(y) = y/5$ and for the second, $f^{-1}(y) = y^{1/5}$

2.3 $E = \{(1,2), (-4, 1/3)\}$. Only the first of these points is 'economically significant'.

2.4 In order, the formulae are

$$\frac{1}{x^4} + 1, \quad \frac{1}{(x^2+1)^2}, \quad \sqrt{x^2+1},\ x+1 \quad \text{and} \quad \sqrt{\frac{1}{x^4} + 1}.$$

2.5 $q^S(p) = 3p - 1$, $q^D(p) = 2 - p$, $p^S(q) = \frac{q+1}{3}$, $p^D(q) = 2 - q$.

2.6 There are no solutions if $-2 < \alpha < 2$. If $\alpha = 2$ or $\alpha = -2$ there is just one solution: when $\alpha = 2$ this is $x = -1$ and when $\alpha = -2$ this is $x = 1$. For all other values of α (i.e., for $\alpha < -2$ and $\alpha > 2$) there are two solutions:

$$\frac{-\alpha + \sqrt{\alpha^2 - 4}}{2} \quad \text{and} \quad \frac{-\alpha - \sqrt{\alpha^2 - 4}}{2}.$$

3.1 $y_t = 10 - 6(1/2)^t.$

3.2 (a) tends to 5, decreasing, as t tends to ∞; (b) tends to 0, oscillating, as t tends to infinity.

3.3 $y_t = -1 + 8(4^t)$; tends to infinity as $t \to \infty$.

3.4 $y^* = 3$ and $y_t = 3 + 3(1/4)^t$, which tends to 3, decreasing, as $t \to \infty$.

3.5 $y_t = \dfrac{100E}{W} + \left(8000 - \dfrac{100E}{W}\right)\left(1 - \dfrac{W}{100}\right)^t.$

3.6 $p^* = \dfrac{3}{2}$, $p_t = \dfrac{3k}{2k-r} + \left(1 - \dfrac{3k}{2k-r}\right)(1 + r - 2k)^t.$

4.1 $y_t = 20\,000 + (P - 20\,000)(1.05)^t.$

4.2 \$21 909.295, \$20 100.27.

4.3 \$12 409.04.
4.4 In all cases, choose the first option.
4.5 (a) \$2160.00, (b) \$2164.86, (c) \$2166.56.

5.1 $p_t = 9 + \left(-\frac{1}{3}\right)^t$, $p_t \to 9$ and $q_t \to 2$ as $t \to \infty$.
5.2 $p_t = 3 - \frac{1}{2}\left(-\frac{5}{4}\right)^t$, $q_t = 4 + 6\left(-\frac{5}{4}\right)^t$.
5.3 Unstable.

6.1 $f'(x) = 3(4x+5)(2x^2+5x+4)^2$ and $g'(x) = (x^2-6x)(x-3)^{-2}$. The function g is not defined when $x = 3$.
6.2 The derivative of $f^{-1}(x)$ is $\frac{1}{5}(x-3)^{-4/5}$.
6.3 $h'(x) = x^2(x^3+8)^{-2/3}$, which is not defined when $x = -2$:

$$f'(x) = \frac{1}{2}\left(x + (x^3+8)^{1/3}\right)^{-1/2}\left(1 + x^2(x^3+8)^{-2/3}\right).$$

6.4 $C'(q) = 20 + \dfrac{2+3q}{2\sqrt{1+q}}$.
6.5 Using the derivative, an approximation to the increase is 107. (The actual change is 116.5.)

7.1 (a) 128, (b) $1/5 = 0.2$, (c) $1/100\,000 = 0.00001$.
7.2 $60\,000(1.06)^{10}e^{5(0.007)} = 152\,480$.
7.3 (a) $30\,000/(1.08)^{10} = 13\,895.80$, (b) $30\,000e^{-10(0.08)} = 13\,479.87$.
7.4 (a) $f'(x) = \dfrac{4x+5}{2x^2+5x+4}$, (b) $g'(x) = 12x^3\exp x + 3x^4\exp x$.
Both derivatives are defined for all values of x.
7.5 $f'(x) = \dfrac{2x^3-2x}{x^4+4x^2+3}$.
7.6 $\dfrac{\cos(\ln x)}{x}$, $\dfrac{2xe^{x^2}}{(\cos(e^{x^2}))^2}$.
7.7 If $X = \ln x$ and $Y = \ln y$, then the relationship is $\frac{1}{4}X + \frac{2}{3}Y = \ln 8$.

8.1 The critical point is 2. It is an inflexion point.
8.2 The maximum occurs at $x = 2$, where the value is 7; the minimum is at $x = 4$, where the value is -1.
8.3 The maximum occurs at $x = 2$, where the value is 9; the minimum is at $x = 4$, where the value is -7.
8.4 The maximum value is 29, at $x = 2$.
8.5 The maximum value is $4/e^2$, obtained when $x = 2$; the minimum is 0, at $x = 0$.
8.6 $P(t) = (500 + 100t)e^{-0.05t}$; sell after 15 years.
8.7 $P(t) = e^{\sqrt{t}}e^{-0.125t}$; sell after 16 years.

9.1 $\varepsilon(p) = \dfrac{4p}{70-4p}$; valid for $0 \le p < \dfrac{70}{4}$; inelastic for $0 \le p < \dfrac{70}{8}$.

9.3 $\varepsilon(p) = \dfrac{3p^3}{2(2+p^3)}$, so the demand is elastic for $p > 4^{1/3}$.

9.5 (i) $p^D(q) = 25 - 0.05q$; (ii) $\Pi(q) = 24q - 0.08q^2$; (iii) $q_m = 150$.

9.6 The maximum profit is 47 633.3, achieved when $q = 36.4839$.

10.1 (a) 196, (b) $\Pi(q) = 2q^3 - 10q^2 - 196$, (c) $q_s = 5$, (d) $q_b = 7$, (e) the firm's supply set is the union of three pieces:

- the points $(0, p)$ for $0 < p < 75$;
- the points (q, p) such that $p = 3q^2 - 20q + 100$ for $75 \le p \le 200$;
- the points $(10, p)$ for $p > 200$.

11.1 $\dfrac{\partial f}{\partial x} = 2xy + y^3$, $\dfrac{\partial f}{\partial y} = x^2 + 3xy^2$,

$\dfrac{\partial^2 f}{\partial x^2} = 2y$, $\dfrac{\partial^2 f}{\partial x \partial y} = 2x + 3y^2 = \dfrac{\partial^2 f}{\partial y \partial x}$, $\dfrac{\partial^2 f}{\partial y^2} = 6xy$.

$\dfrac{\partial g}{\partial x} = \dfrac{10}{3}x^{-1/3}y^{1/4}$, $\dfrac{\partial g}{\partial y} = \dfrac{5}{4}x^{2/3}y^{-3/4}$,

$\dfrac{\partial^2 g}{\partial x^2} = -\dfrac{10}{9}x^{-4/3}y^{1/4}$, $\dfrac{\partial^2 g}{\partial x \partial y} = \dfrac{5}{6}x^{-1/3}y^{-3/4} = \dfrac{\partial^2 g}{\partial y \partial x}$, $\dfrac{\partial^2 g}{\partial y^2} = -\dfrac{15}{16}x^{2/3}y^{-7/4}$.

$\dfrac{\partial h}{\partial x} = 2x(x^2+y^3)^{2/3} + \dfrac{4}{3}x^3(x^2+y^3)^{-1/3}$, $\dfrac{\partial h}{\partial y} = 2x^2y^2(x^2+y^3)^{-1/3}$,

$\dfrac{\partial^2 h}{\partial x^2} = 2(x^2+y^3)^{2/3} + \dfrac{20}{3}x^2(x^2+y^3)^{-1/3} - \dfrac{8}{9}x^4(x^2+y^3)^{-4/3}$,

$\dfrac{\partial^2 h}{\partial x \partial y} = 4xy^2(x^2+y^3)^{-1/3} - \dfrac{4}{3}x^3y^2(x^2+y^3)^{-4/3} = \dfrac{\partial^2 h}{\partial y \partial x}$,

$\dfrac{\partial^2 h}{\partial y^2} = 4x^2y(x^2+y^3)^{-1/3} - 2x^2y^4(x^2+y^3)^{-4/3}$.

11.2 $\dfrac{\partial f}{\partial x} = y^2x^{y^2-1}$, $\dfrac{\partial f}{\partial y} = 2yx^y \ln x$.

11.3 $F'(t) = 9t^2 - 10t - 16$.

11.4 If $Q(t) = q(k(t), l(t))$, then $Q'(t) = 32t^3 + 120t^2 + 100t$.

11.5 $\dfrac{\partial F}{\partial u} = 7u^6 + 10u^4v^2 - 4u^3v^3 + 3u^2v^4 - 4uv^5$,

$\dfrac{\partial F}{\partial v} = 4u^5v - 3u^4v^2 + 4u^3v^3 - 10u^2v^4 - 7v^6$.

11.6 $\dfrac{\partial F}{\partial u} = u^{7/4}v^{3/4}(u^2+v^2)^{-1/2} + \dfrac{3}{4}u^{-1/4}v^{3/4}(u^2+v^2)^{1/2}$,

$\dfrac{\partial F}{\partial v} = u^{3/4}v^{7/4}(u^2+v^2)^{-1/2} + \dfrac{3}{4}u^{3/4}v^{-1/4}(u^2+v^2)^{1/2}$.

12.1 $\dfrac{\partial f}{\partial x} = \alpha A x^{\alpha-1}y^{\beta}$, $\dfrac{\partial f}{\partial y} = \beta A x^{\alpha}y^{\beta-1}$,

$\dfrac{\partial^2 f}{\partial x^2} = \alpha(\alpha-1)Ax^{\alpha-2}y^{\beta}$, $\dfrac{\partial^2 f}{\partial x \partial y} = \alpha\beta A x^{\alpha-1}y^{\beta-1} = \dfrac{\partial^2 f}{\partial y \partial x}$,

$\dfrac{\partial^2 f}{\partial y^2} = \beta(\beta-1)Ax^{\alpha}y^{\beta-2}$. The required gradient is $-4/3$.

12.2 $\dfrac{\partial g}{\partial x} = 2xy^{3/2}$, $\dfrac{\partial g}{\partial y} = \dfrac{3}{2}x^2y^{1/2}$. The required slope is $-8/3$.

12.3 f is homogeneous of degree 1; g is not homogeneous; h is homogeneous of degree 2/9.

12.5 $\dfrac{\partial q}{\partial k} = -\dfrac{\partial g/\partial k}{\partial g/\partial q},\ \dfrac{\partial q}{\partial l} = -\dfrac{\partial g/\partial l}{\partial g/\partial q}.$

$$\frac{\partial q}{\partial k} = -\frac{2q^3k + ql}{3q^2k^2 + kl},\ \frac{\partial q}{\partial l} = -\frac{3l^2 + qk}{3q^2k^2 + kl}.$$

13.1 Maximum value is 10, at the point $(1, 1)$.
13.2 The only critical point is $(20, 20)$, a minimum.
13.4 u has a saddle point at $(0, 0)$ and a minimum at $(-1, 1)$.

v has minima at $(3/2, 3/2)$ and $(3/2, -3/2)$. At $(0, 0)$, $D = 0$ and so the second-order conditions tell us nothing. However, we can observe that $f(x, 0) = x^3$ and this has an inflexion point at 0. It follows that f cannot have a maximum or a minimum at $(0, 0)$. In other words, $(0, 0)$ is a saddle point of f.
13.5 The firm should produce 2 units of X and 4 units of Y.
13.6 The maximum profit is 42, when $x = 1$ and $y = 4$.
13.7 The required production levels are 2 units of X and 2 units of Y.

14.1 (i) $(18, -18, -9)$, (ii) $(5, 22, 25)$, (iii) 0 and 7.
14.2 For Mick, Keith and Charlie, the set is convex. The set is not convex for Bill.
14.3 Mick prefers **r** to **s** and **s** to **t**. Keith prefers **t** to **s** and **s** to **r**. Bill prefers **t** to **s** and **s** to **r**. Charlie prefers **s** to **t** and **t** to **r**.
14.4 Mick will buy 52 glasses of orange juice and no sunflower seeds. Keith and Bill will each buy no orange juice and 260 grams of sunflower seeds. Charlie will buy 26 glasses of orange juice and 130 grams of sunflower seeds.

15.1 $AB = \begin{pmatrix} -41 \\ -50 \end{pmatrix}.$

15.2 $AB = \begin{pmatrix} 6 & 4 & 12 \\ 13 & 11 & 27 \\ 16 & 12 & 32 \end{pmatrix}$ and $BA = \begin{pmatrix} 3 & 7 & 7 \\ 10 & 26 & 23 \\ 9 & 21 & 20 \end{pmatrix}.$

15.3 (i) $AB = \begin{pmatrix} 1 & 4 \\ 1 & 1 \\ -3 & 4 \end{pmatrix}$, (ii) $AC = \begin{pmatrix} 6 & 9 \\ 1 & 5 \\ 6 & 5 \end{pmatrix}$, (iii) $A(B + C) = \begin{pmatrix} 7 & 13 \\ 2 & 6 \\ 3 & 9 \end{pmatrix}.$

15.4 $A^n = \begin{pmatrix} 3^n & (3^n - 1)/2 \\ 0 & 1 \end{pmatrix}.$

15.5 $B^n = \begin{pmatrix} 1 & -n \\ 0 & 1 \end{pmatrix}.$

16.2 $x_1 = -1, x_2 = 5, x_3 = 4.$
16.3 $x_1 = 5/9, x_2 = -4/9, x_3 = 5/9.$
16.4 $x_1 = 1, x_2 = 1, x_3 = 2.$
16.5 $(7, 11, 1) = (1, 4, 7) + (2, 5, -4) + 2(2, 1, -1).$
16.6 $a = 1, b = 1, c = -1.$
16.7 $a = 50, b = -10, c = 20.$

17.1 The rank is 2; general solution is $\mathbf{x} = \begin{pmatrix} 1 \\ 0 \\ 0 \end{pmatrix} + s \begin{pmatrix} 1 \\ -1 \\ 1 \end{pmatrix}, s \in \mathbb{R}.$

17.2 The general solution is $\mathbf{x} = \begin{pmatrix} 37 \\ -43 \\ 0 \end{pmatrix} + s \begin{pmatrix} -9/5 \\ 16/5 \\ 1 \end{pmatrix}, s \in \mathbb{R}.$

17.3 The rank is 2; general solution is $\mathbf{x} = \begin{pmatrix} \frac{9}{2} \\ \frac{29}{2} \\ 0 \\ 0 \end{pmatrix} + s \begin{pmatrix} \frac{1}{2} \\ -\frac{3}{2} \\ 1 \\ 0 \end{pmatrix} + t \begin{pmatrix} \frac{1}{4} \\ -\frac{3}{4} \\ 0 \\ 1 \end{pmatrix}, s, t \in \mathbb{R}.$

17.5 There are no solutions.
17.6 Consistent if $k = -3$. When $k = -3$, $x_1 = x_2 = x_3 = 1$ is the only solution.
17.7 When $c = -16b$, the solution is $x_1 = 6a + 3b$, $x_2 = b - 2a$ and $x_3 = -3a - 4b$.
17.8 One possible arbitrage portfolio is $Y = \begin{pmatrix} 1000 & -1000 & 0 \end{pmatrix}$.

18.1 (a) $B^{-1} = \begin{pmatrix} 10 & -19 & 5 \\ -3 & 7 & -2 \\ 1 & -3 & 1 \end{pmatrix}$, (c) $\mathbf{x} = \begin{pmatrix} 10c_1 - 19c_2 + 5c_3 \\ -3c_1 + 7c_2 - 2c_3 \\ c_1 - 3c_2 + c_3 \end{pmatrix}.$

18.2 $\dfrac{1}{36} \begin{pmatrix} -25 & 38 & -5 \\ -3 & 6 & -3 \\ 7 & -2 & -1 \end{pmatrix}, \dfrac{1}{5} \begin{pmatrix} -1 & 9 & -16 \\ 0 & -5 & 10 \\ -5 & 5 & -5 \end{pmatrix}.$

18.3 To two decimal places, $\begin{pmatrix} 1.72 & 0.78 & 0.63 \\ 0.89 & 1.61 & 0.63 \\ 0.55 & 0.70 & 1.56 \end{pmatrix}.$

18.5 $\dfrac{1}{2} \begin{pmatrix} -3 & 5 & 4 & -4 \\ -4 & 12 & -2 & -2 \\ 3 & -7 & 0 & 2 \\ 3 & -5 & -1 & 2 \end{pmatrix}.$

18.7 The matrix is not invertible.
18.8 $q^* = 10$ and $p^* = 2$.
18.9 $Y^* = 60$ and $r^* = 0.1$.

19.1 $\begin{pmatrix} 0.8 & -0.2 \\ -0.1 & 0.7 \end{pmatrix} \mathbf{x} = \begin{pmatrix} d_1 \\ d_2 \end{pmatrix}, \mathbf{x} = \dfrac{1}{27} \begin{pmatrix} 35d_1 + 10d_2 \\ 5d_1 + 40d_2 \end{pmatrix}.$
19.2 The production (in dollars) of grommets and widgets should be 979.02 and 664.34, respectively.
19.3 The weekly production level (in dollars) of X, Y and Z should be 769.50, 910.55 and 1688.00, respectively.
19.4 To be correct to one decimal place, we need

$$(I - A)^{-1} = I + A + A^2 + A^3 + A^4 = \begin{pmatrix} 1.29 & 0.36 \\ 0.18 & 1.47 \end{pmatrix}.$$

20.1 -108
20.3 4.

20.4 It is invertible if and only if a, b, c are three distinct numbers.
20.5 $x = 4$ and $y = 1$.

21.1 $60(100^{4/3})$, approximately 27 849.53.
21.2 $x = 36$ and $y = 64$.
21.3 The cost function is $C(q) = 2\sqrt{3}q^2$. The optimal production level is $500/\sqrt{3}$, approximately 288.68.
21.4 The supply set contains the points $(q, 5\sqrt{6}q^{3/2})$ for $0 \le q \le L$ and (L, p) for $p > 5\sqrt{6}L^{3/2}$.
21.5 The supply set contains the points $(0, p)$ for $0 \le p \le Z$ and (L, p) for $p > Z$.
21.6 \$200/9.

22.1 35 units of fruit and 7 of chocolate.
22.2 $0.12x_1 + 0.2x_2 = 1.8$, $L = x_1x_2^2 - \lambda(0.12x_1 + 0.2x_2 - 1.8)$. The optimal bundle is $(5, 6)$.
22.3 10 hours should be spent on research and 30 hours on teaching.
22.4 $\alpha = 1/3$, 20 of each.
22.5 $q_1(p_1, p_2, M) = \dfrac{5M}{7p_1}$, $q_2(p_1, p_2, M) = \dfrac{2M}{7p_2}$,

$$V(p_1, p_2, M) = \frac{5^{5/6}2^{1/3}M^{7/6}}{7^{7/6}p_1^{5/6}p_2^{1/3}}, \quad \lambda^* = \frac{1}{6}\,\frac{5^{5/6}2^{1/3}M^{1/6}}{7^{1/6}p_1^{5/6}p_2^{1/3}}.$$

Actual change in indirect utility is 0.419110; the approximation is 0.418628.

22.6 $q_1(p_1, p_2, M) = \dfrac{3M}{5p_1}$, $q_2(p_1, p_2, M) = \dfrac{2M}{5p_2}$, $V(p_1, p_2, M) = \ln\left(\dfrac{108M^5}{3125p_1^3p_2^2}\right)$.

23.1 (a) $y_t = A9^t + B(-2)^t$; (b) $y_t = (Ct + D)(-4)^t$; (c) $y_t = 3^t\left[E\cos\left(\frac{\pi t}{3}\right) + F\sin\left(\frac{\pi t}{3}\right)\right]$.
23.2 $y_t = A(-3)^t + B2^t$; $y_t = 2^t - (-3)^t$.
23.3 $y_t = (Ct + D)4^t$; $y_t = 2t(4^t)$.
23.4 $y_t = \left(\sqrt{27}\right)^t\left[E\cos\left(\frac{\pi t}{6}\right) + F\sin\left(\frac{\pi t}{6}\right)\right]$; $y_t = \left(\sqrt{27}\right)^t\sin\left(\frac{\pi t}{6}\right)$.
23.5 $y_t = 1 + A(-1)^t + B(-11)^t$.
23.6 $y_t = \frac{1}{2} + A3^t + B4^t$; $y_t = \frac{1}{2} + 2(3^t) + 4^t$.
23.7 $y_t = 2 + (Ct + D)5^t$; $y_t = 2 + (t + 1)5^t$.
23.8 $y_t = 3 + 3^t\left[E\cos\left(\frac{\pi t}{3}\right) + F\sin\left(\frac{\pi t}{3}\right)\right]$; $y_t = 3 + 3^t\cos\left(\frac{\pi t}{3}\right)$.
23.10 $y_t - y_{t-1} - 12y_{t-2} = 12$; $y_t = 2(4^t) - (-3)^t - 1$ and $x_t = \frac{3}{2}(4^t) + (-3)^t$.

24.1 (a) $y_t \to \infty$ as $t \to \infty$; (b) $y_t \to \infty$ as $t \to \infty$; (c) in the long run, y_t oscillates increasingly.
24.2 $Y_t = 64 + (\frac{1}{\sqrt{8}})^t\left[\cos\left(\frac{\pi t}{4}\right) + \sin\left(\frac{\pi t}{4}\right)\right]$; in the long run, Y_t oscillates decreasingly to 64.

25.1 $\dfrac{t^2}{2} + t + \ln(t + 3) + c$, $\dfrac{5}{2} + \ln\left(\dfrac{5}{4}\right)$.
25.2 $q^* = 10$, $p^* = 160$, $CS = 50$.
25.3 $2\pi + 2$.
25.4 $\ln\left(\dfrac{3 + \sqrt{10}}{1 + \sqrt{2}}\right)$.

25.5 $\dfrac{14}{3} + 8\ln 2$.

25.6 $\ln(101) - \dfrac{200}{101}$.

26.1 $\dfrac{1}{3}\ln(3t + t^3) + c$, $\dfrac{1}{2}\ln(t-3) - \dfrac{1}{2}\ln(t-1) + c$, $\dfrac{1}{2}e^{t^2} + c$.
(For the first, make the substitution $u = 3t + t^3$; for the second, use the method of partial fractions; for the third, make the substitution $u = t^2$.)

26.2 $\dfrac{1}{3}\ln\left(\dfrac{7}{2}\right)$, $\dfrac{1}{2}\ln\left(\dfrac{28}{27}\right)$, $\dfrac{e-1}{2}$.

26.3 $\dfrac{x^2}{2} + \dfrac{3}{2}\ln(x-1) - \dfrac{1}{2}\ln(x-1) + c$.

26.4 $\dfrac{1}{3}(\sin\theta)^3 + c$, $2\sqrt{x}\ln(x) - 4\sqrt{x} + c$, $\dfrac{1}{3}t^3e^{t^3} - \dfrac{1}{3}e^{t^3} + c$.
(For the first, make the substitution $u = \sin\theta$; for the second, use integration by parts; for the third, make the substitution $u = t^3$ and use integration by parts on the resulting integral.)

26.5 $2\sqrt{t-1} + \ln(\sqrt{t-1} - 1) - \ln(\sqrt{t-1} + 1) + c$.

26.6 $2\ln(2 - \sqrt{x}) - 2\ln(1 - \sqrt{x}) + c$, $\ln(1 + \sqrt{x}) - \ln(1 - \sqrt{x}) + c$.

26.7 $e^2(3e^2 - 1)/4$.

26.8 $(e^4 - 1)/2$.

26.9 $q^* = 5$, $CS = 192\ln\left(\dfrac{3}{2}\right) - 20$.

26.10 $p^* = \dfrac{50}{\ln(20)}$, $CS = 1000\ln\left(\dfrac{\ln(20)}{\ln(10)}\right) - \dfrac{500}{\ln(20)}$.

27.1 $\sqrt{1 - y^2} = \dfrac{1}{2}\ln(1-t) - \dfrac{1}{2}\ln(1+t) + c$.

27.2 $y^3 = 5 + A(t^3 + 1)$, so $y(t) = \left(5 + A(t^3+1)\right)^{1/3}$.

27.3 $y(t) = \sqrt{t^2 + 1}$.

27.4 $y(t) = \dfrac{2}{\sqrt{1 - t^4} + c}$, $y(t) = \dfrac{2}{\sqrt{1 - t^4} + 1}$.

27.5 $\dfrac{dy}{dt} = \dfrac{y^2 - 1}{2yt}$, $y(t) = \sqrt{1 + 15t}$.

27.6 $p(t) = 1 - \dfrac{1}{2}\left(\dfrac{1}{256t + 1}\right)^{1/4}$.

27.7 $q^D(p) = 16/(p+1)$.

27.8 $q^D(p) = 5e^{\sqrt{6}}/e^{\sqrt{p^2+2}}$.

27.9 $K(t) = \left(\dfrac{2}{9}s\alpha t^3 + \dfrac{1}{3}s\beta t^2 + \dfrac{2}{3}s\gamma t + c\right)^{3/2}$, where $s = 1 - b$ and c is a constant.

27.10 $y(t) = \dfrac{2t^6 + 3t^4 + 12}{12(t^2 + 1)}$.

27.11 $f(x) = \dfrac{x+1}{3}\ln\left(\dfrac{x-2}{2(x+1)}\right)$.

27.12 $V(t) = -20 + (A + 20)e^{t/5}$.

28.1 (a) $y(t) = Ae^{3t} + Be^{2t}$; (b) $y(t) = (Ct + D)e^{2t}$;
(c) $y(t) = e^{-t}(E\cos t + F\sin t)$.
28.2 $y^*(t) = \left(-\frac{1}{3}t + \frac{2}{9}\right)e^{3t}$; $y(t) = \left(-\frac{1}{3}t + \frac{2}{9}\right)e^{3t} + Ae^{6t} + Be^{2t}$.
28.3 $y(t) = \frac{1}{2}e^{t} + \frac{1}{2}\cos t + \frac{9}{2}\sin t$.
28.4 $p''(t) - 7p'(t) - 30p(t) = -72$; $p(t) = \frac{12}{5} + 2e^{10t} + e^{-3t}$ and tends to infinity with exponential growth.
28.5 $p''(t) - 14p'(t) + 24p(t) = 44$; $p(t) = \frac{11}{6} + 2e^{12t} - e^{2t}$ and tends to infinity with exponential growth.

29.1 $x_t = 0.8 + 0.2(0.75)^t$ and $y_t = 1.2 - 0.2(0.75)^t$.
29.2 The eigenvalues are $1 - \sqrt{6}$ and $1 + \sqrt{6}$.
29.3 The eigenvalues are 0 and 5. The corresponding eigenvectors are, respectively

$$\begin{pmatrix} -2 \\ 1 \end{pmatrix} \quad \text{and} \quad \begin{pmatrix} 1 \\ 2 \end{pmatrix}.$$

29.4 The eigenvalues are -3 and -1. The corresponding eigenvectors are, respectively

$$\begin{pmatrix} -2 \\ 1 \end{pmatrix} \quad \text{and} \quad \begin{pmatrix} -1 \\ 1 \end{pmatrix}.$$

We can therefore take $P = \begin{pmatrix} -2 & -1 \\ 1 & 1 \end{pmatrix}$ and $D = \begin{pmatrix} -3 & 0 \\ 0 & -1 \end{pmatrix}$.

30.1 We can take

$$P = \begin{pmatrix} 1 & 3 \\ 1 & 1 \end{pmatrix} \quad \text{and} \quad D = \begin{pmatrix} -1 & 0 \\ 0 & 1 \end{pmatrix}.$$

The nth power is

$$A^n = -\frac{1}{2}\begin{pmatrix} (-1)^n - 3 & -3(-1)^n + 3 \\ (-1)^n - 1 & -3(-1)^n + 1 \end{pmatrix}.$$

This means $A^n = I$ if n is even and $A^n = A$ if n is odd.
30.2 The sequences are $x_t = 10(-3)^t - 9(-1)^t$ and $y_t = -5(-3)^t + 9(-1)^t$.
30.3 We can take

$$P = \begin{pmatrix} 1 & 1 \\ -3 & 1 \end{pmatrix} \quad \text{and} \quad D = \begin{pmatrix} 4 & 0 \\ 0 & 8 \end{pmatrix}.$$

The functions are

$$f(t) = \frac{1}{4}e^{4t} + \frac{7}{4}e^{8t} \quad \text{and} \quad g(t) = -\frac{3}{4}e^{4t} + \frac{7}{4}e^{8t}.$$

Index

addition of matrices, 152
adjugate, 218
annuity, 36
anti-derivative, 278
  See also indefinite integral
approximation by derivative, 52
arbitrage portfolio, 156, 179
  and state prices, 180
area under a curve, 277
  and definite integral, 279
assets, 154
associative operation, 14
augmented matrix, 163
auxiliary equation, 253
  of differential equation, 321
  of recurrence equation, 253
average cost, 96
average variable cost, 95
axes, 1

back-substitution, 164
behavioural conditions, 252, 301
behavioural parameters, 196
breakeven point, 96
budget constraint, 139, 239
budget set, 139
  illustrated graphically, 139
bundle, 137
  optimal, 144
  price of, 139
business cycles, 265

calculating matrix inverse, 194
  by determinants, 218
  by row operations, 194
capital, 35, 101
  unit cost, 224
capital growth, 35
chain rule, 55
  for partial differentiation, 104
change of variable in integration, 290
characteristic polynomial, 337
co-factor matrix, 218
Cobb–Douglas function, 67, 103, 230
  homogeneity of, 115
cobweb
  stable, 46
  unstable, 47
cobweb model, 44
  general linear case, 45
coefficients, 161
columns of matrix, 152
complements, 125
completing the square, 15
composite function, 14
  derivative of, 55
composite function rule, 55
composition of functions, 14
  associativity, 14
compounding
  continuous, 64
  interval of, 37, 62
consistent linear system, 174
constant coefficients, 23
constant of integration, 278
constant returns to scale, 115
constrained optimisation, 224
  and tangency, 225
  by elimination of one variable, 224
constraint, 224
consumer, 238
  budget constraint, 239
  theory of, 238
  utility function, 239
consumer surplus, 275
  and definite integral, 281
consumption, 196, 251
continuous compounding, 64, 307
contour, 114, 140
  slope of, 114
convex combination, 142
convex preferences, 143
convex set, 142
coordinates, 1, 12
cosine (cos) function, 68
  derivative, 68

cost, 54
  average, 96
  average variable, 95
  derivation of, 227, 230
  fixed, 73
  variable, 73
coupled differential equations
  change of variable, 352
coupled recurrences
  and matrix powers, 348
  change of variable, 349
  reduction to second order, 335
coupled system
  of differential equations, 352
  of recurrence equations, 334
Cramer's rule, 217
critical point, 74, 126
  first-order conditions, 126
  function of one variable, 74
  function of two variables, 126
  nature of, 74, 131
  second-order conditions, 75, 131
  visualisation, 127

decimal representation, 12
decreasing returns to scale, 115
definite integral, 279
degree of homogeneity, 115
degrees, 68
  conversion to radians, 68
demand, 1
  curve, 1
  elastic, 85
  function, 2, 242
  inelastic, 86
  inverse, 2
  Marshallian, 242
  set, 1
derivative, 52
  d-notation, 52
  and approximation, 52
  and gradient, 74
  definition of, 52
  of a composition, 55
  of a power, 53, 69
  of a product, 53
  of a quotient, 54
  of a sum, 53
  of decreasing function, 74
  of exponential function, 64
  of implicit function, 112
  of increasing function, 74
  of inverse function, 56
  of logarithm, 67
  second, 74
  See also partial derivative
determinant, 212
  and invertibility, 216
  and unique solution, 216
  calculation by row operations, 213
  of 2×2 matrix, 212
  of 3×3 matrix, 212
  of diagonal matrix, 213
  of upper-triangular matrix, 214
diagonal matrix, 339
diagonalisable matrix, 339
diagonalisation, 339
  and eigenvectors, 340
difference equation, 23
differential equation, 302
  See also first-order differential equation
  See also second-order differential equation
differentiation, *see* derivative
discriminant, 16
dot product, 138
dummy variable, 12

echelon form, 164
  leading ones, 164
efficient small firm, 93, 124, 228
  breakeven point, 96
  cost function, 228, 230
  demand set, 93
  inverse supply function, 229
  profit function, 95
  startup point, 95
  supply function, 229
  supply set, 93, 229
eigenvalue, 337
  determining, 337
eigenvector, 337
  determining, 337
elastic demand, 85
elasticity of demand, 85, 311
  and change in revenue, 85
elementary operations, 162
elementary row operations, 163
elements of a set, 11
empty set, 11
entry of matrix, 152
equation
  linear, 15
  quadratic, 15
equilibrium
  in market, 3
  set, 3
  stable, 46
  unstable, 47
equilibrium conditions, 196, 252
equivalent annual rate, 38
Euler's Theorem, 115
excise tax, 4, 82

expenditure, 196
exponential function, 63
  derivative of, 64
  graph of, 63
  properties, 63

factorisation, 15
firm, 93, 224
  capital and labour cost, 224
  efficient small, 93
    (*see also* efficient small firm)
firm producing two goods, 123
first quadrant, 12
first-order conditions, 126
first-order differential equation, 303
  general solution, 304
  initial conditions, 304
  particular solution, 304
  separable, 303
first-order recurrence, 23
  behaviour of solution, 26
  general solution, 25
  solution method, 24
fixed cost, 73
free parameters in solution to linear system, 177
function, 12
  as black box, 12
  composite, 14
  defined implicitly, 110
  diagrammatic representation, 13
  formula, 12
  gradient of, 52
  graph of, 15
  identity, 15
  inverse, 13
function of a function, 14
function of a function rule, 55
function of several variables, 101
function of two variables, 101
  contour, 114
  homogeneous, 115

Gaussian elimination, 164
global
  maximum, 76, 131
  minimum, 76
global optimisation, 76
government spending, 196
gradient of function, 52
graph, 15
  of exponential function, 63
  of logarithm function, 65

homogenous function, 115

identity function, 15
identity matrix, 154
implicit function, 110
  derivative of, 112
income, 196, 251
inconsistent linear system, 174
increasing returns to scale, 115
indefinite integral, 278
indifference curve, 140
indirect utility function, 243
inelastic demand, 86
infinite intervals, 78
inflexion point, 74
initial condition, 24
inner product, 138
integers, 12
integral, 278
  See also definite integral
  See also indefinite integral
integrals
  standard, 280
integrand, 278
integrating factor, 309
integration, 278
  by change of variable, 290
  by partial fractions, 293
  by substitution, 289
integration by parts, 292
interdependent production processes, 204
interest rate, 35
  annual, 35
  equivalent annual, 38
intersection, 3, 11
  symbol, 3
interval of numbers, 76
inverse demand, 2, 13
inverse function, 13
  derivative of, 56
inverse matrix, 191
  2×2 case, 195, 217
  3×3 case, 218
  calculating, 194, 218
  need not exist, 191
  uniqueness of, 191
inverse supply, 3, 13
invertible matrix, 192
investment, 196, 251
IS–LM analysis, 196
isoquant, 114, 224

labour, 101
  wage, 224
Lagrange multiplier, 226
  meaning, 244
Lagrange multiplier method, 226
  first-order conditions, 238
  general formulation, 238
  rationale, 226
Lagrangean, 226, 238

leading ones, 164
Leontief matrix, 206
limit, 26
linear combination, 142
linear equation, 15
linear equations, 161
  augmented matrix, 163
  coefficients, 161
  consistent, 174
  Cramer's rule, 217
  echelon form, 164
  elementary operations, 162
  elementary row operations, 163
  free parameters, 177
  Gaussian elimination, 164
  general solution, 176
  inconsistent, 174, 176
  infinitely many solutions, 175
  leading ones, 164
  matrix form, 161
  rank of, 175
  solution in vector form, 177
  square system, 190
  unique solution, 176, 190
linear recurrence, 23
linear system, *see* linear equations
local maximum, 76, 126
local minimum, 76, 126
logarithm function, 65
  derivative of, 67
  graph of, 65
  notation, 65
  properties, 65
  to base 10, 65

macroeconomics, 196, 251
  dynamics, 252, 265, 301
  equilibrium conditions, 196, 252, 301
main diagonal, 154
marginal cost
  as derivative, 54
  meaning of, 54
marginal product of capital, 103
marginal product of labour, 103
marginal propensity to consume, 302
marginal propensity to invest, 302
marginal rate of substitution, 240
marginal revenue, 87
marginal utility of income, 243
  and Lagrange multiplier, 244
market trends, 319
Markov chain, 334
Marshallian demand function, 242
matrix, 152
  addition, 152
  columns, 152
  determinant, 212
  entry, 152
  identity, 154
  inverse, 191
  invertible, 192
  multiplication, 153
  multiplication by a scalar, 153
  non-singular, 192
  powers, *see* matrix powers
  rows, 152
  size, 152
  square, 154
  transpose, 218
matrix form of linear system, 161
matrix powers, 347
  and coupled recurrences, 348
  via diagonalisation, 347
maximum point, 74
  local, 76, 126
  of one-variable function, 74
  of two-variable function, 126
meaning of Lagrange multiplier, 244
members of a set, 11
minimum point, 74
  local, 76, 126
  of one-variable function, 74
modelling, 1
  the market, 1
money demand, 196
money supply, 196
monopoly, 85, 88, 124
multiplication of matrices, 153
multiplier-accelerator model, 252, 265
  modification of, 268

*n*th root, 62
national economy, 251
  dynamics of, 252
  equilibrium conditions, 252
natural logarithm, 65
  See also logarithm function
natural numbers, 12
non-negative numbers, 12
non-singular matrix, 192
numbers
  integers, 12
  natural, 12
  rational, 62
  real, 12

objective function, 224
optimal bundle, 144
  and tangency to indifference curve, 145
optimisation, 73
  constrained, 224
  global, 76
  of functions of a single variable, 73
  of two-variable functions, 126

on an infinite interval, 78
on an interval, 76
ordered pairs, 12
depiction, 12
oscillatory, 26

partial derivative, 101
chain rule, 104
second, 102
partial fractions, 293
perfect competition, 88, 93
policy parameters, 196
population modelling, 334, 350
portfolio, 155, 179
arbitrage, 156, 179
riskless, 155
positive integers, 12
power rules, 61
powers
derivative of, 69
general definition, 66
integral, 61
of matrix, *see* matrix powers
rational, 62
rules of, 61
powers of a matrix, *see* matrix powers
present value, 35, 82
of annuity, 37
under continuous compounding, 64
price, 1
price adjustment, 305
price ratio, 240
price-taker, 93
principal, 35
product rule for differentiation, 53
production, 251
production function, 101, 224
Cobb–Douglas, 230
isoquant of, 114
returns to scale, 115
profit, 73, 123, 124
maximisation, 87, 123

quadrant
first, 12
quadratic equation, 15
completing the square, 15
discriminant, 16
factorisation, 15
formula for solutions(s), 15
roots, 16
types of solutions of, 16
quantity, 1
quotient rule for differentiation, 54

radians, 68
rank, 175
of linear system, 175
and solution set of linear system, 175
rate of change, 51
average, 51
instantaneous, 51
rational numbers, 62
rational powers, 62
real numbers, 12
decimal representation, 12
interval of, 76
non-negative, 12
recurrence, 23
coupled, 334
first-order, 23
general solution, 25, 253
initial conditions, 24, 253
initial values, 253
linear, 23
second-order, 253
time-independent solution, 24, 256
See also first-order recurrence
See also second-order recurrence
recurrence equation, *see* recurrence
returns matrix, 155, 179
returns to scale, 115
constant, 115
decreasing, 115
increasing, 115
revenue, 73, 124
riskless portfolio, 155
root of equation, 16
row operations, 213
rows of matrix, 152

saddle point, 127
scalar, 137
scalar product, 138
second derivative, 74
second partial derivatives, 102, 132
second-order conditions, 75, 131
second-order differential equation, 303, 320
auxiliary equation, 321
behaviour of solution, 325
general solution, 320
homogeneous, 320
initial values, 321
linear, constant coefficient, 320
non-homogeneous, 323
particular solution, 320, 324
second-order recurrence, 253
auxiliary equation, 253
behaviour of solution, 264
general solution, 253
homogeneous, 253
non-homogeneous, 256
oscillatory solution, 265
particular solution, 256, 268
solution method, 253
separable differential equation, 303

sequence, 23
set, 11
  convex, 142
  elements of, 11
  empty, 11
  intersection, 11
  members of, 11
  subset of, 11
  union, 11
simultaneous equations, 3, 16, 161
sine (sin) function, 68
  derivative, 68
square linear system, 190
square matrix, 154
standard integrals, 280
startup point, 95
state prices, 179
  and arbitrage portfolio, 180
states, 154
stationary point, *see* critical point
subset, 11
substitutes, 124
substitution method
  for definite integration, 291
  for indefinite integration, 289
sum rule for differentiation, 53
supply, 3
  function, 3
  inverse, 3
  set, 3

taking logarithms, 66
tangency condition, 240
tangent (tan) function, 68
tangent to a graph, 52
tax
  excise, 4, 82
  theorem, 8
technology matrix, 205
tending to infinity, 26
tending to zero, 26
theory of the consumer, *see* consumer
theory of the firm, *see* firm
time-independent solution of recurrence
  first-order, 24
  second-order, 256
transpose, 218
trigonometrical function, 68

union, 11
upper triangular matrix, 214
utility, 140
utility function, 140, 239
  contours, *see* indifference curve
  indifference curve, 140
  indirect, 243

variable cost, 73, 95
vector, 137
  addition, 137
  algebraic operations, 137
  as point in space, 137
  convex combination, 142
  dot product, 138
  inner product, 138
  linear combination, 142
  multiplication by a scalar, 137
  scalar product, 138

wage, 224